Internet of Things

Technology, Communications and Computing

The series Internet of Things - Technologies, Communications and Computing publishes new developments and advances in the various areas of the different facets of the Internet of Things. The intent is to cover technology (smart devices, wireless sensors, systems), communications (networks and protocols) and computing (theory, middleware and applications) of the Internet of Things, as embedded in the fields of engineering, computer science, life sciences, as well as the methodologies behind them. The series contains monographs, lecture notes and edited volumes in the Internet of Things research and development area, spanning the areas of wireless sensor networks, autonomic networking, network protocol, agent-based computing, artificial intelligence, self organizing systems, multi-sensor data fusion, smart objects, and hybrid intelligent systems.

**Indexing: *Internet of Things* is covered by Scopus and Ei-Compendex **

Panos Fitsilis
Editor

Building on Smart Cities Skills and Competences

Human factors affecting smart cities development

Springer

Editor
Panos Fitsilis
University of Thessaly
Larissa, Greece

ISSN 2199-1073 ISSN 2199-1081 (electronic)
Internet of Things
ISBN 978-3-030-97817-4 ISBN 978-3-030-97818-1 (eBook)
https://doi.org/10.1007/978-3-030-97818-1

This Springer imprint is published by the registered company Springer Nature Switzerland AG
The registered company address is: Gewerbestrasse 11, 6330 Cham, Switzerland

To my wife Sofia who somehow managed to be supportive once again.

Preface

Smart cities are complex ecosystems that use information and communication technologies for helping their citizens and organizations deal with the challenges of urbanization, safety, resilience, and sustainability. Their creation is a long and laborious transformation that should not be considered mainly as a technical challenge but as a movement to create citizen-centered ecosystems that improve quality of life and stimulate economic activity.

Smart city ecosystems comprise people, organizations and businesses, policies, technologies, legislation, and processes integrated to create the desired outcomes. Within this context, people comprise a vital constituent in these efforts and a highly skilled workforce is key in rendering smart city ecosystems a reality.

Further, cities are important drivers for growth, employment creation, and sustainable green progress. They play a central role in the digital transformation of the economy, achieving sustainable growth in the digital sector and using advanced technology to serve citizens.

Today, more and more smart cities are emerging worldwide creating a sustainable and strongly growing market. According to various market reports, it is expected that this market will reach the size of 1 billion euros worldwide, but the most important fact is that it is expected to change our daily lives.

Although for several years we were working on the development of this new smart ecosystem, quite recently we discovered that the people factor was not considered, sufficiently. We realized that even though billions were invested in technological or urban development, not sufficient effort had been made in training the necessary workforce with the skills to fulfill this vision. The development of a smart city was considered "business as usual" by the IT vendors, or even yet another case of modern technology deployment. However, this is far from true as smart cities transverse every aspect of our lives and our activities.

As we routinely say, we are "living in a software-enabled society and we cannot risk it". And since our lives depend surprisingly so much on the smooth operation of digital services, we need to reassure ourselves and to invent new ways to develop software that empowers us to produce reliable software, with the ability to continuously develop new features and effortlessly deploy it and make it available to our citizens in a matter of seconds.

Therefore, digital transformation requires complementary measures that are not overly directed to support infrastructure investment but are as well aimed at promoting entrepreneurship, improving digital skills, new paradigms for developing transformation strategies and software. All of the above are considered strategically important for medium-sized cities that enable them to be more competitive in the global economy.

This book has been inspired and the ideas are largely originated from the SmartDevOps project, which was funded by Erasmus+ KA2 with Project No.: 601015-EPP-1-2018-1-EL-EPPKA2-SSA and led by the University of Thessaly (https://smartdevops.eu). The SmartDevOps project moved towards this direction and developed three new smart cities professions, an extensive list of smart cities-related competences, a community of practice at the European level, and the Smart Cities Body of Knowledge (SCBoK), which is an attempt to systematically approach the topic of required smart cities competences.

The book is structured in four parts. **Part one** is focused on skills and competences and the first four chapters articulate different perspectives for skills and competences required for smart cities development.

In the chapter "Emerging Smart City Job Roles and Skills for Smart Urban Governance" Theodor Panagiotakopoulos, Omiros Iatrellis, and Achilles Kameas present their research aimed at identifying emerging smart city job roles and skills and providing a mapping among them to steer the development of contemporary lifelong education programs for smart cities' workforce. They conclude that three primary new job roles are required for smart city professionals: the smart city planner, the smart city IT manager, and the smart city IT officer. Moreover, they present a framework with 102 skills classified into four categories (transversal, generic IT, DevOps, and smart city-related skills) and the 42 most important ones for smart cities. Transversal (soft) skills dominated the top ten, while social skills came in first. Finally, three curricula were produced by defining which of these 42 skills are mandatory for each smart city job role. This work was a key outcome of the SmartDevOps ERASMUS+ project.

The second chapter entitled "The co-evolution of the digital transition and appropriate skills at city level" by Lena Tsipouri and Sofia Liarti, analyzes the meaning and the characteristics of smart cities. Subsequently, the chapter is presenting the types of skills needed by the smart cities' stakeholders during the digital transition and how to obtain them. To elaborate on this need, the chapter studies the history and the concept of smart cities as well as the dimensions and characteristics relevant for the labor market. This allows the reader to comprehend, how a city can embark into the virtuous circle of co-evolution, validating the positive relationship between skills, agglomeration, growth, and sustainability.

Subsequently, Paraskevi Tsoutsa and Ioannis Lampropoulos in their chapter "Preparing for future competences: Trends arising through keyword and review analysis" explore the topic of required competences in smart cities, using bibliometric analysis. The analysis produced four distinct clusters of concepts regarding smart cities. Further, they build a visual map of the keyword concepts and finally they present a conceptual atlas of the existing literature.

In the chapter "Building smart city knowledge and competences using problem-based learning in a blended learning environment," Alina Bockshecker, Katharina Ebner, and Stefan Smolnik present an educational concept capable of teaching the transformation of cities towards smart cities. This approach integrates problem-based learning, gamification, and virtual group work. Further, using virtual group work, students acquire theoretical and practical, transferable knowledge and competences concerning diverse and complex concepts of smart cities.

Part two consists of six chapters, and it is focused on smart cities strategy development and on how to manage smart cities projects. Smart cities strategy development is an important process that has to involve all city's stakeholders, and it is central to the city's digital transformation.

The fifth chapter entitled "The dynamic formation of a successful smart city roadmap" by Georgios Siokas and Aggelos Tsakanikas is attempting to capture and present the dynamics behind the municipality's strategic design and implementation of smart initiatives for different types of municipalities.

N. Kishor Narang in the chapter "Sustainable digital transformation of urban landscape through disruptive technologies and standards" enumerates the shifting paradigms and the required skills and competencies for sustainable digital transformation of urban landscape. Further, he presents the standard landscape in smart cities and smart infrastructure domain and how they affect digital transformation.

Chapter seven entitled "Smart cities: Emerging risks and mitigation strategies" by Konstantinos Kirytopoulos, Theofanis Christopoulos, and Emmanuel Dermitzakis is studying how to identify risks and develop appropriate risk response strategies to address the risks that threaten the sustainability of smart cities. This chapter intends to constitute a guide on risks and relevant remedies to practitioners and academics who are involved in the development or operation of smart cities, as well as to highlight the required skills to achieve proper risk management. To achieve this, a systematic literature review is carried out.

The chapter "City resilience and intelligence: Interrelation and reciprocity" by Christos Ziozias and Leonidas Anthopoulos is addressing the question of how to transform a smart city into a resilient one. This chapter presents the major similarities and differences between a smart and a resilient city, as a tool for officials, responsible for planning each city model, to decide and define the proper policies, strategies, and actions. It concludes that research on the smart city or resilient city skills and competencies is in the early stages and there are insufficient findings to compare skills and competencies for a smart or resilient city.

In the chapter "Smart city projects evaluation: A bibliometric approach," Vassilis Gerogiannis and Stella Manika attempt to provide insights into the evaluation of smart city development and deployment projects. More specifically, they assess how these projects successfully contribute to the development of the smart city "intelligence". The performed bibliographic analysis highlighted, that the success of smart city projects, is related more to the projects' technical parameters rather with the social or citizen-centric factors. The results of this study map the major bibliographic trends on the subject of smart cities or projects evaluation or success and then highlight the gap of the citizen-based approach in the evaluation of smart cities.

Chapter ten entitled "Modelling Project Management Complexity in Smart Cities Projects" by Vyron Damasiotis is addressing the issue of project management complexity, especially on smart cities projects. He proposes a 12-dimensional complexity model with 45 corresponding metrics for assessing the complexity of smart cities projects based on project management aspects, technical characteristics of software development, etc. This model can be used for accessing the complexity of smart cities projects.

The **third part** of this book is related to citizen engagement in the development of smart cities. Citizens are engaged through strong collaboration, co-creation, project prioritization, etc.

As such the first chapter of this section by Kleanthis Sitakoulis and Stella Manika is about how citizens are learning to engage in smart cities' development. The chapter initially analyzes the concept of civic engagement and its potential by mapping the dominant bibliographic trends. It then considers the utilization of civic engagement and its integration in the process of developing the intelligence of a city, via a pilot study. Through its two pillars, the chapter highlights and quantifies the role of civic engagement in the effort to achieve the intelligence of a city and concludes by proposing actions to strengthen and redefine this role.

Chapter twelve is presenting how skill development for smart cities can be achieved not only through formal educational programs but also through other initiatives. This chapter, by Judy Pamela Backhouse and Laila al Hadhrami, describes an initiative in Oman to develop a group of **smart city ambassadors** who are experts in certain aspects of smart cities and to use these ambassadors to share knowledge with other stakeholders and contribute their expertise to smart city projects.

Chapter thirteen, "At the root of the smart cities: Smart learning ecosystems to train smart citizens" by Carlo Giovannella, illustrates the relevance that smart learning ecosystems have in the education of smart citizens. In this work, a smart city is people-centered since all citizens must consider themselves as an active agent capable of contributing to the process that leads the ecosystem towards a progressive increase of its smartness, an increase that would not be possible without smart citizens. Then the chapter attempts to define the concept of smartness for citizens and how smart learning ecosystems can contribute towards this direction.

In chapter fourteen entitled "Towards e-deliberation 2.0," Vassilis Tsakanikas, Georgia Rokkou, and Vassilis Triantafyllou present how e-deliberation can become a major tool in increasing citizens' participation in the decision-making process and consequently enhance the democratic process.

Ahmed ElBatanony and Giancarlo Suzzi explore in the chapter "No-code for smart cities" how no-code development platforms could be used to empower smart city citizens with tools to change and improve their city systems. The no-code movement aims to introduce software development tools that require no prior coding or development skills to the general population. Finally, the chapter presents interesting ideas on the development of smart city API able to be utilized by smart citizens.

Finally, the **fourth part** is focusing on innovative technologies that can reshape smart cities ecosystems.

The first chapter of this section is entitled "A big data analytics framework for a smart city: A case study" by Andreas F. Gkontzis, Dimitrios Kalles, Evgenia Paxinou, Rozita Tsoni, and Vassilios S. Verykios. The authors present a case study of a municipality, where many of the existing smart services are not providing added value to the stakeholders, due to the inadequate knowledge and expertise in big data and data analytics technologies. Then a conceptual framework, solving this problem, is presented. This framework can be used for the data integration of the existing and future smart systems of a municipality.

In chapter seventeen entitled "RES-Q: Towards semantic interoperability for risk and disaster management in smart cities," Omiros Iatrellis, Vasileios Kyriatzis, Nicholas Samaras, and Charalampos Dervenis present how the Covid-19 pandemic has imposed new challenges in preserving the goal of developing smart and sustainable cities worldwide while improving urban resilience. They present the RES-Q (RESCUE) semantic model, which includes the needed domain knowledge streams for the smart city crisis management domain.

The chapter "Blockchain for smart cities: Findings from a systematic literature review" by Ifigenia Georgiou, Juan Geoffrey Nell, and Angelika I. Kokkinaki is a systematic literature review where an attempt is made to answer two important research questions about smart cities and blockchains: (i) What was the reason that blockchain was proposed as a potential solution for increased security and trustworthiness? (ii) What blockchain-based applications are being proposed for smart cities? They conclude by highlighting the blockchain challenges, the skills that are needed to implement blockchain for smart cities, and the need to change the current mindset of centralized control and trusted third parties to a more participative engagement model across smart cities.

Finally, chapter nineteen, "Artificial intelligence, big data analytics, and smart cities" by Yiannis Kiouvrekis, Theodor Panagiotakopoulos, Iakovos Ouranos, and Ioannis Filippopoulos explores which digital skills are necessary when dealing with smart cities and especially with artificial intelligence applications. This work concludes that there are urgent needs to create structured education programs covering concepts of AI and Big Data Analytics for smart cities focusing not only on scientists or municipalities employees but on citizens as well.

Together, these contributions shed new light on the numerous technologies that are related to smart cities ecosystems, applications, tools, and platforms.

However, the main contribution of this book is an attempt to shift focus from smart cities technological factors to human and social factors. These factors affect significantly smart city strategy, citizens' engagement and participation, and most importantly the creation of the "smart employees" and "smart citizens."

Finally, we would like to raise awareness of all smart city stakeholders that the development of smart cities is not a technological challenge but is about how we imagine/dream as citizens, our cities in the years to come.

Larissa, Greece Panos Fitsilis

Acknowledgments

This book has been inspired and the ideas have largely originated from SmartDevOps project. The SmartDevOps project was funded by Erasmus+ KA2 with Project No.: 601015-EPP-1-2018-1-EL-EPPKA2-SSA, and it was executed by a consortium of 13 partners led by the University of Thessaly (https://smartdevops.eu).

As such, I would like to thank all project partners and project team members who contributed to the development of this book, as well as all project participants of the SmartDevOps project who helped me improve it and encouraged me to continue this effort during the Covid19 pandemic. Many of the chapters included in this book were initially conceived during these projects and have been included in this book.

Before closing, I would like to thank all reviewers who with their work, comments, and valuable contribution improved greatly the final outcome and of course the advisory board that consisted of:

Prof. Leonidas Anthopoulos, University of Thessaly
Dr. Okan Geray, Smart Dubai
Prof. Vassilios Gerogiannis, University of Thessaly
Prof. Achilles Kameas, Hellenic Open University
Prof. Angelika Kokkinaki, University of Nicosia
Adjunct Prof. Ilpo Laitinen, City of Helsinki, Finland
Associate Prof. Agostino Marengo, University of Bari
Prof. Anastasia Stratigea, National Technical University of Athens

Contents

Part I Skills and Competences for Smart Cities' Development

Emerging Smart City Job Roles and Skills for Smart Urban Governance 3
Theodor Panagiotakopoulos, Omiros Iatrellis, and Achilles Kameas

The Co-Evolution of the Digital Transition and Appropriate Skills at City Level 21
Lena Tsipouri and Sofia Liarti

Preparing for Smart Cities' Future Competences: Trends Arising Through Keyword and Review Analysis 37
Paraskevi Tsoutsa and Ioannis Ch. Lampropoulos

Building Smart City Knowledge and Competences Using Problem-Based Learning in a Blended Learning Environment 53
Alina Bockshecker, Katharina Ebner, and Stefan Smolnik

Part II Strategy and Projects

The Dynamic Formation of a Successful Smart City Roadmap 71
Georgios Siokas and Aggelos Tsakanikas

Sustainable Digital Transformation of Urban Landscape Through Disruptive Technologies and Standards 95
N. Kishor Narang

Smart Cities: Emerging Risks and Mitigation Strategies 123
Konstantinos Kirytopoulos, Theofanis Christopoulos, and Emmanuel Dermitzakis

City Resilience and Smartness: Interrelation and Reciprocity 141
Christos Ziozias and Leonidas Anthopoulos

Smart City Projects Evaluation: A Bibliometric Approach 155
Vassilis C. Gerogiannis and Stella Manika

Modeling Project Management Complexity in Smart Cities' Projects..... 169
Vyron Damasiotis

Part III Citizen Engagement

Learning to Engage Citizens to Make Smarter Cities 187
Kleanthis Sirakoulis and Stella Manika

Developing Smart City Ambassadors in Oman.............................. 201
Judy Backhouse and Laila al Hadhrami

At the Root of the Smart Cities: Smart Learning Ecosystems to Train Smart Citizens.. 217
Carlo Giovannella

Toward E-Deliberation 2.0 229
Vassilis Tsakanikas, Georgia Rokkou, and Vassilis Triantafyllou

No-Code for Smart Cities 247
Ahmed ElBatanony and Giancarlo Succi

Part IV Smart Cities Innovative Technologies

A Big Data Analytics Conceptual Framework for a Smart City: A Case Study 259
Andreas F. Gkontzis, Dimitrios Kalles, Evgenia Paxinou, Rozita Tsoni, and Vassilios S. Verykios

RES-Q: Toward Semantic Interoperability for Risk and Disaster Management in Smart Cities 281
Omiros Iatrellis, Vasileios Kyriatzis, Nicholas Samaras, and Charalampos Dervenis

Blockchain for Smart Cities: Findings from a Systematic Literature Review.. 297
Ifigenia Georgiou, Juan Geoffrey Nell, and Angelika I. Kokkinaki

Artificial Intelligence, Big Data Analytics, and Smart Cities............... 315
Yiannis Kiouvrekis, Theodor Panagiotakopoulos, Iakovos Ouranos, and Ioannis Filippopoulos

Index.. 327

Part I
Skills and Competences for Smart Cities' Development

Emerging Smart City Job Roles and Skills for Smart Urban Governance

Theodor Panagiotakopoulos, Omiros Iatrellis, and Achilles Kameas

1 Introduction

During the last two decades, the smart city concept has been significantly proliferated as an urban development model with the power and potential to address contemporary challenges of cities mainly caused by climate change and urbanization. Major concerns of urban agglomerations include traffic congestions, water and land scarcities, energy shortages, floods, environmental pollution, human health, increased unemployment, and social issues, such as immigration flows and growing inequalities (Nam & Pardo, 2011; Washburn et al., 2009; Caragliu et al., 2011). In this context, smart cities can be seen as a strategy that focuses on implementing advanced technology-driven solutions to deal with the pressing issues facing policy makers today (Viitanen & Kingston, 2014). They aspire to a techno-utopian future where intelligent systems of decision-making and public service provision are expected to deliver on the promise of sustainable urban growth making metropolitan areas livable and prosperous.

In the smart city paradigm, technologies are seen as the means for discovering new forms of urban intelligence, collaboration, and innovation offering a rich variety

T. Panagiotakopoulos (✉)
School of Technology and Science, Hellenic Open University, Patras, Greece

School of Business, University of Nicosia, Nicosia, Cyprus
e-mail: panagiotakopoulos@eap.gr

O. Iatrellis
University of Thessaly, Larissa, Greece
e-mail: iatrellis@hotmail.com

A. Kameas
School of Technology and Science, Hellenic Open University, Patras, Greece
e-mail: kameas@eap.gr

P. Fitsilis (ed.), *Building on Smart Cities Skills and Competences*, Internet of Things,
https://doi.org/10.1007/978-3-030-97818-1_1

of solutions to improve utilities, services, public infrastructures, economy, citizen participation, the environment, and ultimately the quality of living. Going further from the typically focused on technical solutions model however, a more efficient urban metabolism requires a better understanding of other elements also playing a catalytic role in translating smart city imaginaries to effective tangible interventions in urban governance. To this end, Nam and Pardo (Nam & Pardo, 2011) emphasized that beyond technology, human and institutional factors are core components of smart cities, while Kourtit and Nijkamp (Kourtit & Nijkamp, 2012) argued that smart cities rely on a mix of human, social, infrastructural, and entrepreneurial capital to achieve their objectives. Clearly, people comprise a crucial axis for the development and governance of smart cities.

In fact, smart cities are established on the basis of the existence of a range of creative talents able to deliver urban technological innovation (Kourtit & Nijkamp, 2012). The shift toward a knowledge-based economy has converted superior talent, which embodies skills, knowledge creativity, and innovation capacity, into a key ingredient for urban development (Winters, 2011). While a creative and skilled workforce does not guarantee optimal urban performance, its presence plays a determinant role in the success of smart cities (Caragliu et al., 2011). Indeed, growth, economic value, and competitive advantages will increasingly be attained from skilled and knowledgeable people (Dirks et al., 2010).

However, several reports indicate that skilled workers are lagging behind demand. A recent study highlighted that digital illiteracy and lack of specific skills are among the major barriers of incorporating emerging technologies into urban operations (Ubaldi et al., 2019). Moreover, according to a study by the Economic and Social Council of United Nations entitled "Smart cities and infrastructure" that explores smart city trends up to 2030 (UN ECOSOC, 2016), one of the five main challenges that will be encountered in the implementation of smart cities projects relates to skills gaps. Lack of trained workforce and shortage of training funds constitute part of the most prominent challenges in development of smart cities (PwC, 2018). Obviously, addressing the skill readiness of current and prospective smart cities professionals through appropriate educational strategies is a matter of urgency to achieve smart urban governance.

Toward this direction, the starting point would be to identify future professions and, more importantly, the required skillset that the smart city workforce should possess. The literature in defining these skills and especially from a smart city labor market perspective is extremely limited. To fill this gap, the Erasmus+ project Smart-DevOps[1] explored emerging job profiles and skills in smart cities with the ultimate objective of creating Vocational and Educational Training (VET) courses for lifelong training of smart cities' workers. This chapter presents the methodology and findings of our exploratory research and provides a mapping of skills to the identified job profiles to guide lifelong training programs design and development.

[1] https://smartdevops.eu/dev/

Initially, the current framework and existing efforts to determine the skills demand in smart cities are described.

2 Smart Cities: Context and Perspectives

The smart city concept has its contemporary origins in the "smart growth" movement of the 1990s as a reaction to the negative impact of urbanization with objectives that are still relevant (Albino et al., 2015). Despite the steep rise of the global smart cities discourse, there is still not a commonly accepted and consistent definition of the concept among academics and practitioners. Yet, and beyond the points of emphasis in different renders, most smart city definitions have explicit technological dimensions. According to (Dirks & Keeling, 2010), "a smarter city is one that uses technology to transform its core systems and optimize the return from largely finite resources." Or, as Washburn et al. put it, smart city is defined as "the use of smart computing technologies to make the critical infrastructure components and services of a city more intelligent, interconnected, and efficient" (Washburn et al., 2009).

As Gabrys argues (Gabrys, 2014), urban environment infused with digital technologies is not a new development; however, the mantle of sustainability has created a fertile ground for their promotion and accelerated innovative technological urban projects. To achieve their objectives, smart cities have to continuously collect real-time data and intelligently use them to enable well-informed decision-making toward improved public service delivery and optimized resource usage. Thus, smart cities mostly rely on the realization of several technological aspects of pervasive computing, most notably on the Internet of Things (IoT). Everyday objects are enhanced with computational and networking capabilities and seamlessly communicate with each other across heterogeneous networks and IoT platforms to provide information and services (Bibri, 2018).

These views are mostly technology-oriented and consist one of the two major research streams in defining what makes smart cities smart; the other takes a people-oriented path (Guo et al., 2019). In the technology-centered perspective, Information and Communication Technologies (ICTs) are key to realizing smart cities, and the importance of big data analytics, wireless sensor networks, artificial intelligence, machine learning, and cloud computing for optimizing city operation is overemphasized. The people-centered perspective, on the other hand, is based on arguments stressing that technology can't do everything on itself and should not develop and be used in isolation of societal needs and challenges. It brings human and social capital in the foreground considering knowledge, skills, creativity, and innovation capacity as important levers for urban growth.

While the technology-centered view sees smart cities through technocratic lens focusing on ICTs and how they are used to improve urban management and functioning, the people-centered one focuses on policies related to welfare, education, economy, and social participation (which are called soft domains) and how ICTs can enhance them (Kitchin, 2014). To put it simply, the main focus of the latter is shifted from the role of technologies which are now considered as enablers to the role of human and social capital deemed as key drivers of urban development (Lombardi et al., 2012). Intensifying this view, Hollands (Hollands, 2008) highlights that smart cities should "start with people from the human capital side, rather than blindly believing that IT itself can automatically transform and improve cities." Taking both perspectives into account, Caragliu et al. (Caragliu et al., 2011) provided a holistic definition of smart cities, which are based on "investments in human and social capital and traditional (transport) and modern (ICT) communication infrastructures that fuel sustainable economic growth and a high quality of life, with a wise management of natural resources, through participatory government."

Although the driving forces and needs may vary from city to city, smart city applications tend to cluster in specific domains. A comprehensive classification is described by Neirotti et al. (Neirotti et al., 2014) who identified six main application domains: natural resources and energy, transport and mobility, buildings, living, government, and economy and people while differentiating between hard (e.g., buildings, transport, water and energy grids) and soft domains (e.g., culture, education, social inclusion, and public administration). Examples of smart city applications may include cooperative localization in connected and autonomous vehicles to increase road safety (Piperigkos et al., 2020), IoT based smart water applications for water quality assessment, leak detections, and water level monitoring (Panagiotakopoulos et al., 2021); smart homes (Antonopoulos et al., 2015); and ambient assisted living systems enabling older adults to stay at their preferred living environments for longer while continuously receiving healthcare and daily living support services (Panagiotou et al., 2015).

3 Skills Demand in Smart Cities

Digitization of cities and the infusion of digital technologies to all aspects of our everyday life have affected the world of work causing wide changes in the skills that workers need to acquire in order to claim sustainable careers. Identifying the type of skills smart city professionals should develop for smart urban governance will allow enactment of policies and measures to build a workforce capable of efficiently planning, developing, deploying, and managing smart city solutions.

Framed in this context, Markow et al. (Markow et al., 2019) examined the labor market demand for three skill categories: human, business, and digital skills (Fig. 1). They analyzed online job postings in 8 smart cities in the USA with over 600,000 inhabitants and found that at least one of these skills was requested in 70% of the job openings in these smart cities. Human skills were found to be in stronger demand compared to the other categories with communication being more frequently asked for, followed by collaboration and critical thinking. Concerning digital skills, data management and software development had the highest demand, while business process and project management led the business enablers' demand. Moreover, data

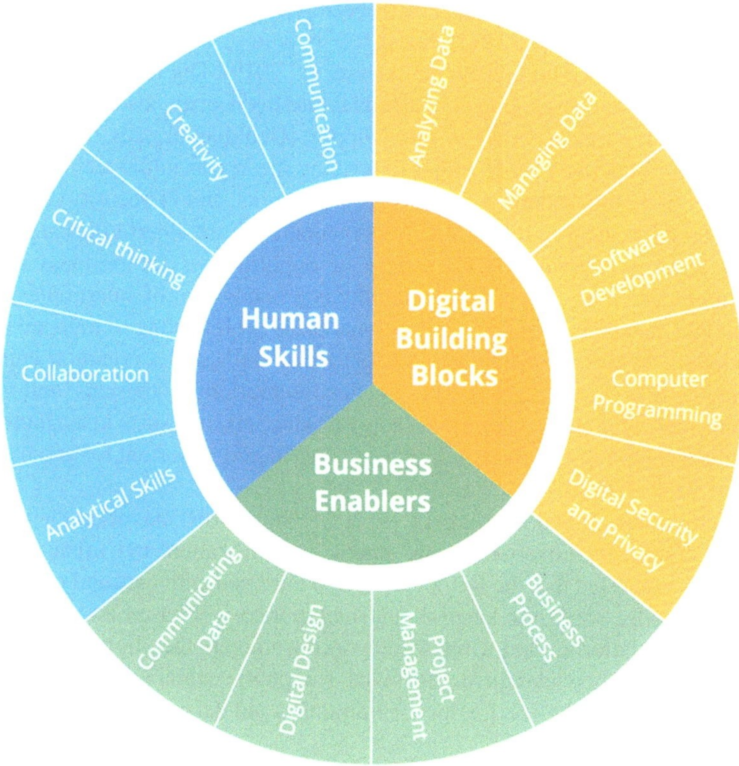

Fig. 1 The new foundation skills of the digital economy (Markow et al., 2019)

analysis met the largest growth (107%) in smart cities' job openings followed by collaboration (98%) and digital security and privacy (98%).

The demand for data management is also reported in (Saunders & Baeck, 2015), where data handling and advanced skills for data specialists are considered important to integrate advanced technologies and new data and knowledge-related processes into traditional organizational workflows. Attempting to explore the skills that smart city managers should possess, an empirical list of 24 skills was administered to Italian smart city managers (Michelucci et al., 2016). Implementing an exploratory factor analysis, the authors identified five groups of skills: (a) city planning skills (e.g., urban innovation and territorial planning), (b) legal skills (e.g., open data management and procurement), (c) soft skills (e.g., empathy and strategic vision), (d) financial skills (e.g., crowd-funding tools and economic principles), and (e) basic skills (e.g., knowledge of foreign languages).

Several smart city skills have been also reported in (Dirks et al., 2010) although not in a structured manner. These skills include digital ones, such as advanced information technology and analytics, as well as soft (or human, or transversal skills), such as creativity and systems thinking. Creativity is probably the most recognized skill in the literature and often been praised for its contribution in urban innovation (Nam & Pardo, 2011; Kourtit & Nijkamp, 2012; Dirks et al., 2010; Albino et al., 2015; Kitchin, 2014; Markow et al., 2019). Digital skills, such as big data analytics, cloud computing, and artificial intelligence, are also often met in the literature (and reported as being in short supply) (Ubaldi et al., 2019), while others can be implicitly derived from studies describing the key enabling technologies for smart cities (e.g., blockchain, wireless sensor networks, and IT security).

Another skill category refers to entrepreneurship, since smart cities are rapidly evolving markets offering a rich variety of opportunities for business activity. Furthermore, as environmental sustainability is a major goal of smart cities, green skills should also be part of every educational program intending to train smart city professionals. While a solid framework for green skills and a common approach to define them have not yet been developed, conservation of landscape and nature, environmental technology, renewable energy production, and waste management are illustrative examples of skills currently being considered important for green jobs in the EU (CEDEFOP, 2018).

In addition, David and McNutt (David & McNutt, 2019) classified the required skills for smart city governance into public administration and urban planning skills. They suggested that the former should be enriched with data and information governance skills, while the latter should incorporate ICT-based monitoring, data analysis, civic platforms, and wireless networking among other digital skills. Going one level lower, they classified the skills for public administration and urban planning of smart cities to technology, community, and data skills.

4 Methodology

Our methodology was divided into three stages including both quantitative and qualitative research methods as depicted in Fig. 2. The first stage aimed at determining emerging job roles for smart city professionals. A roundtable discussion was ran, where each partner of the DevOps project[2] proposed three job roles (this was the project's target), in order to select the three most prevalent. In order to form the job role proposals, we utilized empirical knowledge and information obtained from the literature that falls into the urban studies discipline, which we accessed through the Scopus database. We opted for this approach instead of running a survey with a variety of relevant job roles, since our aim was to identify a specific and very small number of job roles representing the core professionals needed to support smart city projects in a horizontal, cross-sectoral manner and it was our belief that the literature combined with our experience in the field would suffice.

The second stage was the most demanding one and concerned the identification of the emerging skills smart city professionals should have. Initially, we created a

[2] https://smartdevops.eu/dev/who-we-are/

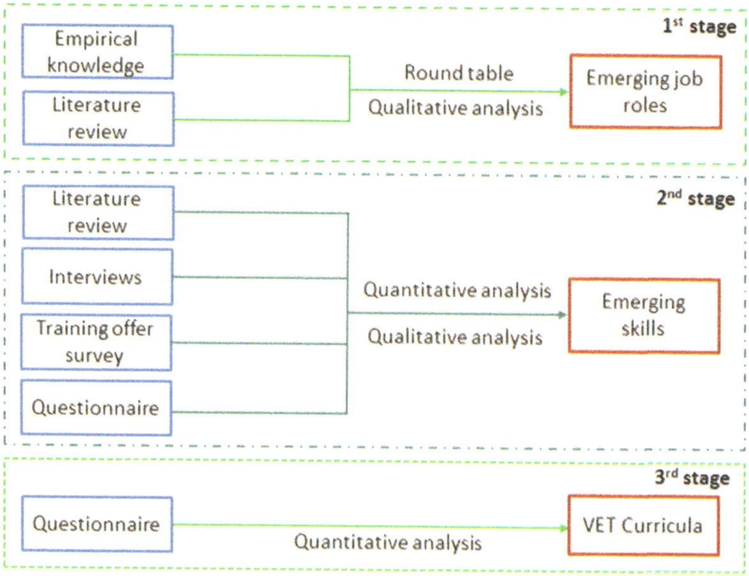

Fig. 2 Research methodology

comprehensive and multidimensional skills framework considering relevant literature (as described in the previous section), courses of European master's programs on smart cities, as well as information generated from interviews with smart city administrators. This multifaceted approach enabled us to incorporate the skills demand and supply aspects in tandem with academics' views. Thus, innovative and efficient training interventions could be then developed on the basis of what is demanded versus what is offered.

The interviewees with smart city officials were 16 in total, from 3 European countries: Greece ($n = 6$), Germany ($n = 5$), and Italy ($n = 5$). Specifically, officials from Trikala, Heraklion, Rome, Florence, Berlin, and Hamburg were asked the following question: What are the most important management and IT skills identified in your "City of the Future"? Qualitative analysis of the collected data revealed various skills we hadn't identify giving us the opportunity to enrich the smart city skills framework. Problem solving, IT management, decision-making, strategic vision, IT quality assurance, IT security, smart city platforms, and user experience were some of the additional skills identified by one or more interviewees.

The final skills framework comprised 102 skills divided into 4 categories: transversal (or soft), generic IT, DevOps, and smart city-specific skills. Since the Smart-DevOps project emphasizes on the DevOps methodology for development and operation of projects and services, we considered an individual category for DevOps skills. We then created an online questionnaire, which was sent to smart city stakeholders to evaluate the importance of these skills for smart city professionals through a 5-point Likert scale (Strongly disagree, Disagree, Neither agree nor disagree, Agree, Strongly Agree) transformed to a ranking from 1 to 5. Brief descriptions were provided for each skill to facilitate better understanding of its meaning, while an open question was included at the end of the questionnaire requesting to report any skills not included in the questionnaire along with a short description. Participants included city officials and employees, self-employed professionals (i.e., engineers), decision makers and IT experts, as well as corporate members of companies operating in areas of smart cities. Despite our intensive efforts to disseminate this questionnaire through several channels, we received 387 responses.

Finally, the third stage focused on mapping the skills identified in the second stage to the job roles identified in the first stage, in order to elicit training curricula and guide the development of appropriate learning material. To this end, we used a questionnaire that included the most important skills, i.e., those with the highest ranking (in mean terms) in the responses given in the previous stage. This questionnaire was administered to 90 people who were asked to indicate whether each skill is mandatory or optional for each job role. Aiming to gain an unbiased rounded view and since this stage had a solid educational interest, respective participants were equally divided among academia, industry, and municipal authorities.

In order to assure that participants in all stages of our research were actually engaged in smart city initiatives and thus were aware of smart city concepts and challenges, we addressed people from cities either included in the Cities in Motion Index (CIMI), 2020 edition,[3] or in the core group of the 100 Intelligent Cities Challenge (ICC).[4] CIMI utilizes 101 indicators that reflect both objective and subjective data to define the smarter cities globally. ICC was launched by the European Commission to support the transition of European cities "towards a more digital, service-oriented and low-carbon economy." Academics were selected based on journal publications found in the Scopus database during the last 5 years under the urban studies discipline. Moreover, all questionnaires included a form asking for personal and professional information notifying that all data would be acquired and used according to the General Data Protection Regulation (EU 2016/679) for research purposes. Respondents could skip this form and proceed with filling in the questionnaires anonymously.

5 Results

5.1 Emerging Smart City Job Roles

Identifying the first two job roles was a straightforward task, since urban planning and management are core processes of cities' governance and smart cities will

[3] https://media.iese.edu/research/pdfs/ST-0542-E.pdf

[4] https://www.intelligentcitieschallenge.eu/

definitely need people capable of undertaking these neuralgic roles. Many articles in the literature advocate for this and highlight the importance of planning and/or management roles for smart cities (e.g., (Viitanen & Kingston, 2014; Kourtit & Nijkamp, 2012; Michelucci et al., 2016; David & McNutt, 2019)). Hence, the Smart City Planner (SCP) and the Smart City IT Manager (SCM) were unanimously recognized as the most fundamental emerging job roles for smart cities. Additionally, the Smart City IT Officer (SCO) was selected as the third job role since technologies are pervasively used to support and enable urban innovation and people with technical expertise are essential to develop, deploy, and operate smart city systems and applications. Descriptions for these job roles are provided in Table 1.

5.2 *Smart City Skills*

In order to determine the most important skills for smart city professionals, we calculated the mean and standard deviation of all responses for each skill. Tables 2, 3, 4 show the 34 skills that received the highest overall ranking grouped by skill category. To these skills were added the six DevOps-related skills (Table 5) forming our basis of reference for the next stage.

Feedback provided via the open question gave us valuable insight for the social aspect of a smart city's sustainability. A lot of comments mentioned the need to either incorporate it in the scope of the skill "smart city sustainability," which was originally focusing on environmental issues, or have it as a new skill. We therefore split the smart city sustainability to two skills, namely, the social sustainability and the green smart city skills. Moreover, the resilience of smart cities emerged as a very important skill for respective managers and planners. Thus, we added a skill referring to smart city resilience reaching a total of 42 skills.

5.3 *Mapping Skills to Job Roles*

The consolidated results of the responses on mapping the 42 skills described in the previous section to the 3 smart city job roles are shown in Tables 6, 7, 8, and 9. The symbol "✓" indicates that a skill was reported as mandatory for a given job role by the majority of the respondents.

6 Discussion and Limitations

The digital transformation of urban places creates new career opportunities either through entirely new occupations or through new forms of existing occupations that

Table 1 Emerging job roles in smart cities

Job role	Description
Smart city planner	A high-level official that is able to bridge the needs that arise from: • Cities' traditional development and operational needs • Smart and sustainable cities' frameworks, best practices, standards, and technologies • Strategic priorities of the city's political leadership A SCP should have overall knowledge of the city's strategic objectives. He/she should be aware of the smart cities' trends and best practices in order to set up the city's strategy and strategic plan for the implementation of the city's vision. He/she should define specific key performance indicators (KPIs) for monitoring the progress of the implementation of the city's strategic plan
Smart city IT manager	A SCM can be defined as an ICT consultant with responsibilities that include: • Setting objectives and strategies for the IT department of the municipalities • Deciding and implementing suitable technological solutions to support all internal operations and optimize their strategic benefits • Designing and customizing the IT systems, frameworks, and platforms to improve citizen experience • Planning the implementation of new IT systems and providing systematic guidance to IT professionals and other staff within the organization • Making procurement decision for technological equipment and software as well as establish partnerships with IT providers • Managing the technological infrastructure (networks and computer systems) of the smart city to ensure their performance • Managing IT-related projects A SCM should have strong IT background to consistently keep up-to-date with recent advancements in technologies for smart cities. Apart from technical skills and experience, he/she must possess public administration skills and should be able to align smart cities' strategical objectives with IT development, deployment, and operation strategy, so that IT services maximize the value offered to citizens
Smart city IT officer	A SCO is an IT expert that should be able to: • Analyze urban organizational data • Determine information system requirements and define project objectives • Apply software development process, development environments, tools, and techniques • Make recommendations for necessary IT system components, e.g., hardware, software, and networking systems • Design, implement, and deploy new IT systems and services • Operate IT systems and services • Provide support and training to various types of users

result from the incorporation of new skills. In this dynamically changing career ecosystem, we argue that smart city planners, IT managers, and IT officers comprise three primary emerging job roles. As reported in (Michelucci et al., 2016), dedicated

Table 2 Highest rated transversal (soft) skills

Skill	Mean ± SD	Overall rank
Social skills	4.61 ± 0.54	1
Project and process management	4.53 ± 0.61	2
Strategic vision & strategy development	4.47 ± 0.64	4
Leadership	4.41 ± 0.73	6
Ability to work in a team	4.40 ± 0.77	7
Decision-making and problem solving	4.38 ± 0.78	8
Stakeholder management	4.35 ± 0.81	10
Knowledge management	4.29 ± 0.74	13
Creativity	4.23 ± 0.79	17
Ambiguity tolerance	4.11 ± 0.93	21
Intercultural skills	4.06 ± 0.77	23
Motivation to learn	4.04 ± 0.79	25
Advanced presentation skills	3.95 ± 0.83	28
Design thinking	3.94 ± 0.74	29
Emotional intelligence	3.86 ± 0.84	31
Entrepreneurial thinking	3.76 ± 0.68	33

Table 3 Highest rated generic IT skills

Skill	Mean ± SD	Overall rank
Internet of Things	4.31 ± 0.65	12
Software development	4.28 ± 0.76	14
GIS technologies/spatial data analysis	4.27 ± 0.82	16
Big data analytics	4.21 ± 0.86	18
Cloud computing	4.17 ± 0.72	19
IT security	4.08 ± 0.84	22
System and software architecture	4.01 ± 0.71	26
Artificial intelligence and machine learning	3.99 ± 0.67	27
IT quality assurance	3.92 ± 0.71	30

departments undertaking the design and implementation of smart city projects have started to establish in cities led by smart city managers. However, considering that these managers should have a strong IT expertise, we explicitly added this dimension in the respective job role title. SCOs would be professionals working in these departments possessing key technical skills and undertaking cross-sectoral responsibilities. A rich variety of other job roles have been also indicated, but were not included in the final trio due to being domain-specific (e.g., smart mobility operator), or partially fulfilled by the primary ones (e.g., chief innovation officer) or simply considered less important for the smart city operation (e.g., smart city community manager).

Concerning emerging smart city skills, transversal skills were rated as the most important. This observation agrees with the findings of (Markow et al., 2019; Michelucci et al., 2016), which, to the best of our knowledge, are the most

Table 4 Highest rated smart city-specific skills

Skill	Mean ± SD	Overall rank
Smart city context, policies, and operating procedures	4.51 ± 0.66	3
Urban management	4.45 ± 0.71	5
Smart cities' business models and financial management	4.36 ± 0.74	9
Smart city sustainability	4.32 ± 0.73	11
Digital urban infrastructures and services	4.28 ± 0.79	15
Citizen- and community-driven design	4.16 ± 0.80	20
Urban simulation/digital twins	4.05 ± 0.77	24
Smart city procurement	3.78 ± 0.76	32
Smart city standards and legal issues	3.73 ± 0.79	34

Table 5 DevOps skills

Skill	Mean ± SD	Overall rank
Using build, deployment, and monitoring tools	3.61 ± 0.76	42
Repository management	3.46 ± 0.86	51
Code analysis and continuous testing tools	3.37 ± 0.84	56
Continuous integration	3.28 ± 0.80	62
DevOps basic concepts, culture, and practices	3.17 ± 0.80	68
Configuration management	3.03 ± 0.80	78

Table 6 Transversal skills of smart city job roles

Skill	SCP	SCM	SCO
Creativity	✓	✓	✓
Entrepreneurial thinking	✓	✓	✓
Ability to work in a team	✓	✓	✓
Social skills	✓	✓	✓
Ambiguity tolerance	✓	✓	
Motivation to learn	✓	✓	✓
Emotional intelligence	✓	✓	✓
Strategic vision & strategy development	✓	✓	
Intercultural skills	✓	✓	✓
Design thinking	✓	✓	✓
Decision-making and problem solving	✓	✓	✓
Leadership	✓	✓	
Stakeholder management	✓	✓	
Knowledge management	✓	✓	
Project and process management	✓	✓	✓
Advanced presentation skills	✓	✓	✓

systematic studies in relevant literature, utilizing however a limited skill pool. Specifically, seven of the top ten rated skills are transversal, with the rest being smart city-specific skills. This could be justified by the fact that most of the participants represented cities (e.g., Pescara, Ioannina, and Hamburg) that were in early stages

Table 7 Generic IT skills of smart city job roles

Skill	SCP	SCM	SCO
Software development		✓	✓
IT quality assurance		✓	✓
IT security	✓	✓	✓
System and software architecture	✓	✓	✓
Cloud computing	✓	✓	✓
Internet of Things	✓	✓	✓
Big data analytics	✓	✓	✓
Artificial intelligence and machine learning		✓	✓
GIS technologies/spatial data analysis	✓	✓	✓

Table 8 DevOps skills of smart city job roles

Skill	SCP	SCM	SCO
DevOps basic concepts, culture, and practices	✓	✓	✓
Repository management			✓
Continuous integration			✓
Configuration management			✓
Using build, deployment, and monitoring tools			✓
Code analysis and continuous testing tools			✓

Table 9 Smart city-related skills of smart city job roles

Skill	SCP	SCM	SCO
Digital urban infrastructures and services	✓	✓	✓
Urban simulation/digital twins	✓	✓	✓
Smart cities' business models and financial management	✓	✓	
Smart city context, policies, and operating procedures	✓	✓	✓
Social sustainability	✓	✓	✓
Smart city standards and legal issues	✓	✓	
Urban management	✓	✓	
Citizen- and community-driven design		✓	✓
Smart city procurement	✓	✓	
Smart city resilience	✓	✓	✓
Green smart cities	✓	✓	✓

of transition to smart cities establishing their roadmaps and reform policies, and technical skills will be required in later phases.

The process of transforming a city into a smart city is complex and depends on various factors that need to coevolve such as built infrastructure, natural environment, business models, cultural practices, citizens' needs and preferences, technological know-how, and available resources (Carvalho, 2015). Thus, translating aspiration into reality involves changes of a socio-technical nature, and therefore social skills are particularly important. This is also verified by our findings, where social skills topped the ranked list of skills as shown in Table 2. Taking as well open answers into account, we could argue that this is because, on the one hand, having the capacity to resolve conflicts, align asymmetrical interests between the supply and the demand sides of smart city systems, and smoothen tensions stemming from the strive to interconnect disaggregated urban infrastructures and services is crucial to substantiate smart city visions. On the other hand, since smart cities are a fairly new concept, social skills are needed to communicate the objectives and benefits of smart city technologies to citizens, in order to foster acceptance and empower participation.

Looking at the skills of each job role, it is obvious that the SCMs' skillset is the most demanding one including many skills from all categories. SCPs are mostly oriented toward transversal and smart city-related skills, while IT skills should be at a basic level compared to SCMs and SCOs. The latter should also own skills from all categories with a clear focus however on IT skills, which become more specialized if we look in lower positions of the ranked skills list (e.g., mobile development, computer vision, etc.)

One limitation of our study is the low response rate. We cannot accurately calculate it, as the questionnaire was also distributed centrally through smart city associations and technical chambers; however, we estimate it below 5%. This may be due to the time period we conducted our research that coincided with the first phase of the pandemic (late winter–spring 2020) where strict restrictive measures disrupted daily routines and workflows. Workshops, info days, and targeted promotion of the questionnaire could potentially increase the response rate. Another limitation concerns the small percentage of respondents that provided personal information in the skills assessment questionnaire. To overcome this, we could provide a small text on the skill assessment questionnaire highlighting the contribution of personal information in analyzing the findings from different perspectives. However, the careful and balanced selection of the participants did not introduce bias in our results.

7 Conclusions

Smart cities are complex systems of systems, involving many different domains and infrastructures and organizations and activities. It is imperative that all these are integrated and work together effectively for a city to become smart. Therefore, the transition of cities to smart cities requires experts with different knowledge and expertise to examine their complex needs from different points of view. Simultaneously, technology and society are constantly evolving and require new capacities to underpin technologically enhanced urban management and growth. In this dynamically changing landscape, the types of professions and skills required for the transition to and governance of smart cities have to be continuously redefined and updated.

Our work addresses the question regarding the need of a new group of specialized professionals for smart urban governance. Furthermore, we address the pressing need to define the required skills smart cities' workforce should possess, in order to cope with the challenging responsibilities of future urban governance models. We concluded that three primary job roles are needed: the smart city planner, the smart city IT manager, and the smart city IT officer. Concerning emerging smart city skills, we explored the importance of 102 skills classified into 4 categories, transversal (soft), generic IT, DevOps, and smart city related, concluding in a ranked list of 42 skills. This list underlines the dominance of transversal skills with the highest ranked technical skill lying in the 12th place. Finally, we mapped these 42 skills to the previously identified job roles laying the foundations for competence-based training programs design and development.

Our findings contribute to understanding the evolution of professions and skills demand in smart cities. By doing so, critical guidance is offered to various stakeholders to harness the value of future smart city professions and skills. People that intend to pursue smart city careers will gain valuable insights into the required skills of different job roles and look for appropriate learning opportunities to develop them. Businesses can seek for solutions to offer tailored training programs to their workers, in order to increase their competitiveness. Finally, education providers can create short, flexible competence-based training programs and contemporary educational contents on the basis of the identified skills and individual needs.

From our perspective, we will exploit the findings of this study to design educational programs for smart cities. Our future work will focus on the development and delivery of such programs for upskilling and reskilling smart city professionals. We intend to apply several strategies, such as Massive Open Online Courses (MOOCs), specialization courses through blending learning models, and work-based learning (Iatrellis et al., 2021), with a strong emphasis on support of diverse learning paths.

Acknowledgments This work was supported by the Erasmus+ KA2 under the project "DevOps competences for Smart Cities" (Project No.: 601015-EPP-1-2018-1-EL-EPPKA2-SSA). This paper reflects the views only of the authors, and the Commission cannot be held responsible for any use which may be made of the information contained therein.

References

Albino, V., Berardi, U., & Dangelico, R. M. (2015). Smart cities: Definitions, dimensions, performance, and initiatives. *Journal of Urban Technology, 22*(1), 3–21.

Antonopoulos, C. P., Panagiotakopoulos, T. C., Panagiotou, C., Touliatos, G., Koubias, S. A., Kameas, A., & Voros, N. S. (2015). On developing a novel versatile framework for heterogeneous home monitoring WSN networks. *EAI Endorsed Transactions on Pervasive Health and Technology, 1*(1), e4.

Bibri, S. E. (2018). The IoT for smart sustainable cities of the future: An analytical framework for sensor-based big data applications for environmental sustainability. *Sustainable Cities and Society, 38*, 230–253.

Caragliu, A., Del Bo, C., & Nijkamp, P. (2011). Smart cities in Europe. *Journal of Urban Technology, 18*(2), 65–82.

Carvalho, L. (2015). Smart cities from scratch? A socio-technical perspective. *Cambridge Journal of Regions, Economy and Society, 8*(1), 43–60.

CEDEFOP. (2018). *Skills for green jobs 2018 update. European synthesis report*. Retrieved August 24, 2021, from https://www.cedefop.europa.eu/files/3078_en.pdf
David, N., & McNutt, J. (2019). Building a workforce for smart city governance: Challenges and opportunities for the planning and administrative professions. In *Informatics* (Vol. 6, No. 4, p. 47). Multidisciplinary Digital Publishing Institute.
Dirks, S., Gurdgiev, C., & Keeling, M. (2010). *Smarter cities for smarter growth: How cities can optimize their systems for the talent-based economy*. IBM Institute for business Value.
Dirks, S., & Keeling, M. (2010). A vision of smarter cities: How cities can lead the way into. Ministry of Science, Technology and Innovation Aras 1-7, Blok C4 & C5, Kompleks C, Pusat Pentadbiran Kerajaan Persekutuan, 62662 Putrajaya, Wilayah Persekutuan.
Gabrys, J. (2014). Programming environments: Environmentality and citizen sensing in the smart city. *Environment and Planning D: Society and Space, 32*(1), 30–48.
Guo, Y. M., Huang, Z. L., Guo, J., Li, H., Guo, X. R., & Nkeli, M. J. (2019). Bibliometric analysis on smart cities research. *Sustainability, 11*(13), 3606.
Hollands, R. G. (2008). Will the real smart city please stand up? Intelligent, progressive or entrepreneurial? *City, 12*(3), 303–320.
Iatrellis, O., Panagiotakopoulos, T., Gerogiannis, V. C., Fitsilis, P., & Kameas, A. (2021). Cloud computing and semantic web technologies for ubiquitous management of smart cities-related competences. *Education and Information Technologies, 26*(2), 2143–2164.
Kitchin, R. (2014). The real-time city? Big data and smart urbanism. *GeoJournal, 79*(1), 1–14.
Kourtit, K., & Nijkamp, P. (2012). Smart cities in the innovation age. *Innovation: The European Journal of Social Science Research, 25*(2), 93–95.
Lombardi, P., Giordano, S., Farouh, H., & Yousef, W. (2012). Modelling the smart city performance. *Innovation: The European Journal of Social Science Research, 25*(2), 137–149.
Markow, W., Hughes, D., & Walsh, M. (2019). *Future skills, future cities: New foundational skills in smart cities*. Business-Higher Education Forum.
Michelucci, F. V., De Marco, A., & Tanda, A. (2016). Defining the role of the Smart-City manager: An analysis of responsibilities and skills. *Journal of Urban Technology, 23*(3), 23–42.
Nam, T., & Pardo, T. A. (2011). Conceptualizing smart city with dimensions of technology, people, and institutions. In *Proceedings of the 12th annual international digital government research conference: digital government innovation in challenging times* (pp. 282–291).
Neirotti, P., De Marco, A., Cagliano, A. C., Mangano, G., & Scorrano, F. (2014). Current trends in smart city initiatives: Some stylised facts. *Cities, 38*, 25–36.
Panagiotakopoulos, T., Vlachos, P., Bakalakos, T., Kanavos, A., & Kameas, A. (2021). A FIWARE-based IoT framework for Smart water distribution management. In *The 12th international conference on information, intelligence, systems and applications (IISA)*, IEEE.
Panagiotou, C., Panagiotakopoulos, T., & Kameas, A. (2015). A multi: Modal decision making system for an ambient assisted living environment. In *Proceedings of the 8th ACM international conference on pervasive technologies related to assistive environments* (pp. 1–8).
Piperigkos, N., Lalos, A. S., & Berberidis, K. (2020). Graph based cooperative localization for connected and semi-autonomous vehicles. In *2020 IEEE 25th International workshop on computer aided modeling and design of communication links and networks (CAMAD)* (pp. 1–6). IEEE.
PwC (2018). Retrieved April 15, 2021, from https://www.pwc.com/gx/en/sustainability/assets/creating-the-smart-cities-of-the-future.pdf
Saunders, T., & Baeck, P. (2015). *Rethinking smart cities from the ground up*. Nesta.
Ubaldi, B., Le Fevre, E. M., Petrucci, E., Marchionni, P., Biancalana, C., Hiltunen, N., et al. (2019). *State of the art in the use of emerging technologies in the public sector*. OECD Publishing.
UN ECOSOC (2016). *Smart cities and infrastructure*. Retrieved August 12, 2021, from http://unctad.org/meetings/en/SessionalDocuments/ecn162016d2_en.pdf.
Viitanen, J., & Kingston, R. (2014). Smart cities and green growth: Outsourcing democratic and environmental resilience to the global technology sector. *Environment and Planning A, 46*(4), 803–819.

Washburn, D., Sindhu, U., Balaouras, S., Dines, R. A., Hayes, N., & Nelson, L. E. (2009). Helping CIOs understand "smart city" initiatives. *Growth, 17*(2), 1–17.

Winters, J. V. (2011). Why are smart cities growing? Who moves and who stays. *Journal of Regional Science, 51*(2), 253–270.

The Co-Evolution of the Digital Transition and Appropriate Skills at City Level

Lena Tsipouri and Sofia Liarti

1 Introduction

The world population living in cities is expected to reach 68% by 2050 (United Nations Secretary-General, 2020) because urban areas offer the best infrastructure and job opportunities. This *agglomeration effect*, the cost savings arising from urban agglomeration, however, has also a flip side: diseconomies of scale manifested in traffic congestion, air pollution, unemployment, and social exclusion to name just the most important ones. The advent of information and communication technologies (ICTs) has raised hopes that the consequences of the diseconomies could be mitigated: that smart tools will help overcome or at least minimise the inconveniences. An effective and flourishing labour market marshalling the proper skills is a necessary (yet not sufficient) condition to capture ICT benefits. Local human resources define the path (and pace) towards the digital transition affecting the development of digital infrastructure, business investments, public interventions/governance, and an active civil society.

Because all stakeholders can contribute, an ecosystem network must be developed that will involve all of them: citizens, organisations, institutions, governments, universities, companies, experts, research centres, and non-profit organisations (Berrone & Ricart, 2020). A smart city is both a pole of attraction for new and better skills and, at the same time, also the outcome of skilled people's performance: soft and hard skills in business and the public sector, researchers, and the digital literacy of citizens together create (or not) a smart city. One may think of a chicken-and-egg situation or conversely examine how skill development/attraction can co-evolve with the digital transition. Pioneer cities, which were already developed before the ICT

L. Tsipouri (✉) · S. Liarti
Department of Economic Sciences, National and Kapodistrian University of Athens, Athens, Greece
e-mail: tsipouri@econ.uoa.gr; sliarti@econ.uoa.gr

P. Fitsilis (ed.), *Building on Smart Cities Skills and Competences*, Internet of Things,
https://doi.org/10.1007/978-3-030-97818-1_2

revolution, attracted skills because they offered better working conditions and higher job satisfaction. In a virtuous circle, the newly attracted skills nurtured the next steps of digitalisation. Under this perspective lagging regions seemed doomed to further fall behind. This is a pessimistic view that the virtuous circle of the winners will reinforce a vicious circle for lagging cities to become lasting losers: you need skills for the digital transition, but if you are not sufficiently digitised, the best people will out-migrate depriving the city of origin from catching up opportunities. Yet, a more optimistic view is that smart cities and skills co-evolve in a process of an interacting, mutually reinforcing change. The co-evolution is not a closed system; on the contrary agglomeration forces interfere and positively or negatively affect progress. Skills can be developed everywhere because competent people exist everywhere. Assuming skills and smartness can change, grow, or decline together, then effective public policy can accelerate digital transition, catching up and occasionally even leapfrogging.

The target of this paper is to shed light on this co-evolution process by suggesting types of skills needed by the actors involved in the digital transition and how to obtain them. To make our point in the rest of the paper, we study the history and the concept of smart cities as well as the dimensions and characteristics relevant for the labour market. This allows us to suggest how a city can embark into the virtuous circle of co-evolution, validating the positive relationship between skills, agglomeration, growth, and sustainability.

2 The Notion *Smart City* and Its Evolution

Smart is the predominant name (Manville et al., 2014) in the array of similar terms, digital/smart/intelligent city, used to describe how ICTs contribute to the operation of cities, strengthening effectiveness, improving competitiveness, and providing new ways in which cities can tackle problems (Urenio, 2016). As interacting or subsequent generations of the terms digital/smart/intelligent cities appear in the literature, they are sometimes used as synonyms, while in other cases there are clear views on what distinguishes them. For instance, as far as their contents are concerned, a digital city labels the use of ICT in urban areas, while a smart city labels the environmental quality in cities. As far as their nature and relationship with the government is concerned, a digital city is a free trend arising from the daily use of smart and digital devices by citizens, and it triggers the local governments to provide e-services which gradually transform the city. On the other hand, a smart city is a political trend, led by institutions internationally for implementing adequate initiatives and improving the environmental quality in cities (Cocchia, 2014).

To make things more complex, the notion of intelligent cities appears, with four main characteristics: ‘a creative population and developed knowledge-intensive activities or clusters of such activities; embedded institutions and routines for cooperation in knowledge creation allowing to acquire, adapt, and advance knowledge and know-how; a developed broadband infrastructure, digital spaces, e-services, and online knowledge management tools; and a proven ability to innovate, manage and resolve problems that appear for the first time, since the capacity to innovate and to manage uncertainty are the critical factors for measuring intelligence' (Komninos, 2008). A digital city is not necessarily an intelligent one, but an intelligent city has digital components. Intelligent is a city with the ability of supporting learning, technological development, and innovation procedures (Nam & Pardo, 2011).

Still, the label smart city is a fuzzy concept and is used in ways that are not always consistent (Hollands, 2008; Nam & Pardo, 2011). Both the definition and content of smart and digital cities are addressed differently in the literature. According to Cocchia (2014), smart cities are the results of a triple combination: the EU source, focusing on the environmental requirements; the digital source, based on the previous experiences of digital cities; and the cultural source, that is, the human and social capital able to build the smart community. According to the European Commission (European Commission, 2020b), smart are the cities where 'traditional networks and services are made more efficient with the use of digital and telecommunication technologies for the benefit of its inhabitants and business'. In another definition, Mora and Bolici (2017) consider that 'smart cities are urban areas in which information and communication technologies are used as a tool for providing a solution to the multi-faceted problems that limit their [the cities'] sustainable development in social, economic, and/or environmental terms'. Egger and Skowron (2018) use a somewhat different approach noting that a really smart city uses technology for making better decisions about policy and its citizens.

Historically, ICT applications in cities were developed long before any of the above terms were coined. The application of digital technologies led first to the term *digital cities*. A definition by Komninos (Komninos, 2008) stated that 'a Digital City denotes an area that combines broadband communication infrastructure with flexible, service-oriented computing systems. These new digital infrastructures seek to ensure better services for citizens, consumers and business in a specific area'. Yovanof and Hazapis (Yovanof & Hazapis, 2009) added open industry standards and innovative services to meet the needs of governments and their employees, citizens, and businesses and included 'the goal of a digital city ... to create an environment for information sharing, collaboration, inter-operability & seamless experience for all its inhabitants anywhere in the city'. Couclelis (2004) pointed at social, cultural, political, ideological, and also theoretical dimensions for digital cities.

The first *smart city* is reported in the literature in 1994, when the city of Amsterdam was characterised as a virtual digital city (Rommes et al., 1999), thanks to the development of ICT-based projects introducing new applications, services, devices, and technological infrastructures (Mora & Bolici, 2017). Large ICT companies, like Cisco and IBM, immediately identified the opportunity to use their capabilities applying sensors, networks, and analytics to help cities to be more efficient, i.e. smart (Verdict, 2020). At the time, the term 'smart city' was not frequently used. Bibliometric analysis shows that smart city research established itself as a new area of scientific enquiry in 2009, and since then, it has been fast-growing, arousing strong interest from an expanding scientific community of researchers (Mora & Bolici, 2017). It was not until 2010 when the European

Commission made use of the term 'smart city' in its *Europe 2020 Strategy* that the term proliferated (Cocchia, 2014). Yet, more recently the European Commission adopted the term intelligent cities for its most recent programme supporting over 100 European cities.

Rather than trying to distinguish and fine-tune the terminology between digital, smart, and intelligent, we focus hereafter on the characteristics of smart cities 'that leverage digitalisation and engage stakeholders to improve people's well-being and build more inclusive, sustainable and resilient societies' (OECD, 2020) because we consider them as the practical aspects which can help derive lessons for the necessary skills to create smart cities and maintain their dynamism.

3 Smart City Features

The definitions indicate that smart cities are dynamic entities, and we need to know who contributes to their creation and evolution, what needs to be created and constantly upgraded, and how. Public policy is fundamental and so is interoperability between actors and systems. The need for interaction with the higher levels of public administration and the private sector makes interoperability a major challenge which can be addressed by the European Interoperability Framework, an Interoperability Framework for Smart Cities and Communities (EIF4SCC[1]) helping cities prepare for the new era of interoperable interactions.

3.1 Actors and Interactions

The actors (public sector, business, and citizens-civil society) are the driving force behind smartness. The indivisibility of the physical infrastructure, social characteristics, and resistance to change and global competition give municipal authorities a pivotal role for the level of smartness and the speed of transition.

The collaborative nature of smart cities: As Baccarne et al. (2014) have mentioned, both the quadruple helix model for innovation and living labs are linked to the 'collaborative nature of smart cities'. Quadruple helix model refers to collaboration between universities, government, industry, and citizens. Such collaborations facilitate the exchange of ideas and technologies (Etzkowitz, 2008). On the other hand, 'Living Labs are user-centered, open innovation ecosystems where research and innovation processes are combined at the same time within a public-private-people partnership' (4P's) (Wikipedia, n.d.).

[1] https://joinup.ec.europa.eu/collection/nifo-national-interoperability-framework-observatory/news/connecting-eif-smart-cities-communities-eif4scc

The city is not isolated. Several dimensions, like standards and interoperability needs, are designed and imposed at the national or occasionally regional level. Funding sources may come from local, regional, national, or even international sources. Such links apply to all actors. Proximity arguments justify the importance of the city level, but the resources of higher governance levels call for interaction and cooperation.

The domains of intervention: The pervasive nature of ICT makes it ubiquitous for all city roles and functions. Alexopoulos et al. (2019) have pointed out and defined the main axes in a more practical approach: ICT infrastructure; environment; transportation, mobility; e-government; safety and security; economic development; energy, sustainable development; waste management and water resources; and health and tourism, culture. These domains are confirmed broadly everywhere in the literature using slightly different levels of aggregating or distinguishing individual dimensions (Batty et al., 2012; Giffinger et al., 2007; Lombardi et al., 2012; Negre & Rosenthal-Sabroux, 2014).

3.2 The Components of Smart Cities

Smart Governance: According to Manville et al. (2014), smart governance means a 'joined up within-city and across-city governance, including services and interactions which link and, where relevant, integrate public, private, civil and European Community organisations so the city can function efficiently and effectively as one organism'. Smart governance includes participation in decision-making processes, transparent governance (use of open data), public and social services, and political strategies and perspectives (Batty et al., 2012; Giffinger et al., 2007; Kolokytha et al., 2015; Negre & Rosenthal-Sabroux, 2014; Vanolo, 2014).

Smart Economy: Smart economy refers to the city's competitiveness and includes innovative spirit, entrepreneurship, trademarks, productivity, flexibility of labour markets, international embeddedness, and ability to transform (Batty et al., 2012; Kolokytha et al., 2015).

Smart Mobility: Smart mobility refers to ICT-supported and ICT-integrated transport and logistics systems (Manville et al., 2014). Smart mobility includes factors such as local accessibility, international accessibility, availability of ICT infrastructure, and sustainable, innovative, and safe transport systems (Batty et al., 2012; Giffinger et al., 2007; Kolokytha et al., 2015). Facilitating mobility for people with disabilities helps smart cities increase their humane face.

Smart Environment: Smart environment refers to the efficient use of ICT for natural resources (Ismagilova et al., 2019) and includes factors such as attractiveness of natural conditions, pollution, environmental protection, and sustainable resource management (Batty et al., 2012; Dameri & Rosenthal-Sabroux, 2014; Kolokytha et al., 2015). According to Manville et al. (2014), smart environment includes also smart energy including renewables.

Smart Living: Smart living refers to the use of ICT for enhancing the citizens' quality of life and includes factors such as cultural and educational facilities, health conditions, individual safety, housing quality, touristic attractivity, and social cohesion (Batty et al., 2012; Dameri & Rosenthal-Sabroux, 2014; Kolokytha et al., 2015). Public safety is an area of concern in growing urbanisation (Breetzke & Flowerday, 2016; Cilliers & Flowerday, 2017) addressing the diseconomies of scale.

Finally, *smart people* is a transversal dimension in the city characteristics: smart people refer to the social and human capital and include factors such as the level of qualifications, affinity to lifelong learning, social and ethnic plurality, flexibility, creativity, cosmopolitanism/open mindedness, and participation in public life (Batty et al., 2012; Dameri & Rosenthal-Sabroux, 2014; Kolokytha et al., 2015).

3.3 Benchmarking Smart Cities

Understanding the features is important not only for appreciating the nature of smart cities but also for measuring their progress and benchmark them against each other. As cities differ it is useful to have a grid of parameters that could be adjusted in every city (Cavada et al., 2014).

Boyd Cohen (2014) has used these characteristics for attributes and indicators that help analyse smart cities' trajectory and performance. Measuring performance is essential to ensure effectiveness. With the COVID-19 crisis severely crunching municipal budgets, it is more critical than ever to devise cost-effective solutions to deliver public services. Assessing smart city performance also helps ground policy intervention in solid evidence by guiding decision makers, both at national and local levels, in setting realistic targets, understanding where cities stand vis-à-vis their objectives, tracking progress, and adjusting policy interventions for greater efficiency and effectiveness. Ultimately, smart city measurement enhances accountability and helps citizens monitor how governments deliver on their commitments (OECD, 2020).

A recent literature review of smart city indicators identifies as many as 1152 different smart city indicators (Petrova-Antonova & Ilieva, 2018). For example, the indicator framework for sustainable, resilient, and smart cities, called 'Sustainable development in communities—indicators for smart cities' developed by the International Organization for Standardization (ISO), has 85 indicators. Another example lies in the 91 key performance indicators (KPIs) for Smart Sustainable Cities (SSC), developed by the United for Smart Sustainable Cities (U4SSC), a UN initiative co-ordinated by the International Telecommunication Union (ITU), UNECE (United Nations Economic Commission for Europe), and UN Habitat (OECD, 2020). CITYKeys has also developed a measurement framework on the performance of smart cities targeted at European cities and includes 75 indicators (OECD, 2020). Skills play a key role in benchmarks, through indicators on education, on inclusion, and on jobs. Conversely, the IESE Business School's Cities in Motion Index (CIMI), which refers to the importance of making cities smart, only indirectly includes indicators specific to smart characteristics (Berrone & Ricart, 2020).

4 Skills for Smart Cities

The main goal of any city should be to improve its human capital. A city with smart governance must be capable of attracting and retaining talent, creating plans to improve education, and promoting both creativity and research (Berrone & Ricart, 2020). The degree of how smart a city is or plans to evolve determines its current and future skill needs.

The academic literature is mostly concerned with the key question of labour displacements, namely, whether the transition will lead to massive unemployment and a redundant labour force (Harari, 2018) or whether the need of more and different professions will overcompensate for the displaced middle skills (Pissarides, 2018). While this dispute is still in foresight debates and there is no agreement where the scale will tilt, one issue for the labour market is undebatable: there will be a major need for new skills, continuous upskilling, and reskilling. Specialists with appropriate qualifications will be needed involving the active participation of both producers and consumers in the development of 'smart' technologies (Avdeeva et al., 2019).

Individual future needs will be derived from labour market analyses taking into consideration status, and future plans. Cities may wish to be part of the big tech production, be pioneers in using smart technologies across the field, or select priorities. Certain skills are necessary across the board.

4.1 Type of Skills Needed: Proficient Employees for Smart Cities

The approach to the definition of digital skills has shifted from a technical orientation towards a wider perspective that considers content-related or higher-order skills (Claro et al., 2012). The results show that twenty-first-century skills are broader than digital skills, not necessarily always underpinned by ICT. Seven core skills are suggested in the literature: technical, information management, communication, collaboration, creativity, critical thinking, and problem-solving. Five contextual skills were also suggested: ethical awareness, cultural awareness, flexibility, self-direction, and lifelong learning (van Laar et al., 2017).

The Partnership for twenty-first Century Skills (Partnership for 21st Century Skills, 2007) is a joint government-corporate organisation which lists three types of skills: learning skills (creativity and innovation; critical thinking and problem-solving; communication and collaboration), literacy skills (information literacy;

media literacy; ICT literacy), and life skills (flexibility and adaptability; initiative and self-direction; social and cross-cultural skills; productivity and accountability; leadership and responsibility). Another initiative is the international research project Assessment and Teaching of twenty-first Century Skills (ATC21S). The ATC21S project resulted in ten skills grouped into four categories: ways of thinking (creativity and innovation; critical thinking, problem-solving, and decision-making; learning to learn and metacognition), ways of working (communication; collaboration), tools for working (information literacy; ICT literacy), and living in the world (citizenship; life and career skills; personal and social responsibility) (Binkley et al., 2012). The Organisation for Economic Co-operation and Development (OECD), for example, has categorised twenty-first-century skills as information, communication, and ethics and social impact (Ananiadou & Claro, 2009). Digital skills themselves include both basic skills necessary to use the Internet, and skills required to comprehend and use online content should be accounted for (van Laar et al., 2020).

Yet often a simplistic myth dominates the discussion of smart cities suggesting that it is the lack of ICT skills that hold progress back. This myth is nurtured by visible and measured skills shortages: there remains a shortage of ICT specialists on the labour market. During 2018, 57% of enterprises that recruited or tried to recruit ICT specialists reported difficulties in filling such vacancies. It was experienced by 64% of large enterprises and 56% of SMEs (European Commission, 2020a). In 2018, some 9.1 million people worked as ICT specialists across the EU (European Commission, 2020a).

Without in the slightest arguing against the need for more skilled ICT people, yet based on the discussion of the broader needs, we suggest that a lot more than that is needed for cities not only to proceed in the digital transition but to remain smart throughout the coming years, where technologies will evolve more rapidly than they can be diffused and absorbed. We thus suggest, for simplification reasons, three main types of skills.

ICT skills are crucial for developing the algorithms, platforms, and tools needed to create and advance the digital economy and society. The main challenge for the ICT skills needed is the very rapid technological progress that makes knowledge obsolete very quickly and calls for continuous upskilling. ICTs are hard skills. Information technology is an effective activity, as an option, necessitating both the development and use of a technological product, and the increase of the level of knowledge and skills in the field of information technology by specialists in all areas and types of modern activities (Cong et al., 2016; Kupriyanovsky et al., 2016; Namiot et al., 2017). Digital creation skills include e.g. computational thinking and coding, entrepreneurship and systems thinking, information architecting, as well as a risk-informed perception of data privacy and security. The challenges of delivering such a skillset are many, from designing a twenty-first-century curriculum to ensuring fair access to technology for people of all abilities, race, gender, age, and class (Tryfonas & Crick, 2018). ICT skills will be both generic (hardware and software engineering) and also very specific such as cybersecurity and data privacy. It will be used for ICT infrastructure, algorithms, and specific tools either using open-access programmes or creating tailor-made solutions. Creating ICT skills is connected to the commitment to developing policies to improve youth ICT skills (Picatoste et al., 2018).

Hardware and software ICT skills are not only needed for the business sector, i.e. producing, but also for municipal employees giving them the opportunity to understand, prioritise, and select smart solutions for their respective cities. Smart governance and smart economy need people with ICT skills, constantly upgrading to be able to produce or select the tools to create, buy, or adapt.

All science, technology, engineering, and mathematics (STEM) will be affected by the digital transition to adapt to changing systems of energy, mobility, waste treatment, environmental engineering, the automotive industry, doctors, and health workers as well as ball business activities will need to be at least knowledgeable in basic digital skills, yet also increasingly in more advanced knowledge to design and implement the numerical tools for their work. Workforces need to be capable of continuously adapting to shifting job requirements related to new skill-intensive technologies (Levy & Murnane, 2004). As workplaces have become more complex and supported by ICT, more jobs require technical skills (van Laar et al., 2020).

Smart economy, smart mobility, and smart environment need STEM in parallel with ICT skills.

Social Sciences and Humanities (SSH) will be equally affected because they will be called upon to address the emerging legal (data privacy and protection; intellectual property, competition), business (cost-benefit, risk management) and societal challenges associated with the change. Forensic readiness (Tan et al., 2021) for instance requires more SSH than STEM. SSH will take responsibility for developing the necessary *soft skills*, their value being recognised before the digital transition, but becoming increasingly important now. Soft skills include deep learning skills; critical thinking skills; acquisition of entrepreneurial skills; innovation and creativity methods and techniques; teamworking and virtual teamworking; and last but not least strategic thinking. The digital transition without soft skills risks turning into a technocratic, inhuman society.

4.2 The Stakeholders Involved

In a city every activity and every person may be both a contributor and a beneficiary of the digital society. In this sense digital skills are needed for everyone, yet they differ subject to the stakeholder community and the role of individuals in it.

Public sector: For sustainable smart governance, it is necessary to continuously improve digital services, their architectural design, their development following the open platform approach and the principles of flexible development software, as well as the adoption of standard data and interfaces. Important elements are also the integration of systems around universal (horizontal) support services and their modernisation procedures related to procurement and implementation of ICT projects. The interventions have a purpose further upgrading the public administration, contributing to increase competitiveness, productivity, and investment, as well as citizen participation. *This implies abandoning bureaucratic approaches and involving skills and resources that are external to the traditional administrative apparatus. It also implies strong governance capacity.* The municipal competence and skills needed are, however, different as they depend on the level of financial and governance autonomy of the city.

Skills are needed for the regulatory framework (to effectively guide the market ensuring competition, innovation, and consumer protection), for public procurement (infrastructure for connectivity and software acquisition/development), and for operating/maintaining the various applications at national, regional, and city level. The likelihood is that cities will need more project level skills, whether innovative or more conventional projects rather than technological skills for new tools, privacy, safety, and cybersecurity issues surrounding the novel and emerging technologies for smart cities looming large. But skills will be needed to design effective strategies and enhance the policy capacity to accelerate smart city development; skills will also be needed to ensure participation in large, national infrastructure projects.

The recent study by Tan et al. (2021) revealing a disproportionate emphasis on analytical capacity as opposed to operational capacity and political capacity points towards the direction that cities are still not investing in project management skills, including soft skills.

There is also another skills dimension at city level: *local higher education establishments and research centres* may play a significant role in smart city development, not only through educating the public and the business sector but also through supporting municipalities with contracts complementing lacking skills in the rest of the public sector. In this sense, supporting research at local level is expected to generate indirect benefits for the digital transition.

Civil society: The target of smart cities is to benefit society. There is a chicken-and-egg situation because digitally literate citizens require smart services, while smart services need digitally literate citizens to use them. Yet, of the 85% of citizens using the Internet in 2019, prior to the COVID-19 crisis, only 58% possessed the level of *at least basic digital skills*. Digital skills are the backbone of the digital society, without which one cannot fully benefit from digital technologies. While the current crisis may be having the positive impact of increasing the number of Internet users, the development of digital skills does not come automatically with increased usage (European Commission, 2020a).

Increasing demand for smart cities/digital services means high demand for ICT-proficient employees, but not only. The human capital dimension of the DESI has two sub-dimensions covering 'Internet user skills' and 'advanced skills and development'. The former draws on the European Commission's Digital Skills Indicator, calculated based on the number and complexity of activities involving the use of digital devices and the Internet. The latter includes indicators on ICT specialists and ICT graduates (European Commission, 2020a).

'Technology-pushed' solutions have often failed to engage the citizens and the public authorities themselves, who didn't take ownership of the 'smart' services experimented in this way. A claim for democracy, innovation, and participation is becoming increasingly pressing, establishing the need to 'listen and talk to the streets' and ultimately changing the governance paradigm. These challenges call for a transformation in the way citizens work, live, play, and build their future. Cities are only smart when they manage to take full advantage of the human capital of their citizens (Oliveira, 2016).

Critical authors claim that 'By promoting the skills and capabilities of communities, technologies can be developed and appropriated based on the need in the local context. In other words, it is not the technologies that smart cities need to give significant attention; they need more to focus upon enabling citizens to enhance their capabilities, who then utilise their skills and capabilities to invent and promote the usage of technology while addressing their own problems' (Kummitha & Crutzen, 2017).

Business sector: The business sector is the provider of tools and solutions for smart cities. Yet, industrial policies are mostly conceived and implemented at the national level (at least in Europe), and both hardware and software are produced and sold massively. At city level larger local businesses may be important, as in the case of researchers, incentivising, supporting, and complementing local authorities, while smaller companies can undertake the necessary adaption of mass production solutions and develop local products, but more importantly they can create and organise the platforms needed by the public sector at local level.

A grid like that on Table 1 can be used by the cities to plot the real skills' needs at local level. We claim that the minimum requirements for the co-evolution with smartness are high ICT skills and medium STEM and soft skills for the public administration and more than basic skills for civil society. The blank cells are filled tailor-made to local productive capabilities, size, and ambitions.

4.3 Understanding and Obtaining Skills

Our analysis above suggests that knowledge creation, knowledge adaptation, and knowledge diffusion skills are necessary for smart city development. For the municipalities soft skills are at least as important as technical skills, yet somehow neglected; and civil society skills play as important a role as public sector skills, and they are often neglected as well.

So, the crucial question is: how can the necessary skills be well anticipated, developed/attracted, and maintained in a city keeping in mind that digital knowledge itself is in constant evolution? Skills are created in several ways:

- Formal educational system, which can offer the necessary technical skills, which need constant update with on-the-job training; it is also recently in the process of emphasising soft skills, and the advantage of soft skills is that they do not become obsolete. Formally educated people are quite mobile, easy to attract or lose.
- Self-educated through open or payable Internet courses, some of which are very high quality. There are several caveats with this mode: for one most people do not have the self-discipline to go through the whole array of courses necessary

Table 1 Minimum skill requirements for smart cities (the number of starts denotes the level of proficiency needed)

	Business sector		Public administration			
	Creation	Applications	Research	System design/purchasing	System operation	Civil society
ICT skills			***	***	**	**
Other STEM				**	**	
Soft skills				***	***	

to develop a set of skills, but also, the public sector in particular, is still skewed towards respecting formal qualifications more than real skills.
- New type of schooling like the Ecole 24 model,[2] which is experimental and expensive but may prove a very appropriate mode for future market needs.
- Lifelong learning through reskilling and upskilling.

The OECD states that the countries with well-established vocational and educational training (VET) and apprenticeship programmes have been more effective in holding the line on youth unemployment (OECD, 2016), while the European Commission has put significant emphasis on the Pact for Skills that promotes joint action to maximise the impact of investing in improving existing skills (upskilling) and training in new skills (reskilling). It calls on industry, employers, social partners, chambers of commerce, public authorities, education and training providers, and employment agencies to work together and make a clear commitment to invest in training for all working-age people across the Union. The Pact for Skills is accompanied by a Charter outlining a shared vision from industry, social partners, vocational and educational training (VET) providers, and national, regional, and local authorities as regards quality training. Its main objective is to mobilise resources and incentivise all relevant stakeholders to take real action to upskill and reskill the workforce, by pooling efforts and setting up partnerships supporting green and digital transitions as well as local and regional growth strategies (European Commission, 2020b).

5 Concluding Remarks

Smart cities apply ICT technologies to promote smart governance, smart economy, smart mobility, smart environment, smart living, and smart people. They are dynamic organisms, which need to embark on a successful, evolutionary transition, constantly upgrade, and occasionally reinvent themselves. ICT skills are a necessary but by far not a sufficient condition for this. ICT skill shortage being the main barrier for the development of smart cities is a myth. Soft skills like management, deep learning, teamworking, and critical, innovative, and strategic thinking are at least as important. The digital transition without soft skills will not only be ineffective, but it will also ultimately turn the city into a technocratic, inhuman society.

Agglomeration effects, omnipresent in urban economics, apply fully in the case of the digital transformation attracting educated workers. Skills and agglomeration are connected to a virtuous circle (or vicious if either one is missing). As a consequence, it is critical for municipalities to organise accordingly, hire, and promote employees with a combination of technical and soft skills that will enable the city to procure and operate tools adapted to their needs at a reasonable cost. An

[2] https://42.fr/en/homepage/

integral part of this is a strategy development, tailor-made to local characteristics, and a sine qua non condition is the education and involvement of civil society. This may imply a new mindset for the city, greater ambitions, and higher salaries than in the past. Such an approach is likely to create resistance to change and discourage the willingness to adjust. Yet, if they really want to become smart, cities must mobilise the means, learn from good practices to improve local skills, and attract external skills wherever needed. Opportunities exist; they only need to be grasped, with whatever challenges and risks this implies. Cities that shy in front of these challenges may be doomed to be constantly falling behind.

Acknowledgements We would like to acknowledge the European Commission programme Intelligent Cities Challenge (ICC) which inspired us to write this chapter.

References

Alexopoulos, C., Pereira, G.V., Charalabidis, Y. & Madrid, L. (2019). A taxonomy of smart cities initiatives. In *Proceedings of the 12th international conference on theory and practice of electronic governance* (pp. 281–290). doi: https://doi.org/10.1145/3326365.3326402

Ananiadou, K., & Claro, M. (2009). *21st century skills and competences for new millennium learners in OECD countries (OECD education working papers No. 41)*. OECD Publishing.

Avdeeva, E., Davydova, T., Skripnikova, N., & Kochetova, L. (2019). Human resource development in the implementation of the concept of "smart cities". *E3S Web of Conference* (pp. 110). doi: https://doi.org/10.1051/e3sconf/201911002139

Baccarne, B., Schuurman, D., Mechant, P., & De Marez, L. (2014). The role of urban living labs in a smart city. In *XXV ISPIM innovation conference, proceedings*.

Batty, M., Axhausen, K. W., Giannotti, F., Pozdnoukhov, A., Bazzani, A., Wachowicz, M., Ouzounis, G., & Portugali, Y. (2012). Smart cities of the future. *The European Physical Journal Special Topics, 214*(1), 481–518. https://doi.org/10.1140/epjst/e2012-01703-3

Berrone, P., & Ricart, J. E. (2020). *IESE cities in motion index*. Retrieved October 10, 2021, from https://media.iese.edu/research/pdfs/ST-0542-E.pdf

Binkley, M., Erstad, O., Herman, J., Raizen, S., Ripley, M., Miller-Ricci, M., & Rumble, M. (2012). Defining twenty-first century skills. In P. Griffin & E. Care (Eds.), *Assessment and teaching of 21st century skills: Methods and approach* (pp. 17–66). Springer. https://doi.org/10.1007/978-94-007-2324-5_2

Breetzke, T., & Flowerday, S. V. (2016). The usability of IVRs for smart city crowdsourcing in developing cities. *The Electronic Journal of Information Systems in Developing Countries, 73*(1), 1–14.

Cavada, M., Hunt, D. V., & Rogers, C. D. (2014). Smart cities: Contradicting definitions and unclear measures. In *World Sustainability Forum*. MDPI AG (pp. 1–12).

Cilliers, L., & Flowerday, S. (2017). Factors that influence the usability of a participatory IVR crowdsourcing system in a smart city. *South African Computer Journal, 29*(3), 16–30. https://doi.org/10.18489/sacj.v29i3.422

Claro, M., Preiss, D. D., San Martín, E., Jara, I., Hinostroza, J. E., Valenzuela, S., Cortes, F., & Nussbaum, M. (2012). Assessment of 21st century ICT skills in Chile: Test design and results from high school level students. *Computers & Education, 59*(3), 1042–1053. https://doi.org/10.1016/j.compedu.2012.04.004

Cocchia, A. (2014). Smart and digital city: A systematic literature review. In R. Dameri & C. Rosenthal-Sabroux (Eds.), *Smart city. Progress in IS*. Springer. https://doi.org/10.1007/978-3-319-06160-3_2

Cohen, B. (2014). *The smartest cities in the world 2015: Methodology*. Retrieved April 9, 2021, from https://www.fastcompany.com/90691393/viral-tiktok-videos-gen-z-lessons

Cong, C., Chen, X. G., & Yang, Q. (2016). *Proceedings of the 2016 3rd international conference on management, education technology and sports science (METSS 2016)* (Vol. 25, pp. 64–69).

Couclelis, H. (2004). The construction of the digital city. *Environment and Planning B: Planning and Design, 31*(1), 5–19. https://doi.org/10.1068/b1299

Dameri, R. P., & Rosenthal-Sabroux, C. (Eds.). (2014). *Smart city: How to create public and economic value with high technology in urban space*. Springer.

Egger, W.D., & Skowron, J. (2018). *Forces of change: Smart cities*. Deloitte. Retrieved April 9, 2021, from https://www2.deloitte.com/us/en/insights/focus/smart-city/overview.html

Etzkowitz, H. (2008). *The triple helix: University–industry–government. Implications for policy and evaluation*. Swedish Institute for Studies in Education and Research.

European Commission. (2020a). *Digital economy and society index (DESI) 2020, Thematic chapters*. Retrieved September 10, 2021, from https://digital-strategy.ec.europa.eu/en/policies/desi

European Commission. (2020b). *The pact for skills: Mobilising all partners to invest in skills*. Retrieved September 10, 2021, from https://ec.europa.eu/social/main.jsp?langId=en&catId=1517&furtherNews=yes&newsId=9827

Giffinger, R., Fertner, C., Kramar, H., & Meijers, E. (2007). *City-ranking of European medium-sized cities* (pp. 1–12). Centre of Regional Science.

Harari, Y. N. (2018). *21 lessons for the 21st century*. Random House.

Hollands, R. G. (2008). Will the real smart city please stand up? Intelligent, progressive or entrepreneurial? *City, 12*(3), 303–320. https://doi.org/10.1080/13604810802479126

Ismagilova, E., Hughes, L., Dwivedi, Y. K., & Raman, K. R. (2019). Smart cities: Advances in research—An information systems perspective. *International Journal of Information Management, 47*, 88–100.

Kolokytha, E., Kolokythas, G., Valsamidis, S., & Florou, G. (2015). The contribution of the open data to the development of the smart cities. *Scientific Bulletin—Economic Sciences, 14*(2), 3–16.

Komninos, N. (2008). *Intelligent cities and globalization of innovation networks*. Routledge.

Kummitha, R. K. R., & Crutzen, N. (2017). How do we understand smart cities? An evolutionary perspective. *Cities, 67*, 43–52. https://doi.org/10.1016/j.cities.2017.04.010

Kupriyanovsky, V., Utkin, N., Nikolaev, D., Yartsev, D., Namiot, D., & Sinyagov, S. (2016). On localization of British standards for smart cities. *International Journal of Open Information Technologies, 4*(7), 13–21.

Levy, F., & Murnane, R. J. (2004). *The new division of labor: How computers are creating the next job market*. Princetown University Press.

Lombardi, P., Giordano, S., Farouh, H., & Yousef, W. (2012). Modelling the smart city performance. *Innovation: The European Journal of Social Science Research, 25*(2), 137–149. https://doi.org/10.1080/13511610.2012.660325

Manville, C., Cochrane, G. Cave, J., Millard, J., Pederson, J. K., Thaarup, R. K., Liebe, A., Wissner, M., Massink, R., & Kotterink, B. (2014). *Mapping smart cities in the EU. European Parliament—Directorate-General for Internal Policies*. Retrieved September 10, 2021, from https://www.europarl.europa.eu/Mapping_Smart_Cities_in_the_EU

Mora, L., & Bolici, R. (2017). How to become a smart city: Learning from Amsterdam. In A. Bisello, D. Vettorato, R. Stephens, & P. Elisei (Eds.), *Smart and sustainable planning for cities and regions. Results of SSPCR 2015. Green energy and technology*. Springer. https://doi.org/10.1007/978-3-319-44899-2_15

Nam, T., & Pardo, T. A. (2011). Conceptualizing smart city with dimensions of technology, people and institutions. In *Proceedings of the 12th annual international digital government research conference: Digital government innovation in challenging times* (pp. 282–291). ACM. https://doi.org/10.1145/2037556.2037602

Namiot, D., Kupriyanovsky, V., Samorodov, A., Karasev, O., Zamolodchikov, D., & Fedorova, N. (2017). Smart cities and education in digital economy. *International Journal of Open Information Technologies, 5*(3), 56–71.

Negre, E., & Rosenthal-Sabroux, C. (2014). Recommendations to improve the smartness of a city. In R. P. Dameri & C. Rosenthal-Sabroux (Eds.), *Smart city. How to create public and economic value with high technology in urban space* (pp. 101–115). Springer. https://doi.org/10.1007/978-3-319-06160-3_5

OECD. (2016). Education at a glance 2016: OECD Indicators, OECD Publishing, . Retrieved September 10, 2021, from https://www.oecd-ilibrary.org/docserver/eag-2016

OECD. (2020). *Measuring smart cities' performance. Do smart cities benefit everyone? 2nd OECD roundtable on smart cities and inclusive growth*. Retrieved September 10, 2021, from https://www.oecd.org/cfe/cities/Smart-cities-measurement-framework-scoping.pdf

Oliveira, A. D. (2016). The human smart cities manifesto: A global perspective. In G. Concilio & F. Rizzo (Eds.), *Human smart cities: Rethinking the interplay between design and planning* (pp. 197–202). Springer. https://doi.org/10.1007/978-3-319-33024-2

Partnership for 21st Century Skills. (2007). *Framework for 21st-century learning*. Retrieved from http://www.p21.org/documents/P21_Framework_Definitions.pdf

Petrova-Antonova, D. & Ilieva, S. (2018). Smart cities evaluation—A survey of performance and sustainability indicators. doi: https://doi.org/10.1109/SEAA.2018.00084

Picatoste, J., Pérez-Ortiz, L., Ruesga-Benito, S. M., & Novo-Corti, I. (2018). Smart cities for wellbeing: Youth employment and their skills on computers. *Journal of Science and Technology Policy Management, 9*(2), 227–241. https://doi.org/10.1108/JSTPM-04-2017-0014

Pissarides, C.A. (2018). The impact of AI on the future of work: Technology and jobs in the digital era. *Royal Society British Academy event: AI and the Future of Work*. Retrieved September 30, 2021, from https://personal.lse.ac.uk/ghatak/Chris1.pdf

Rommes, E., Oost, E. V., & Oudshoorn, N. (1999). Gender in the design of the digital city of Amsterdam. *Information Communication and Society, 2*(4), 476–495. https://doi.org/10.1080/136911899359510

Tan, S.Y., Taeihagh, A., & Li, L. (2021). Designing effective strategies to enhance policy capacity in Smart City development. *5th International Conference on Public Policy (ICPP5) July 5th–9th, 2021–Barcelona, Spain*.

Tryfonas, T., & Crick, T. (2018). Public policy and skills for smart cities: The UK outlook. In PETRA '18: *Proceedings of the 11th Pervasive Technologies Related to Assistive Environments Conference*, pp. 116–117. Association for Computing Machinery (ACM).

United Nations Secretary-General. (2020). *Secretary-General's message on World Cities Day*. Retrieved April 9, 2021, from https://www.un.org/sg/en/content/secretary-generals-message-world-cities-day-scroll-down-for-french-version

Urenio. (2016). *Digital and intelligent cities: Concepts examples and Core technologies*. Retrieved April 9, 2021, from http://www.urenio.org/el/wp-content/uploads/2008/10/Digital__Intelligent_Cities_Concepts_Examples_and_Core_Technologies.pdf

van Laar, E., van Deursen, A. J., van Dijk, J. A., & de Haan, J. (2017). The relation between 21st-century skills and digital skills: A systematic literature review. *Computers in Human Behavior, 72*, 577–588. https://doi.org/10.1016/j.chb.2017.03.010

van Laar, E., van Deursen, A. J., van Dijk, J. A., & de Haan, J. (2020). *Determinants of 21st-century skills for workers skills: A systematic literature review* (pp. 1–14). SAGE Open. https://doi.org/10.1177/2158244019900176

Vanolo, A. (2014). Smartmentality: The smart city as disciplinary strategy. *Urban Studies, 51*(5), 883–898. https://doi.org/10.1177/0042098013494427

Verdict. (2020). *History of smart cities: Timeline*. Retrieved April 9, 2021, from https://www.verdict.co.uk/smart-cities-timeline/

Wikipedia. (n.d.). *Living lab*. Retrieved August, 29, 2021, from https://en.wikipedia.org/wiki/Living_lab

Yovanof, G. S., & Hazapis, G. N. (2009). An architectural framework and enabling wireless technologies for digital cities & intelligent urban environments. *Wireless Personal Communications, 49*(3), 445–463. https://doi.org/10.1007/s11277-009-9693-4

Preparing for Smart Cities' Future Competences: Trends Arising Through Keyword and Review Analysis

Paraskevi Tsoutsa and Ioannis Ch. Lampropoulos

1 Introduction

People nowadays require key competences in their daily operations and transactions, given the rapidity of recent technological breakthroughs and their influence in all fields. The demand for new essential competences in a lifelong context has raised high expectations for education and lifelong learning while also broadening the range of competences required.

The concepts "competence" and "skill" are often used as synonyms and in the literature, there are numerous definitions available for these words. We adopt the definition of Cedefop (2015); a competence is the "ability to apply learning outcomes adequately in a defined context (education, work, personal, or professional development) or the ability to use knowledge, skills and personal, social and/or methodological abilities, in work or study situations for professional and personal development."

The most important skills are influenced by the current business environment, the employment market, the rising information economy, the enhanced globalization, and task automation (Cedefop, 2019; Kotak & O'Neill, 2021; Semeijn & Nikolova, 2021; Tsoutsa et al., 2018). Gaining competences plays an exponential function in anticipating the global labor patterns of the near future in this disruptive landscape (Iliescu, 2021; Fitsilis et al., 2018). In education, rapid social, economic, and technological changes have prompted the development of a flexible and educable workforce, resulting in the redefinition of educational approaches and competences that learners should acquire at various levels of their education in order to become

P. Tsoutsa (✉) · I. C. Lampropoulos
University of Thessaly, Thessaly, Greece
e-mail: ptsoutsa@uth.gr; ilampropoulos@uth.gr

P. Fitsilis (ed.), *Building on Smart Cities Skills and Competences*, Internet of Things,
https://doi.org/10.1007/978-3-030-97818-1_3

competent citizens and significant contributors to their countries' development (EC, 2006).

This research was conducted to extort and present recent data on competency issues and trends in a dynamic, unpredictable, complex, and ambiguous ecosystem such as smart cities that reveal the points of convergence among diverse domains and research disciplines. We analyze the concepts that derive from the keywords that the literature production contains about skills and competences. Keywords analysis is staged to visualize the research patterns in domains. We present a co-occurrence map for author keywords and additional keywords, and the results are segmented into sections, where each section is discussed. Web of Science (WOS), which provides access to multiple databases that provide comprehensive data for many different academic disciplines, was used to collect publications on the topic. Our goal was to get an overview of the authors' keywords used, and we hoped to identify subtopics by using VOSviewer (VoS) (Van Eck & Waltman, 2010). The terms competences (or synonyms) and smart city (or similar terms) had to appear in the topic, title, abstract, or keywords of last decade's articles. By the keywords were identified using VoS, we selected 110 keywords. The map revealed many subtopics and some relatively new concepts related to key competences in selected studies. Another objective of this research is to determine which are the primary domains where skills are researched, as well as any future trends that may emerge from these domains. As a result, we continued by analyzing the competence review papers from the previous 2 years and present our findings.

This work contributes to the theoretical understanding on the topic of competences by giving a keyword map of the competence notion. The co-occurrence analysis of related phrases will at the very least provide a clearer knowledge of how these terms relate to competences, whereas the study of literature reviews will characterize the current and future dynamics that competences may have in many areas.

The rest of this paper is structured as follows: In Sect. 2 related work is presented, Sect. 3 describes the method which was followed and the procedure for searching, identifying, and selecting the articles, Sect. 4 provides the results, and the last section concludes the paper.

2 Related Work on Competences

Competence has a multidisciplinary history, which is evident today as many disciplines are interested in its definition and application. For the last decade, the competence topic is an engaging one, generating interest among researchers.

As the amount of research on intercultural competence (ICC) has increased over the last two decades, a thorough evaluation of the various literature and its growth process is doing in research by Peng et al. (2020). The study conducted a bibliometric analysis for the knowledge domain of intercultural competence. Through analysis, they provide empirical observation from multiple perspectives in intercultural competence research. According to their results, there is an emphasis and future direction of research for scholars, which is conducive to the expansion of the researchers' ideas and the in-depth exploration in pedagogy, management, medicine, and neurology.

In Antera (2021), the many interpretations of the concept of professional competence are analyzed and studied in relation to vocational teaching. The researcher discovers similar concept attributes as well as nearby ideas related with professional competence using a conceptual analysis approach, which follows the data collecting procedure of a systematic literature study. According to her findings, only a few studies provide adequate professional competence concept definitions. Furthermore, the scholars agree on the major characteristics of professional competence, such as the contextual and developmental nature of professional competence. Since complex concepts like the one under consideration can generate misunderstanding, the author recommended that their use should be accompanied by a description of its multiple meanings.

Zait (2017), in his study, attempt to identify the main necessary competences for smart cities' development regarding civilizational competences and their effect in smart cities' development. Civilizational competences, soft skills, or human-related characteristics of cities highly influenced by culture (at national, regional, organizational, and individual levels), according to their findings, are critical for the development of smart, competitive cities. They group the civilizational competences into four categories, enterprise culture, discursive culture, civic culture, and daily culture and argue that in order to develop smart cities, we must first define them, then determine their antecedents or influencing factors, and last measure them.

The literature of competences encompasses many surveys, yet all emphasize specific areas and methodologies used which provides a distinction between each context the concept is used, although the research is all part about the same notion. This study is the first survey that focuses on competences in whatever area of the smart city domain, covering papers published in journals indexed by the Web of Science database after 2010.

3 Research Method

Bibliometric is a statistical technique for analyzing bibliographic data from articles and books, such as titles, keywords, authors, and cited references. It is used to measure the productivity of institutions and countries, as well as define current trends and forecast future research foci. The term bibliometric was first coined by Alan Pritchard in his paper "Statistical Bibliography or Bibliometrics" (Pritchard, 1969). For this study it used the bibliometric technique to "analyze and illustrate the literature on the research topics of 'Skills' and 'Competences'."

3.1 Data Source and Search Strategy

3.1.1 Data Source

The data source was built using secondary data by journal and conference articles that were identified by searching the online database Web of Science (WOS) for the time span from 2010 to 2021. The choice of this database is due to the global reputation this instrument has, as it represents a main source for finding publications with the greatest impact while providing data for bibliometric analysis. The analysis tool used in the study was VOSviewer.

3.1.2 Procedure for Searching, Identifying, and Selecting Articles

An advanced search was conducted for the retrieval of data, and the inclusion criteria for the selection of relevant articles that were taken into consideration for the performance of this research are:

(i) Published after 2010 and including August 2021 in order to extract the most current research and trends in this field.
(ii) Contain the specified search descriptors (skill, competences, smart cities) and their synonyms (e.g., intelligent cities) either in title, keywords, or abstract.
(iii) Are related to the field of human skill and competences.
(iv) Are related to skills in professional development at any stage of the education system or lifelong learning.

The following exclusion criteria were applied: (i) studies in competences aside from humans; (ii) unpublished data or not published in conference papers, book chapters, and journal articles; (iii) studies with animals; (iv) studies including participants with disabilities, diseases, or disorders; and (v) not written in English language.

3.1.3 Search Summary

A total of 152 relevant articles were identified in the database using the aforementioned search strategy as it is depicted in Fig. 1. By excluding duplicates, the total number of articles was reduced to 146, and these are selected to be included in the research.

The data retrieved from this first phase studies were first interconnected and metrically presented using the VoS software. Following that, at a second phase of the research, only review papers were selected published between January 2019 and August 2021, and an analysis was conducted as it is presented in Sect. 4.2.

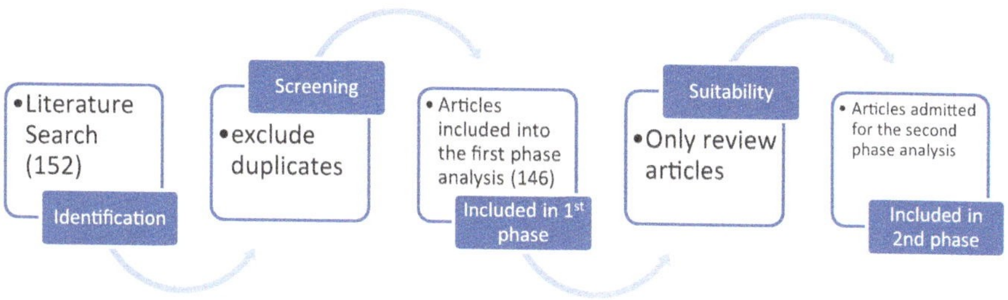

Fig. 1 Flow diagram for selection of articles

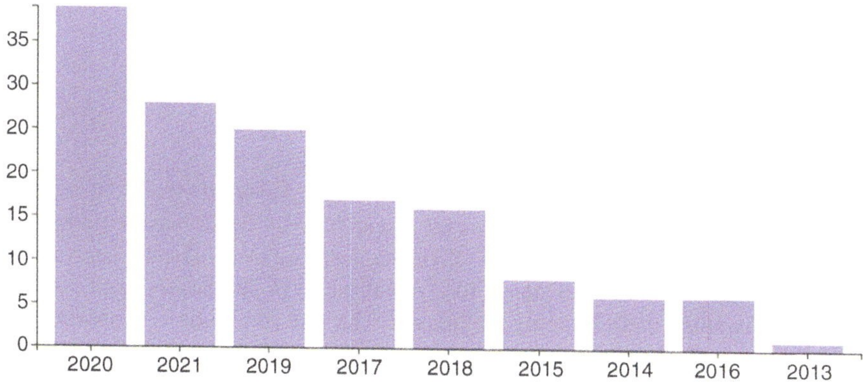

Fig. 2 Annual scientific production per year (Source: WOS)

3.1.4 Data Extraction

The information that were extracted from each article included: the names of the first three authors, year of publication, title, keywords, type of study, domain of research, and abstract. All retrieved data were entered in a spreadsheet and merged into a single file.

Then, data were imported into the VoS software, they were analyzed and bibliometric products (maps, graphs, spreadsheets) were generated. In the analysis and interpretation of developed studies, descriptive, quantitative, and correlational techniques have been combined with the semantic application of keyword study through the VoS software. Content analysis was performed to reduce the amount of data and identify category clusters (Patton, 1990). Through the analysis, a comparative analysis of articles and concepts has been carried out, and their visual representation has been conducted.

It is clear that there is a growing interest in the publications on the topic over time, since we see an evolution from the first articles appeared in 2013 (Fig. 2).

To best serve the research aims, we present in the next section how we examine and determine from the dedicated literature the intellectual correlations of the competency concept. This will allow us to examine the existing ties between the notion of competence and other significant concepts, as well as emphasizing the most crucial relationships that we have uncovered between them.

4 Data Analysis Presentation and Discussion

The findings are presented in two stages. In Sect. 4.1 we show the network graph produced by the analysis of keywords, thereby responding to the first objective of the study. In Sect. 4.2, we discuss on the analysis that was done on review studies on competences over the previous 2 years.

4.1 Keyword Clustering and Classification

This section covers the aspect of the analysis where keywords from the articles are clustered to perhaps unveil dominant research areas which have emerged in the past studies. We have used the VOSviewer software to exhibit the analysis of keywords required. The VOS stands for "visualization of similarities" and uses its exclusive mapping and clustering techniques. The size of a node is proportionate to the number of occurrences of a keyword, and the link between nodes represents the number of co-occurrences of these keywords. The higher the number of co-occurrences between two keywords, the lesser is the distance between them. A thesauruses file was developed and used to eliminate the problem of acronyms, plurals, dashes, etc. in the keywords.

The resulting map is highlighted in Fig. 3, which shows that after mapping and clustering operations, there were four decisively formed distinct clusters from 110 clustered keywords. The figure displays the co-occurrence for the most cited and strongly correlated keywords. The circle color shows the thematic cluster each keyword belongs to, and the size of nodes indicates the frequency of occurrence. The curves between the nodes represent their co-occurrence in the same publication. The shorter the distance between two nodes, the larger the number of co-occurrence of the two keywords. The distinct clusters and the keywords each one contains indicate the trends for the study of related terms used by researchers in their discipline. The keywords in each resulted group are semantically linked to target the research papers that are also related to more terms of this group, rather than just focusing on targeted papers simply containing the specific keyword:

- Cluster 1—Smart city (in red): The "smart city" term registers the most substantial values here with the terms learning, system, environment, technology, data, and economy to follow. Other important keywords to highlight in this cluster are network, platform, IoT, and communication technologies. It is the cluster with the largest number of terms and occurrences, which are 25 and 389, respectively.

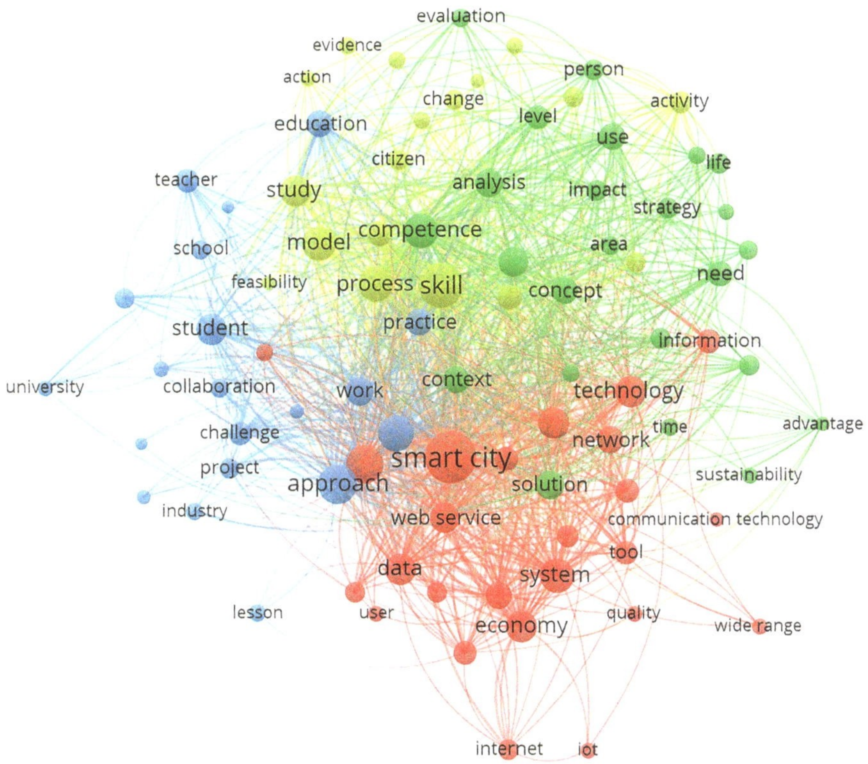

Fig. 3 Graph of clustered keywords and themes

- Cluster 2—Knowledge (in green): The term competence stands out in this cluster strongly related to the knowledge and analysis terms that follow. The knowledge term also gives the name of the cluster. Although it is the 16th most encountered term across all analyzed documents, it is behind "competence" in this cluster. More keywords to highlight here are community, management, services, and performance sustainability and advantage.
- Cluster 3–Education (in blue): In this cluster we meet terms regarding the education like learning, teacher, student, and curriculum. We found related articles proposing innovating practices with the use of ICT or different digital courses that are useful for improving the learning-teaching process of students in the context of smart city.
- Cluster 4—Skill (in yellow): The term skill as synonym to the term competence in cluster 1 counts 46 appearances contrary to 26. It is the second most encountered term across all analyzed documents behind "smart city." It is strongly related to terms study, process, activity, experience, and assessment that are also present in this cluster. It is also related to the competence and study concepts from different clusters. Although this cluster has less terms than all other ones, it totally has more occurrences than cluster 3.

Table 1 Top ten keywords for each theme

	Label	Oc	L	LS		Label	Oc	L	LS
Cluster 1: Smart city	Smart city	59	86	987	Cluster 3: Education	Approach	32	84	541
	Learning	30	85	507	Development	29	85	541	
	System	27	84	438	Student	22	82	390	
	Environment	22	85	436	Work	18	82	313	
	Technology	22	82	395	Education	16	82	344	
	Data	21	82	344	Practice	15	77	262	
	Economy	21	82	358	Challenge	12	76	239	
	Web service	19	82	370	Teacher	12	75	250	
	Framework	15	82	280	Project	10	71	198	
	Network	15	78	243	Curriculum	9	63	160	
Cluster 2: Knowledge	Competence	26	85	465	Cluster 4: Skill	Skill	46	86	723
	Knowledge	20	83	358	Process	28	84	537	
	Analysis	17	83	329	Model	25	82	404	
	Solution	17	83	332	Study	22	84	367	
	Concept	16	81	323	Experience	14	78	238	
	Context	16	82	273	Research	14	82	258	
	Use	15	81	308	Activity	11	68	201	
	Need	14	80	299	Assessment	10	67	157	
	Level	13	78	250	Opportunity	10	74	189	
	Implementation	10	75	211	Change	9	74	174	

Oc Occurrences, *L* Links, *LS* Link strength

Table 1 presents for each cluster the top ten most frequently used phrases, listed from most frequently used to least frequently used assigned by VOSviewer under each cluster, as well as the occurrences, links, and link strength value for each of the terms.

In Fig. 4 we see the most co-occurred terms in all studies.

In Fig. 5, we can observe the representation of the density overview of the clusters, broadcasting the most visited concepts in the literature, correlated with the competence and skill concept. By default, colors range from blue to green to yellow. The terms "smart city," "skill," "competence," "economy," "system," "study," "student," and "technology" have the most visible hallows on the map, and this is in alignment with their leading clusters' positions and highest values in their cluster when it comes to the occurrences. It is also interesting to note the appropriation between items belonging to different clusters. In this respect, we notice how "competence" (cluster 2) is very close to "skill" (cluster 4), since they are frequently used interchangeably and are considered equal, which is very close to "education" (cluster 3).

The density of yellow circles determines the most prominent relationships by strongly correlated keywords which lead to sets of concepts that are studied together

Fig. 4 Most cited terms

and could reveal specific competences. By combining Fig. 3 and Fig. 4, we can observe representations of networks of correlated concepts that could abstractly describe competences to be researched in smart city domains. One such network contains the concepts smart city, web services, data, and users. Further research will enable us to validate that the co-occurrence of concepts leads to the need of developing composite competences. For example, the correlation of concepts such as: smart city, network, management, system, and economy could allow us to conclude that a competence for digital economy management may be needed.

4.2 Most Reviewed Domains

While the first specified aim of the research has been reached in the previous section, a complementary analysis process was conducted to achieve the second defined objective. Our intention was to highlight the areas of interest which is studied by researchers in the field of competences independent of the smart city, in order this

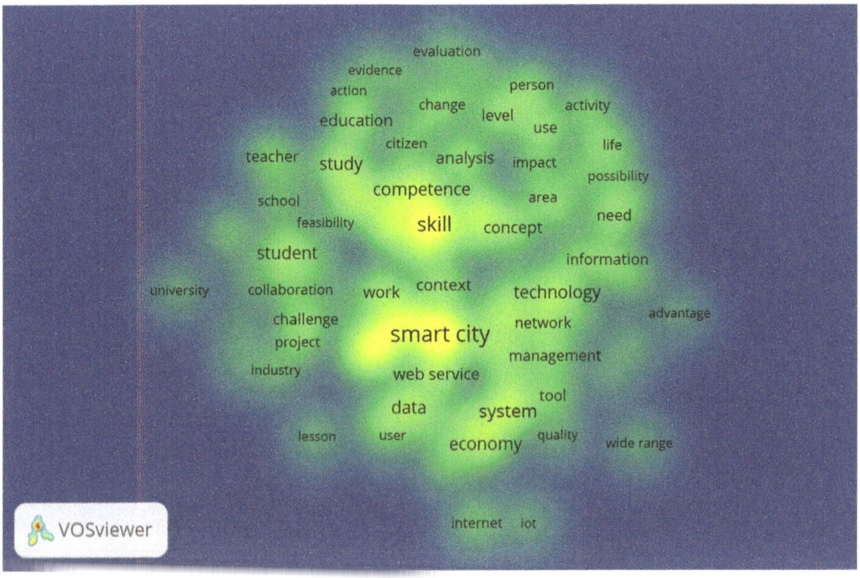

Fig. 5 Density visualization of the clustered keywords

to be an inspiration for exploration in the most specific field of the smart city. Undoubtedly, it is clear that the majority of the resulting domains are relevant to the field of interest of which is smart city competences. More precisely, in order to gain understanding of the different domains, the terms skill and competences are reviewed; we conducted an analysis of the literature review articles published between January 2020 and August 2021, without taking into account the filter for the smart city domain. Each review was critically appraised by the authors of the paper. By analyzing only the review articles, it is observed that the majority of the articles are concerned with the health field and especially the university education of the health executives which is based both on the level of digital capacity and on the adjustment of the programs of studies. Especially during the situation of Covid-19 pandemic, it seems that the areas in which most research has been recorded can contribute to the scientific community for further research so that, by each subject of education, the scientific community can offer all the necessary directions that will be used by government agencies for the maximum efficiency of human capacity per object of activity.

Figure 6 shows the research areas where it is observed that the fields of health and education are the sciences that are particularly active in the study of skills in these areas.

The advent of Covid-19 in connection with the rapid growth of technology has led to digital capability as one of the burning issues (Zhao et al., 2021). Digital competence is one of the main skills in the education system in order to acquire the appropriate knowledge, attitudes, and skills in a safe, appropriate, and decisive way

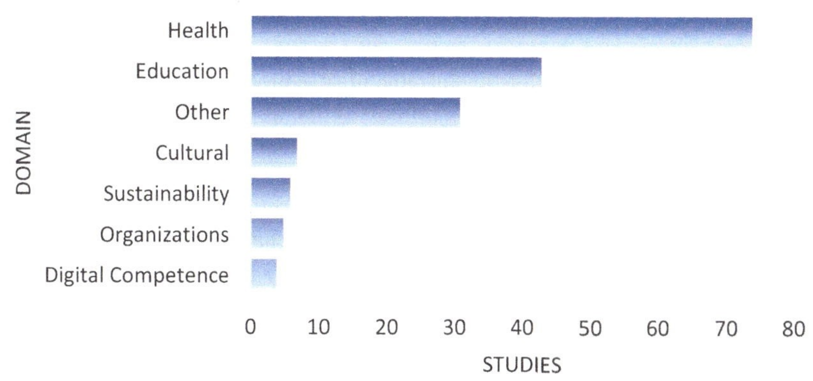

Fig. 6 Areas of research for competence review studies

that will contribute to both education and the professional sectors (Esteve-Mon et al., 2020). In the field of health sciences, Jimenez et al. (2021) urge the scientific community to find alternative ways of providing high levels of public health care, while Antera (2021) shows the need for a complex tool that will facilitate the design of knowledge, intelligence, performance, and skills that will allow measurement and study.

At the university level, according to Pedromo and Gonzalez-Martinez (2020), most of the recorded research is in Spain, while in Latin American countries, a small range of work is presented, although digital skills in higher education show a growing trend due to the demand for information technology and communications (Ocana-Fernandez et al., 2020). The research of Fernandez Batancro et al. (2020) emphasize the importance of digital competence and the lack of teacher training in information communication technologies. University professors, in particular, must be highly skilled in order to face the new digital society and its new challenges (Esteve-Mon et al., 2020), while Kiat Bong and Chen (2021) argue that digital access must be student-free. Regarding the professional competence of teachers, the work of Antera (2021) shows a gap in the study of professional competence at a collective level, while Ingrid (2021) emphasizes the gap and at the same time the importance of socio-emotional skills in education and professional success. Socio-emotional skills are important for health and well-being in social relationships and life in general.

Tarraga-Minguez et al. (2021) highlight a significant gap in teacher training programs in digital competence. Curricula need to be adjusted to find possible ways of teaching and assessment which will reduce the gap between school, society, and work (Gonzalez-Salamanca et al., 2020; Tsoutsa et al., 2022a). For the curricula, Galleli et al. (2020) propose a revision of the curricula in order to harmonize them with the organizational framework that connects higher education with the labor market.

Particular emphasis is placed on research and sustainability where it is a highly developing field for viability and skills, with teachers playing an important role in their transformational skills for sustainability education. Teachers perform a particular task that requires different abilities, skills, as well as critical knowledge of factors related to environmental injustice either in society or within the school environment (Corres et al., 2020). The role of teachers in sustainable development aims at the development of tomorrow's citizens who will have to take responsible action in the future in order to solve the problem (Chen & Liu, 2020). Capacity for environmental sustainability according to Dzhengiz and Niesten (2020) is also determined by managers who if they have the necessary knowledge and develop environmental skills can improve environmental sustainability.

Additionally, the study of the selected articles identifies various gaps that need to be further explored in order to be able to develop specific strategies in the abilities of people, aiming at a process of effective interaction between people and the environment. Liu et al. (2020) propose further research for a systematic evaluation of the effectiveness of the framework and the investigation of innovative educational interventions that will ensure that the training of both students and professionals will be timely and effective. Javier and Purifivacion (2020) and Salmon et al. (2020) also support the importance of further research training and moreover, the need to update educational programs from the initial stages of the educational process, so that no stage of the educational system is excluded.

Minarevic and Tokic Zec (2021) emphasize the importance of further study in the attitudes of students and teachers by comparing practices with different countries being implemented around the world. Schoon (2021) in the context of exploring socio-emotional abilities emphasizes the importance of holistic evaluation in all cultures in order to reflect the formation, development, and possible changes that need to be made. Regarding the integration of technology in teacher training programs, Fernandez-Batanero et al. (2020) emphasize the importance of reviewing curricula in order for future professionals to be fully trained.

In the context of sustainability, Galleli et al. (2020) propose to conduct theoretical studies that will study human skills related to sustainability by incorporating management models that will combine strategy, organizational, and human skills.

In conclusion, the evaluation of the study articles on skills results in interesting analyses without emphasizing the gap in the literature by the researchers and the proposal of specific studies that will contribute significantly to the field of skills development.

5 Conclusions and Future Work

The current growth trends predict a large increase in the number of publications on smart cities' competences. It is particularly important to benefit from such a great number of research articles about skills and omnipotences and gather useful data. The goal of this research was twofold, firstly to address through bibliometric analysis the emerging of different issues and future trends in smart city competences and secondly to go over recent competency reviews on the field and reveal the most important thesis and motivate relevant research in smart cities' competences.

The Web of Science core collection database was used as the data source, and VoS was applied for analysis of the metadata of the retrieved articles. The tendency of research in the field of skills and competences with the analysis of scientific collaboration network and keyword co-occurrence is depicted in networked graphs by using cluster analysis. Through the analysis of keywords, they identified four clusters of themes covered by the research study about competences in smart cities. With this work we depict a different aspect of the skill and competence researched bibliography in the field of smart cities. We have developed a network and a map which are based on the keywords given by the authors of all relevant research articles with the aforementioned concepts during the last decade. These graphs primarily justify the main concepts that are at the heart of the relevant research, and they can be used by researchers to find additional search terms or to limit the scope of their topic. Moreover, they can conclude other research interests that scientists research in different disciplines for the same subject.

To reach the second indicated goal, a complementary analysis technique was used. According to a survey of reviews, the most investigated domains in terms of skills are health and education. Special mention is given to the fields of education and digital skills as well as the study of intercultural behaviors. In addition to the education sector, which seems to have occupied the scientific community, the health sciences sector also has a wide range of literature studies, which is of research interest in emerging emergencies such as Covid-19.

As future work we plan to investigate a larger set of concerns and challenges that arise in the field of continuous professional development, which many sources emphasize (Cedefop, 2015; Fitsilis & Kokkinaki, 2021) and the importance of having a strategy to implement a unified continuum of professional development by exploiting appropriate tools (Tsoutsa et al., 2022b) and technologies.

Acknowledgments This work was supported by the Erasmus+ KA2 under the project "DevOps competences for Smart Cities" (Project No.: 601015-EPP-1-2018-1-EL-EPPKA2-SSA).[1] This paper reflects the views only of the authors, and the Commission cannot be held responsible for any use which may be made of the information contained therein.

References

Antera, S. (2021). Professional competence of vocational teachers: A conceptual review. *Vocations and Learning*. https://doi.org/10.1007/s12186-021-09271-7

Cedefop. (2015). *Skills, qualifications and jobs in the EU: The making of a perfect match? Cedefop reference series No. 103*. Office for the Official Publications of the European Union. Retrieved from http://www.cedefop.europa.eu/en/publications-and-resources/publications/3072

[1] SmartDevOps project w1ebsite is https://smartdevops.eu

Cedefop. (2019). *Skills for green jobs: 2018 update*. Retrieved from https://www.cedefop.europa.eu/en/publications-and-resources/publications/3078

Chen, S.-Y., & Liu, S.-Y. (2020). Developing students' action competence for a sustainable future: A review of educational research. *MDPI, 12*(4).

Corres, A., Rieckmann, M., Espasa, A., & Ruiz-Mallen, I. (2020). Review educator competences in sustainability education: A systematic review of frameworks. *MDPI, 12*(23).

Dzhengiz, T., & Niesten, E. (2020). Competences for environmental sustainability: A systematic review on the impact of absorptive capacity and capabilities. *Journal of Business Ethics, 162*, 881–906.

EC. (2006). *Key competences for lifelong learning /online/*. Retrieved from http://eur-lex.europa.eu/LexUriServ/LexUriServ.do?uri=CELEX:32006H0962:EN:HTML

Esteve-Mon, F. M., Llopis-Nebot, M. A., & Adell-Segura, J. (2020). Digital teaching competence of university teachers: A systematic review of the literature. *IEEE Revista Iberoamericana de Tecnologias del Aprendizaje, 15*(4), 399–406.

Fernandez-Batanero, J. M., Montenegro-Rueda, M. M., Fernandez-Cecero, J., & Garcia-Martinez, I. (2020). Digital competences for teacher professional development. Systematic review. *European Journal of Teacher Education*. https://doi.org/10.1080/02619768.2020.1827389

Fitsilis, P., Kokkinaki, A. (2021). *Smart cities body of knowledge*. Retrieved from https://smartdevops.eu/scbok/.

Fitsilis, P., Tsoutsa, P., & Gerogiannis, V. (2018). Industry 4.0: Required personnel competences. Industry 4.0, 3(3), 130-133. In *3rd International scientific conference on industry 4.0*. Varna.

Galleli, B., Hourneaux, J., & Munck, L. (2020). Sustainability and human competences: A systematic literature review. *Benchmarking: An International Journal, 27*(7).

Gonzalez-Salamanca, J. C., Lucia Agudelo, O., & Salinas, J. (2020). Key competences, education for sustainable development and strategies for the development of 21st century skills. A systematic literature review. *MDPI, 12*(24).

Iliescu, A. N. (2021). Conceptual atlas of the knowmad literature: Visual mapping with VOSviewer. *Management Dynamics in the Knowledge Economy, 9*(3), 379–392.

Ingrid, S. (2021). Towards an integrative taxonomy of social-emotional competences. *Frontiers in Psychology, 12*.

Javier, C., & Purifivacion, C. (2020). From competences to key competences in early childhood education. Comparative and upgrading of the competences in the curriculum. *Propósitos y Representaciones, 8*(1).

Jimenez, C. L., Lopez Poyato, M., Casado Montanes, I., Guix-Comellas, E. M., & Fablrellas, N. (2021). Paediatric nursing clinical competences in primary healthcare: A systematic review. *Journal of Advanced Nursing, 77*(6), 2662–2679.

Kiat Bong, W., & Chen, W. (2021). Increasing faculty's competence in digital accessibility for inclusive education: A systematic literature review. *International Journal of Inclusive Education*. https://doi.org/10.1080/13603116.2021.1937344

Kotak, A., & O'Neill, K. (2021). *Work in progress: Emerging smart city occupations*. Information and Communications Technology Council (ICTC).

Liu, J., Gill, E., & Li, S. (2020). Revisiting cultural competence. *The Clinical Teacher, 18*(2), 191–197.

Minarevic, V., & Tokic Zec, R. (2021). Intercultural competences in initial teacher education—Comparative analysis. *Croatian Journal of Education, 22*(4).

Ocana-Fernandez, Y., Valenzuela-Fernandez, L., & Morillo-Flores, J. (2020). The digital competence in the university teacher. *Propositos Representaciones, 8*(1).

Patton, M. Q. (1990). *Qualitative evaluation and research methods* (2nd ed.). Sage Publications.

Pedromo, B., & Gonzalez-Martinez, O. A. (2020). Digital competences in faculties: A systematic review. *EDMETIC, 9*(2), 92–115.

Peng, R. Z., Zhu, C., & Wu, W. P. (2020). Visualizing the knowledge domain of intercultural competence research: A bibliometric analysis. *International Journal of Intercultural Relations, 74*, 58–68.

Pritchard, A. (1969). Statistical bibliography or bibliometrics. *Journal of Documentation, 25*(4), 348–349.
Salmon, M., Ahmad, G., & Saleem, I. (2020). The concept of competence: A thematic review and discussion. *European Journal of Training and Development, 44*(6/7), 717–742.
Schoon, I. (2021). Towards an integrative taxonomy of social-emotional competences. *Frontiers in Psychology, 12*.
Semeijn, J. H., & Nikolova, I. (2021). Career challenges in smart cities: A sociotechnical systems view on sustainable careers. *Human Relations, 74*(5), 656–677.
Tarraga-Minguez, R., Suarez-Guerrero, C., & Sanz-Cervera, P. (2021). Digital teaching competence evaluation of pre-service teachers in Spain: A review study. *IEEE Revista Iberoamericana de Tecnologias del Aprendizaje, 16*(1).
Tsoutsa, P., Damasiotis, V., Tsifora, E. (2022a). Accounting students perceptions from internship that trigger adaptations in training after Pandemia. Challenges and emerging strategies for global networking post COVID-19, January 2022, ISBN: 9781799888567.
Tsoutsa, P., Fitsilis, P., & Iatrellis, O. (2022b). Towards an ontology for smart city competences. In *Proceedings of the 25th pan-Hellenic conference on informatics (PCI 2021)*, ACM.
Tsoutsa, P., Fitsilis, P., & Ragos, O. (2018). Services simulation in industry 4.0: A comparison of simulators. In *3rd International scientific conference on industry 4.0*. Varna.
Van Eck, N. J., & Waltman, L. (2010). Software survey: VOSviewer, a computer program for bibliometric mapping. *Scientometrics, 84*(2), 523–538.
Zait, A. (2017). Exploring the role of civilizational competences for smart cities' development. Transforming government: People, process and policy.
Zhao, Y., Llorente, A. M. P., & Cruz Sanchez Gomez, M. (2021). Digital competence in higher education research: A systematic literature review. *Computers & Education, 168*.

Building Smart City Knowledge and Competences Using Problem-Based Learning in a Blended Learning Environment

Alina Bockshecker, Katharina Ebner, and Stefan Smolnik

1 Introduction

With roughly 70% of the world's humans living in cities by 2050 (United Nations, 2018), dramatic and far-reaching changes to cities are expected. The cities of the future will have to cope with a strong increase in traffic, waste, energy consumption, noise, and pollution. By using modern information and communication technologies (ICT), smart cities offer a promising perspective to handle the challenges induced by the aforementioned urbanization (Giffinger et al., 2007). Following the work of Giffinger et al. (2007), we consider a city as smart if it is able to deliver outstanding, future-oriented performance and services in the six domains smart governance, smart economy, smart people, smart mobility, smart environment, and smart living. In addition, a smart city is characterized by involving relevant stakeholders into the necessary change and decision processes (Jaekel & Bronnert, 2013; Marrone & Hammerle, 2018). This involvement requires two fundamental things from the stakeholders. First, a high amount of engagement is necessary to have the relevant stakeholders actually involved (Ebner et al., 2019). Second, the stakeholders need a specific set of competences concerning the different domains of a smart city to actually and meaningfully contribute to the transformation from cities to smart cities (Zakirova et al., 2021; Baltac, 2019). The proposed chapter will deal with this important second aspect.

In this chapter, we introduce an education concept that addresses the digital transformation of cities to smart cities from both a conceptual and a didactical education perspective. Our key argument is that change processes in cities are not just a matter of content but especially a matter of debate, discussion, and

A. Bockshecker · K. Ebner (✉) · S. Smolnik
Chair of Business Information Systems, University of Hagen, Hagen, Germany
e-mail: alina.bockshecker@fernuni-hagen.de; katharina.ebner@fernuni-hagen.de; stefan.smolnik@fernuni-hagen.de

P. Fitsilis (ed.), *Building on Smart Cities Skills and Competences*, Internet of Things,
https://doi.org/10.1007/978-3-030-97818-1_4

negotiation: most of the changes smart cities bring into cities will, at first, stimulate discussions, and concerns, and, if the concerns are not addressed, finally, even result in resistance. Stakeholders acting in such processes therefore need competences that allow them to quickly dive into the conceptual dimensions of specific activities while at the same time being able to quickly internalize and understand the positions of other stakeholders. Finally, stakeholders need competences allowing them to work constructively toward meaningful, socially accepted solutions for smart city problems that are both feasible and financeable.

Our insights base on the learning unit "Digitalization of Cities and Traffic" of the module "Digital Transformation" that we offer since 2019 at the University of Hagen, a German distance-teaching university, for both bachelor and master students. We will introduce the specific problem-based learning concept used in this learning unit, which centers around a virtual group work, in which eight students take over different roles and work on solutions for real-world-inspired complex smart city cases. The cases address very different topics in the six smart city domains, such as car and e-scooter sharing, autonomous vehicles, digital schools and technostress, human-robot co-working, chatbots as governmental service agents, smart buildings, new uses of existing places (e.g., old airports or shopping malls), new forms of shopping (e.g., using beacons, location-based services, augmented reality), package drones, etc. The roles in every case represent typical stakeholders, such as mayor, entrepreneurs, and environmentalists, archetypically representing positive, negative, or neutral positions toward the smart city topics of each problem situation. In line with problem-based learning (Müller Werder, 2013; Schmidt, 1983), the cases and roles leave the students with contradictory and sometimes seemingly irreconcilable views on the topics. Based on the conflicting positions, the students are ought to design a solution or find a compromise in the sense of a concrete project or process proposal for the fictitious smart city "Neuhagen."

The challenges of any distance-learning setting involve, first, to assure that relevant competences are taught comparably intensively and qualitatively as in presence teaching settings (Freeman & Urbaczewski, 2019) and, second, to secure the persistence of the learned contents (Xu & Jaggars, 2013). Beyond the addressed smart city contents, we will therefore also introduce the didactic mechanisms to handle these challenges of distance learning and teaching as well as introduce the technical environment we use (e.g., virtual collaboration rooms like Zoom, Jitsi, Moodle, etc.) to connect with the students and to facilitate the virtual group work. We also present the results of different evaluations of the setup (based on exam results, activity statistics, structured and verbal feedback), reflect on the learnings in every semester since the course started, and describe how we have been and still are incorporating the feedback.

Following, we present the relevant background of our education concept, involving smart cities and required smart city competences as well as problem-based learning in Sect. 2. The design requirements underlying our concept are presented in Sect. 3. In Sect. 4, we introduce the applied process for designing the education concept, and present the concept in Sect. 5. In particular, we outline the module structure and introduce the smart city problem situations. We finish this chapter with an evaluation (Sect. 6) and a conclusion (Sect. 7) discussing the transferability of the concept to other smart city education contexts.

2 Problem-Based Learning as a Concept for Teaching the Contents of Smart Cities and Required Competences

Urbanization and digitization of the city and traffic are associated with many problems. Various smart city initiatives are built on the idea of improving the situation of a city as a whole—but often at the expense of individuals who have to adapt their (e.g., consumption or mobility) behavior. Thus, one of the most wicked challenges of smart cities is to reconcile citizens' needs, concerns, approvals, and disapprovals with the initiatives and technologies. Apart from the observation that the digitalization of cities leads to an increasing interconnection of people, tasks, and technology, thus requiring large investments in information technology (IT) and information systems (IS) infrastructure, smart cities require smart citizens. As a result, a culture of lifelong learning (Phoenix, 2002) is inevitable, as citizens need to understand the technologies involved, their potential advantages and disadvantages, their impact on their personal environment (which can be both positive and negative), and different coping strategies for dealing with the consequences posed by the technologies (Selwyn, 2021).

Making decisions in these contexts of digital transformation and smart cities requires a number of additional competences that are only partially considered in current education concepts such as communication, collaboration, critical thinking, and creativity. Zularnaen et al. (2019) also refer to such competences as "twenty-first century skills," thus highlighting their rather novel character compared to traditional expected competence sets (such as technical skills; see also Baltac, 2019). A thorough overview of relevant smart city competences is suggested by Fitsilis et al. (2021), who differentiate 5 different knowledge areas with 32 relevant skills and competences for smart city decision-makers. Moreover, since the required competences develop constantly and are best acquired in practical contexts, decision-makers must be empowered to acquire the necessary competences themselves, integrated with their regular professional activities (Zakirova et al., 2021).

Hence, with its complexity and diversity, the transformation of cities toward smart cities and the numerous relevant competences confront lecturers with challenges in preparing students well for future problem situations in their working lives and as engaged citizens. For this reason, it is required to teach students the basics of smart cities in high quality while at the same time enabling them to work independently and intensively on wicked problem situations and develop suitable solutions in these contexts on their own. This is why the education concept presented below employs a problem-based group work, in which lecturers "instead (...) explaining a principle, defining a concept, or guiding students through procedures, [lecturers] assign problems that will force students to inductively discover explanations, definitions, and processes" (Weiss, 2003). In our concept, the students are supposed to independently identify relevant learning contents based on a given problem situation, determine and close their own knowledge and competence gaps, and work out a possible solution. This solution is not defined a priori but depends on the individual interpretations and the knowledge level of the students. Through problem-based learning, comprehensive and transferable knowledge and competences are acquired as well as effective problem-solving skills can be developed.

A main difference to traditional education concepts is that the acquired knowledge is not queried during problem-solving but is created as part of it (Müller Werder, 2013; Weiss, 2003; Hung et al., 2008). The problem situation generates cognitive and emotional conflicts as well as interactions between the students. These conflicts and interactions provide the stimulus for learning and enhance the motivation as well as the willingness to learn. It also promotes long-term cognitive anchoring of learned content, an expansion of problem-solving skills, higher-order thinking, and the acquisition of self-directed and lifelong learning skills (Hung et al., 2008; Nobaew, 2018). An important aspect of lifelong learning is virtuality and persistence. Students need to be able to learn contents in their own pace and setting. To address this, the education concept presented here in one learning unit is designed as a blended leaning module for distance learning that can be adapted for different application scenarios in a smart city and for different competences to be learned.

The seven-step (Fig. 1) represents the didactic process structure of problem-based learning (Müller Werder, 2013; Schmidt, 1983); it serves as an orientation and structuring of the problem-based virtual group work for the students. The evaluation of the own solution and procedure in the group represents an eighth step, which is detached from the content-related problem-solving in group work. The evaluation is of great importance as a reflection step, since the students critically evaluate their own learning strategies as well as the group process. They also cognitively process the learned contents and the developed solution, which is central for learning persistence (Xu & Jaggars, 2013).

3 Required Design Requirements

The design requirements (DRs) of the education concept for the "Digitalization of Cities and Traffic" learning unit presented here were identified in design processes. Due to the blended learning study model at the University of Hagen, which consists of a combination of online and face-to-face teaching, the module is designed for distance learning, i.e., minimal physical presence of the students as well as learning from a distance. Students often complete their studies while working full- or part-time, so flexibility and optionality of group work are important criteria for the success of the education concept. In addition, the students are located over the entire German-speaking area (i.e., Germany, Switzerland, Austria, Liechtenstein),

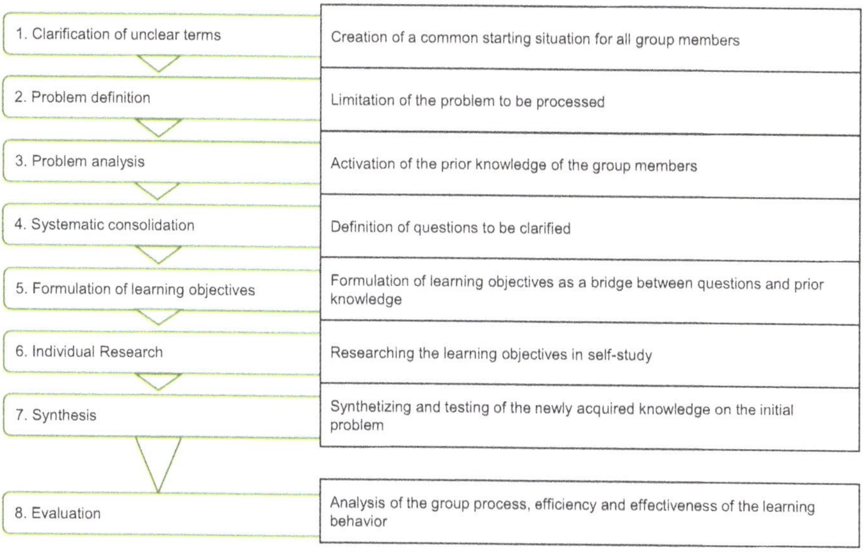

Fig. 1 Steps of the problem-based education concept in virtual group work (based on Müller Werder, 2013; Schmidt, 1983)

some even internationally. The ongoing Covid-19 situation forced many education providers to shift their teaching endeavors to a virtual room. We are therefore confident that our didactical setup is of interest for many lecturers teaching smart cities. Furthermore, the virtual environment supports lifelong learning about smart city concepts for additional interested parties, especially smart city stakeholders. Furthermore, it should be noticed that the whole setup can be transferred to non-virtual settings almost without effort.

The module and the virtual group work in one learning unit must be feasible with few resources. In our case, more than 220 students have to be supervised by 2 research assistants and 1 student assistant. In order to achieve sustainable study and learning success, especially in a distance-learning setting, the education concept should be motivating for the students, e.g., by integrating game elements ("gamification"), which promote both content-related and organizational engagement. Finally, with every problem situation, students are expected to acquire a particular set of smart city competences. Since it is hardly possible to teach all types of competences within one problem situation, it is important to carefully define a well-aligned set of smart city competences involving technical, organizational, as well as soft skills. Altogether, we identified seven DRs that guided our didactical concept development:

- DR1 (Virtuality): Physical meetings of the students are to be kept optional for the processing of the group work.
- DR2 (Flexibility): The group work must allow for time and space flexibility.

- DR3 (Optionality): It must be possible to complete the module successfully without the virtual group work, i.e., students can consciously decide not to participate in the virtual group work.
- DR4a (Legal feasibility): The legal framework of the examination must be adhered to, and assessment-relevant components of the group work must be clearly defined.
- DR4b (Scalability): The group work must be scalable, i.e., it must be realizable even with very large numbers of students and few personnel and financial resources.
- DR5 (Persistence and quality): The didactical concept should foster learning persistence and support the quality of learning.
- DR6 (Competences and skills): The group work should explicitly address the building of a consistent set of smart city competences and skills.

4 Design Process of the Education Concept

We followed a structured design process in order to address all DRs and therefore employed the ADDIE[1] process (see Fig. 2) according to Branch (2009). The design process model involves a cyclical development of the virtual education concept based on the concept of problem-based learning.

The first phase of the ADDIE process focuses on the **analysis** of the module and group work objectives (teaching the basics as well as enabling students to independently familiarize themselves with new problem situations in the context of smart cities and digital transformation), the target group (middle-aged students of business administration, economics, and business informatics, most of whom are employed), and the available resources for implementing the education concept (initially available conference technology (Adobe Connect) and learning management system (Moodle) as well as two research assistants and one student assistant). In the **design** phase, the education concept with clear links between content and learning objectives and introduction to content- and problem-based learning are to be defined. Furthermore, the learning objectives for five thematic learning units within the module as well as the examination components and achievements are to be determined. In the following phase of **development**, the learning resources (e.g., problem situations for group work, methodological introductions, reference material) have to be created, and the required resources as well as the organization of the virtual group work have to be designed more concretely. In the **implementation** phase, the new learning resources are used, and the developed education concept is implemented. In this phase, the supervisors encourage the students as learning guides and contact persons. In the **evaluation** phase, the achievement of objectives is checked, and feedback from the students is compared with the formulated

[1] Abbreviation for analysis, design, development, implementation, and evaluation.

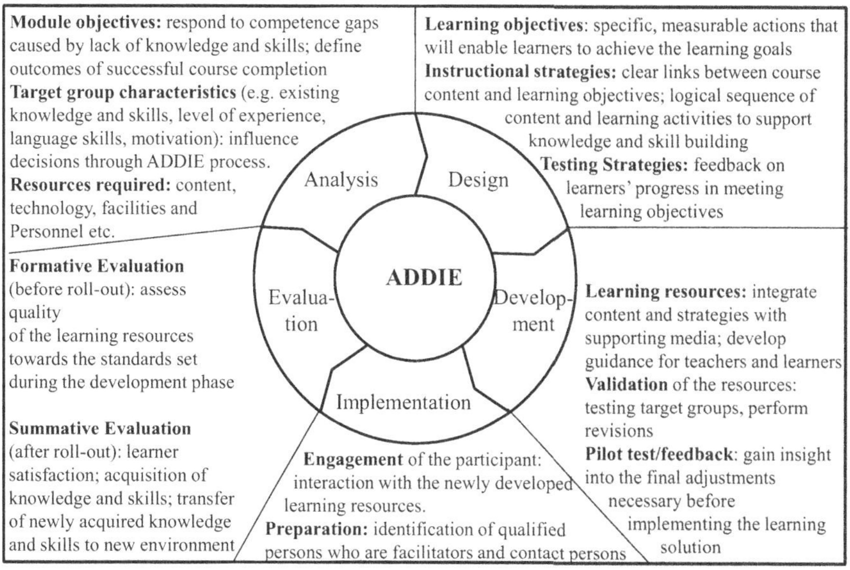

Fig. 2 Phases of the ADDIE process for the development of the education concept (based on Branch, 2009; Obsidian Learning, 2021)

Table 1 Design cycles and major decisions

Design cycle	Major decisions
1	• Development of module content • Design of problem situations • Design of problem-based learning concept • Technical platform (Moodle, Adobe Connect) • Timely and organizational implementation of virtual group work for each semester
2	• Enhanced technical implementation (Zoom) • Tool of automated group formation • Options to improve organizing high numbers of students
3 (ongoing)	• Design of gamification elements • Technical options for further learning support

requirements; if necessary, these represent the basis for the next design process as modified requirements.

In total, we went through two design cycles so far and are currently working on the third cycle. During and after each cycle, we gathered feedback from the students. In Table 1, we summarize the three design cycles along with their major decisions.

5 Design of a Virtual Problem-Based Education Concept for Smart Cities

5.1 Module Structure and Examination

The "Digital Transformation" module is divided into five learning units,[2] each requiring 60 hours of work during the semester. The virtual group work is integrated in the third learning unit to provide students with the required skills and competences to contribute to the transformation process of cities toward smart cities. In the group work, 20 out of 100 module points can be achieved. The remaining 80 module points are part of the module's final written examination. Consequently, participation in the virtual group work is voluntary for the students insofar as the module can also be passed without the virtual group work; however, the 20 points to be achieved through this problem-based group work cannot be compensated (DR3). The virtual group work is divided into three grading domains, with the individual performance outweighing the group performance in total:

1. **Individual performance I**: elaboration of one role and the respective argumentative structure as part of an elevator pitch lasting 90 s (Denning & Dew, 2012) (equivalent to approximately half a page of continuous text).
2. **Group performance**: design of a one-page handout and slides for the final presentation as well as development of a solution for the problem situation in the form of a 3-minute wrap-up.
3. **Individual performance II**: writing of a brief individual (critical) reflection of the group work (maximum two pages; step 8 in Fig. 1).

The semester starts with a virtual meeting to welcome the students and explain the module structure and the timeline of the group work. Following this, the students are able to register themselves for one of the groups via the learning platform Moodle. After a second meeting, the kick-off meeting for the group work phase, the groups will get in touch and agree on three preferences regarding the available problem situations, which they submit. The groups are then automatically assigned by a self-developed VBA (Visual Basic for Application) script to the problem situations in line with their preferences. For synchronous coordination, collaboration, and joint work, each group is provided with a Jitsi room. The main group work phase lasts 6 weeks, in which the students work on the problem situations following the seven-step. A number of support forums (between students), FAQs, and instruction documents are intended to support the students in their most frequent questions. Beyond that, a weekly consultation hour is offered for questions not addressed in the support materials. During this time, the students also prepare

[2] (1) Concepts and technologies of digital transformation, (2) changed value creation through digitization, (3) digitization of cities and traffic, (4) digitization of the financial sector, and (5) use and success of information systems in the age of digitization.

an interim and final presentation, as well as a handout for the other students. The final presentations take place on 4 to 5 evenings via Zoom. The reflection paper (see Fig. 1; step 8) is submitted by the students in PDF format via the learning platform. Students will be notified of the achieved points via the learning platform before the exam.

With this particular group work setting, several of our DRs are addressed. The students are required to playfully design an archetypical role and the respective argumentations with concepts of smart cities and digital transformation. By doing so, we intend to support longer and more intense engagement with the course and learning contents as well as with the module as a whole (DR5). This involves weighing up individual arguments, holding substantive discussions with fellow students, and debating solutions. In addition, compared to other design options, this education concept also enables large numbers of participants to be supervised and examined (DR4b) as well as the virtual implementation and realization (DR1).

5.2 *Context of the Problem Situations*

In the thematic context of the learning unit "Digitization of Cities and Transport," the socio-technical interconnections of the digital transformation are especially pronounced. Based on the six smart city domains of Giffinger et al. (2007; Giffinger & Haindlmaier, 2010), complex problem situations are developed, each with at least eight different roles of the city development committee of the fictitious city "Neuhagen" (e.g., Marc Mayor, Eric Entrepreneur, Bert Background Information). The roles represent archetypical positions with regard to the respective problem situation and address contradictory, complex, and diverse concepts of the transformation of cities toward smart cities. Fifteen problem situations are currently available. Table 2 provides an overview of six problem situations. Within each problem situation, different smart city concepts and challenges are addressed, thus stimulating the learning of different smart city competences. Each problem situation focuses on a consistent set of smart city competences (DR6), e.g., decision-making and problem-solving, teamwork, specific technologies, stakeholder management and citizen engagement, smart city management and planning competences, entrepreneurial competences, resilience, and sustainability competences.

The problem situations are based on current technological developments as well as real problems discussed in media and press in recent years. Figure 3 shows an exemplary problem situation (upper part) and the structure of a role developed as part of the group solution (bottom part). Figure 4 illustrates one possible solution approach.

Table 2 Overview over six exemplary problem situations

Smart governance: design and use options for an unused area (former airport) in the city **Addressed concepts**: *smart shopping, value co-creation, ambient-assisted living, smart home, smart waste, smart lighting, smart buildings, virtual and augmented reality, beacons, location-based services* **Possible solution**: a smart city area concept with shopping facilities and social smart housing for all generations and needs enriched with smart technologies and solutions
Smart economy: advantages and disadvantages of the use of industrial robots **Addressed concepts**: *digital transformation, human-machine collaboration, industrial robots, IT threats, IT security, hacker attacks, lifelong learning* **Possible solution**: collaboration of humans and robots; robots not as substitute for employees but a supplement; further training and training on the job for employees addressing fears and challenges but also highlighting the chances and potentials
Smart people: use of digital media and tablets in the classroom at school **Addressed concepts**: *blended learning, lifelong learning, technostress, media and digital literacy, bring your own device (BYOD), IT security, data security, applications, big data, digital schools* **Possible solution**: pilot use of tablets in some classes to test potentials and identify challenges in combination with a media concept for teachers; implementation of education apps in the classroom; maintenance and servicing concept for the technology
Smart mobility: problematic parking situation in "Neuhagen" **Addressed concepts**: *smart parking, more advanced systems, soil sensors in streets, computer vision, lidar sensors in buildings, artificial intelligence for forecasting, autonomous vehicles, e-scooters, e-bikes, smart people, environmental awareness, sharing concepts* **Possible solution**: use of smart mobility systems, smart parking, reduction of individual traffic through mobility as a service, ride-sharing services, increased the attractiveness of public transport
Smart environment: use of smart technologies in a city hall building and the associated new construction or modernization of the building **Addressed concepts**: *sensors, actuators, preconditions for smart buildings, information and communication technologies (ICT), cyberphysical systems, wearables, e-health, data security, chatbots as governmental service agents* **Possible solution**: modernization of the existing building to create the prerequisites for a smart building (such as connectivity of personal wearables and the room climate control) in supplement with a new building/extension
Smart living: use of smart home technologies and privacy issues **Addressed concepts**: *smart home technology, internet of things (IoT), open and closed systems, transmission standards, information and communication technologies (ICT), big data, data security, ambient-assisted living systems, freemium, pay-per-use, smart service, cyberphysical systems* **Possible solution**: "Neuhagen" as a model city for smart homes can be subsidized with public funding as part of the smart city initiative in order to reach the environmental requirements; smart home technology implementation to support elderly or disabled people

6 Evaluation and Next Steps

Student feedback on the implementation of problem-based learning and learning goal achievement at the end of the 2019/2020 winter semester (first semester of implementation) was very positive. The module evaluation of 28 students also showed a very good average score of 1.43 for the module (rating of 1–5, from very

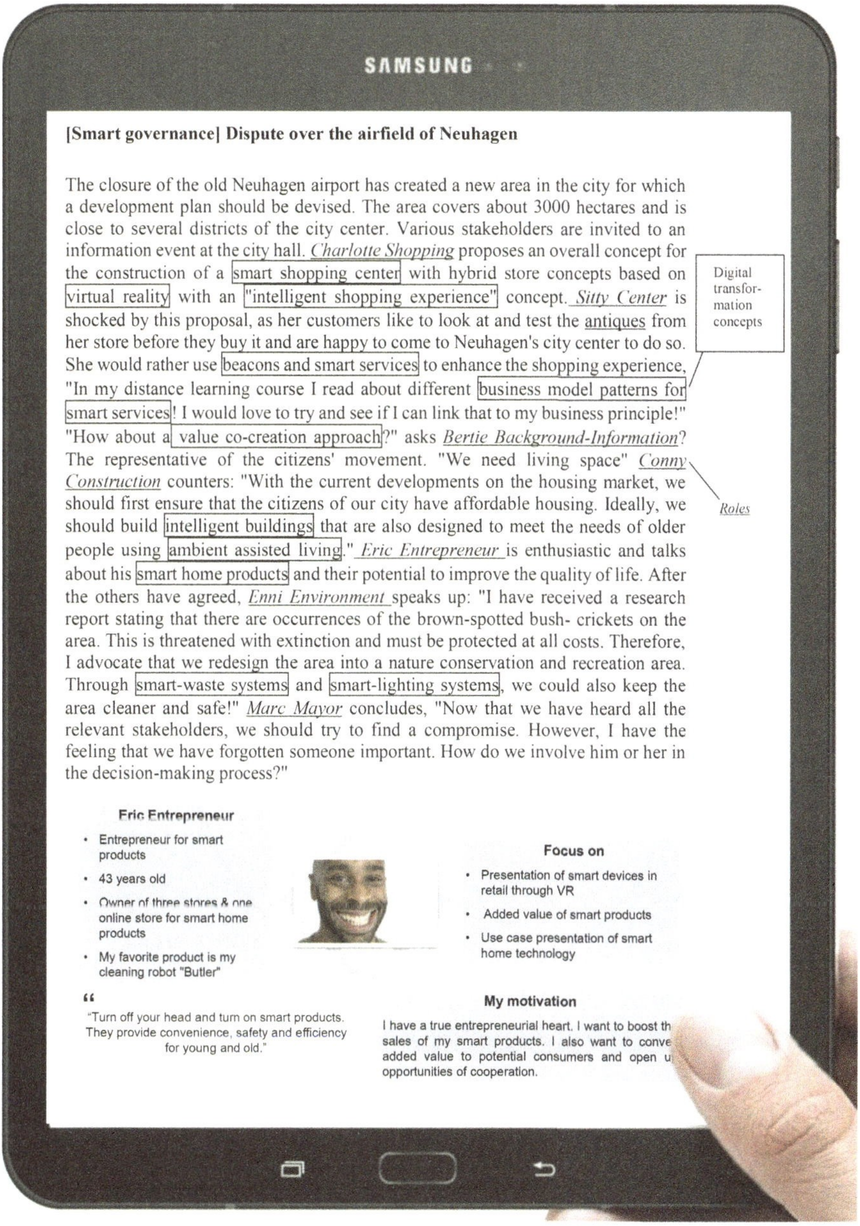

Fig. 3 Exemplary problem situation (concepts of the transformation of cities in boxes and roles underlined) and design of a role

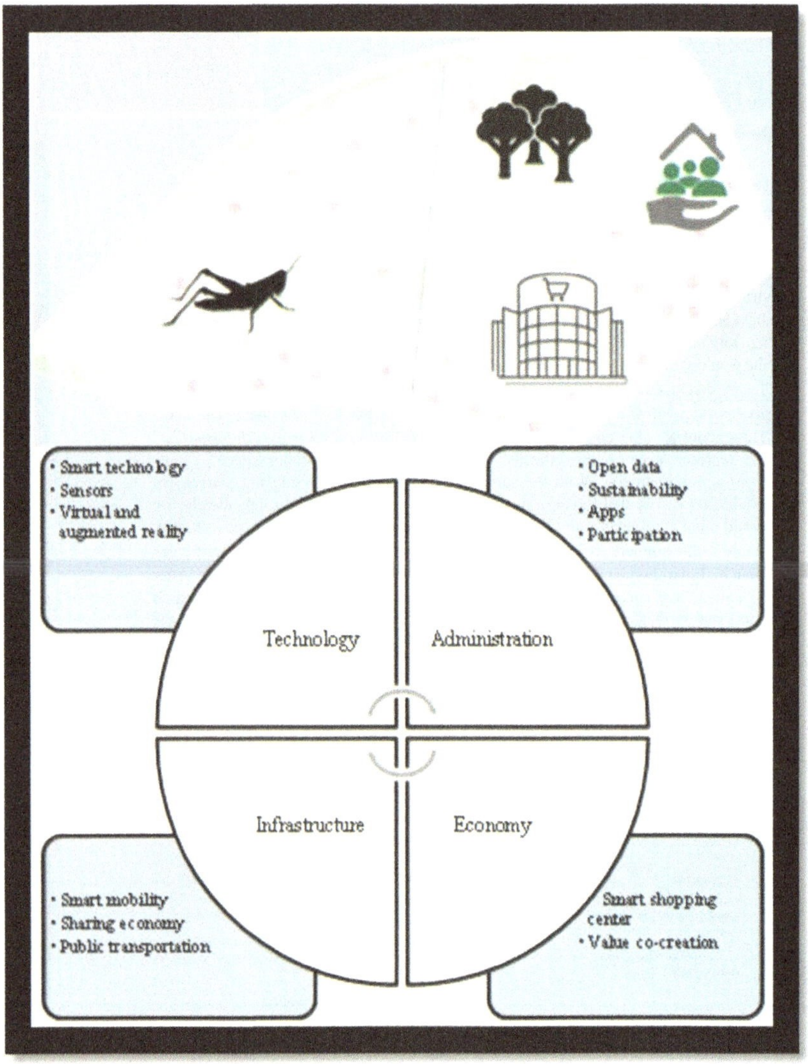

Fig. 4 Possible solution of the problem situation

good to poor). The students further highlighted the high quality of the group work and felt persistence of the learned contents as well as trained competences. However, students pointed out some aspects to be improved: the technical implementation of the virtual group work (change from Adobe Connect to Jitsi and Zoom, DR1, DR2), the time span of the group work (extension of the group work from 4 to 6 weeks, DR2), and further material on the requirements of the virtual group work and problem-based learning (in the form of notes). The notes on the one hand communicated concrete expectations regarding the interim and final presentations and on the other hand supplemented the content of the group work (especially formulation of further problem situations and roles). In addition, a student assistant was assigned for organizational support in future semesters.

In the following semesters, there were only few suggestions for improvement by the students, although the increasing number of participants in the group work (WS 2019/2020, 95; SS 2020, 150; WS 2020/2021, 229; as well as in SS 2021 208 students) resulted in a need to optimize the supervision and realization of the group work. The efficient design of group presentations and the supervision of large numbers of students are thus at the forefront of current further developments. It is currently being analyzed how the virtual group work's supervision can be further improved in terms of efficiency, which aspects of supervision and support can be (further) automated, and at which point individual support for students is definitely required. Evaluating the module results for the last three semesters (SS 2020, WS 2021/22, and SS 2021), students participating in the virtual group work achieve around four points more in the exam than students who do not participate in the group work, leading on average to a better grade.[3]

In addition, a holistic gamification concept is developed to create further incentives for students to actively shape their learning process not only during the group work, in which problem-based learning is anchored, but throughout the whole semester (DR5). For example, a holistic gamification-based education concept requires knowledge of the different motives and goals of the students for the successful completion of the module. The different student motives and goals determine the individual learning progress. For example, a level system with conditional unlocking of certain learning elements and contents is intended to support the students in the structured, step-by-step execution of the group work based on the seven-step. After the fourth step (systematic knowledge deepening), the students receive, e.g., supporting materials for the formulation of learning objectives, which have to be worked out in the fifth step. The level unlocking is complemented with motivational messages such as "Now you are ready for the intermediate presentation!" and "Congratulations. You have completed all the steps of problem-based learning. You are ready for the final presentation!" Conditional unlocking also supports supervisors in that questions about the next steps in the process, as well as the preparatory work students are doing, are more guided and accompanied (e.g., for the interim presentation). If the students receive all information at once at the beginning, there is a high risk that students are overwhelmed by the number of things to consider and remember. As a result, they will not remember relevant information when they need it. Badges such as "Bug Hunter" for identifying errors in the course materials should also provide students with an extrinsic incentive to help improve the course material. Finally, students will also be able to embed the faces of their roles as avatars in the learning platform. The different motives and goals of the students are taken into account in the final concept

[3] In the German system, grades range from 1.0 (with honors) to 5.0 (fail) with steps of 0.3 and 0.7, e.g., 1.3 or 1.7. In the exam, the grades are distributed in five-point steps.

through different gamification elements, so that as far as possible all students receive gamified support according to their preferences. For example, badges to be achieved provide an incentive for the "achiever" player type to actively participate in the gamified learning environment, while for "explorers" new discovery opportunities in the learning platform provide greater incentives (Anschütz et al., 2020). The gamification concept is currently being conceptualized.

7 Transferability of the Approach and Outlook

The presented virtual education concept represents a design option for problem-based learning in a virtual form and can be transferred as well as adapted if necessary to other courses. Other forms of designing problem-based learning can be found, for example, in Hung et al. (2008) and for German-speaking lecturers in Müller Werder (2013). Especially due to the current Covid-19 protection measures, the virtual format also offers potentials for education concepts in presence, since the individual steps of the problem-based education concept do not necessarily have to be gone through in presence but can be transferred to a virtual or hybrid format relatively easily.

A problem-based education concept such as the one presented in this chapter requires constant support for the students from supervisors. This is particularly the case in a virtual implementation, since students often lack the exchange with fellow students. In such an (virtual) education concept, however, communication and interaction are in particular essential for successful completion of the group work. The organizational effort of group work can be reduced for large numbers of students by automated group selection activities as well as by consolidated information transfer of the groups to the chair. Nevertheless, a certain effort remains, especially due to the evaluation of the individual examination performances and coordination with the supervisors. In addition, due to problems with the students' collaboration in groups as well as discontinuation of group work, the intervention of the supervisors is needed throughout the group work phase.

For the virtual group work, it was necessary for the students to familiarize themselves intensively with the topics of the entire module (this was also confirmed by the students in the module evaluation). We also observed an unusually high participation rate of students in the module's final exam (compared to other modules). While in other modules many students often withdraw shortly before the exam, the group work seems to have a positive effect on the exam participation rate. The difference between the number of group work participants and exam participants has been 15–20 students in each exam during the last three semesters. While we have to admit that part of this persistence is probably related to the fact that it is not possible to carry over the points achieved from group work into following semesters[4] (legal restriction), feedback from students still indicates a

[4] Students would have to repeat the group work exercise in the next semester.

positive impact of the group work on the willingness to take the exam. With values between 35 and 50% in the last four semesters, the activity rate, which illustrates the relationship between enrollment numbers and exam numbers, is higher than the average activity rate at the faculty (around 30%). The average exam grade of 2.3 is also higher than the average exam grade of other modules in the faculty (around 2.6).

Overall, the implementation of the virtual education concept based on problem-based learning is an enrichment of the distance-learning program at the University of Hagen. The exchange among the students and with the supervisors is valuable for the further development of the education concept. For example, due to tight contact with the students, opportunities arose for the implementation of specific guest lectures (e.g., a virtual smart factory tour) and the revision of learning contents. Another advantage of the concept is that the learning content and smart city competences are not only learned for the exam but are processed more substantially and cognitively, thus being present over a longer period. The complex problem situations and group discussion, involving transfer, conflicts, and emotional reactions, make the learned knowledge and competences applicable and accessible in everyday situations (Weiss, 2003; Hung et al., 2008). Further observation and analysis will nevertheless have to address how the transfer of such an education concept, which is also more complex for the students, affects other modules and the success of the students when several modules use these concepts in parallel.

The described education concept for the module "Digital Transformation" at the University of Hagen as well as the development along the ADDIE design process support intensive and qualitative teaching and learning as well as high persistence of the learning content. For university education concepts, but especially for the accompaniment of change processes in cities toward smart cities, it is not only about content and concepts but above all about debates, discussions, and negotiations about different possibilities which are addressed through the problem-based learning approach. With our virtual education concept, we provide future city stakeholders with competences that allow them to work constructively toward meaningful, socially accepted solutions that are both feasible and fundable.

References

Anschütz, C., Ebner, K., & Smolnik, S. (2020). Spielerisch zum Ziel: Initiale Designprinzipien für die nachhaltige Gestaltung von Smart-Mobility-Apps auf Basis einer Marktanalyse. *Internationale Tagung Wirtschaftsinformatik (WI 2020)*, 538–553.

Baltac, V. (2019). Smart cities—A view of societal aspects. *Smart Cities, 2*, 538–548. https://doi.org/10.3390/smartcities2040033

Branch, R. M. (2009). *Instructional design: The ADDIE approach*. Springer.

Denning, P. J., & Dew, N. (2012). The myth of the elevator pitch. *Communications of the ACM, 55*, 38–40. https://doi.org/10.1145/2184319.2184333

Ebner, K., Mattes, P., & Smolnik, S. (2019). Are you responsible for traffic congestion? A systematic review of the socio-technical perspective of smart mobility services. In *Proceedings of the 52nd annual hawaii international conference on system sciences (HICSS)*. IEEE Computer Society Press.

Fitsilis, P., Anthopoulos, L., Tsoutsa, P., et al. (2021). *Competences for resilient smart cities officers: A survey: Technical report.*

Freeman, L., & Urbaczewski, A. (2019). Critical success factors for online education: Longitudinal results on program satisfaction. *Communications of the Association for Information Systems (CAIS), 44*, 630–645. https://doi.org/10.17705/1CAIS.04430

Giffinger, R., Fertner, C., Kramar, H., et al. (2007). Ranking of European medium-sized cities. http://www.smart-cities.eu/download/city_ranking_final.pdf

Giffinger, R., & Haindlmaier, G. (2010). Smart cities ranking: An effective instrument for the positioning of the cities. *ACE: Architecture, City and Environment, 4.*

Hung, W., Jonassen, D. H., & Liu, R. (2008). Problem-based learning. *Handbook of Research on Educational Communications and Technology, 3*, 485–506.

Jaekel, M., & Bronnert, K. (2013). *Die digitale Evolution moderner Großstädte: Apps-basierte innovative Geschäftsmodelle für neue Urbanität*. Springer Vieweg.

Marrone, M., & Hammerle, M. (2018). Smart cities: A review and analysis of stakeholders' literature. *Business & Information Systems Engineering, 60*, 197–213. https://doi.org/10.1007/s12599-018-0535-3

Müller Werder, C. (2013). Problem-based Learning erfolgreich gestalten. In H. Bachmann (Ed.), *Hochschullehre variantenreich gestalten: Ansätze, Methoden und Beispiele rund um Kompetenzorientierung* (pp. 50–77). Hep Verlag.

Nobaew, B. (2018). The strategies of implementing role-play within problem-based learning approach to improve learning efficiency in higher education. *Educational Access and Excellence, 1*, 82.

Obsidian Learning. (2021). *The ADDIE model is a five-phase approach to building effective learning solutions*. Retrieved from https://www.obsidianlearning.com/projects/addie-model/

Phoenix, D. A. (2002). A culture of lifelong learning? *Journal of Biological Education, 37*, 4–5. https://doi.org/10.1080/00219266.2002.9655837

Schmidt, H. G. (1983). Problem-based learning: Rationale and description. *Medical Education, 17*, 11–16.

Selwyn, N. (2021). *Embracing a culture of lifelong learning:(re) imagining the futures of lifelong learning: Some sociotechnical tensions*. UNESCO Institute for Lifelong Learning.

United Nations. (2018). World *Urbanization prospects: The 2018 revision*. Retrieved July 3, 2019, from https://population.un.org/wup/Publications/Files/WUP2018-PressRelease.pdf

Weiss, R. E. (2003). Designing problems to promote higher-order thinking. *New Directions for Teaching and Learning*, 25–31. https://doi.org/10.1002/tl.109

Xu, D., & Jaggars, S. S. (2013). The impact of online learning on students' course outcomes: Evidence from a large community and technical college system. *Economics of Education Review, 37*, 46–57. https://doi.org/10.1016/j.econedurev.2013.08.001

Zakirova, A. B., Koshanova, D. K., & Akhayeva, Z. (2021). Identifying competencies in designing the educational program «Smart City». *Bulletin Series of Physics & Mathematical Sciences, 73*, 174–177. https://doi.org/10.51889/2021-1.1728-7901.25

Zulkarnaen, R. H., Setiawan, W., Rusdiana, D., et al. (2019). Smart city design in learning science to grow 21 st century skills of elementary school student. *Journal of Physics: Conference Series, 1157*, 22021. https://doi.org/10.1088/1742-6596/1157/2/022021

Part II
Strategy and Projects

The Dynamic Formation of a Successful Smart City Roadmap

Georgios Siokas and Aggelos Tsakanikas

1 Introduction

As cities grow, both territorially and demographically, the challenges of social and environmental sustainability exacerbate (Yigitcanlar et al., 2018). For example, the environmental impact of cities is continuously growing, as they are responsible for more than 70% of global greenhouse gas emissions (Satterthwaite, 2008). To cope with this challenge, policymakers seek to modernize infrastructure, improve urban planning and utilize new innovative technologies. By allowing the urban environments to gradually transform into smart and sustainable cities, possibly unlocking their potentials in reaching the desired results of sustainability (Kummitha, 2018).

1.1 Challenges in the Smart Environment

According to the literature, an urban ecosystem can be distinct into three main dimensions: (i) technology, such as software, hardware and platforms; (ii) human, such as education, innovation and creativity; and (iii) institution, such as government, regulations and policies (Nam & Pardo, 2011). Each factor consists of numerous aspects of urban life. In order to form a smart urban ecosystem, a city needs to have an isomeric and parallel development of all three dimensions (Al-Hader et al., 2009; Heaton & Parlikad, 2019; Zanella et al., 2014). During the last years with the economic crisis, cities focused on promoting economic growth, foreign investment and job creation, while their digital transformation faced

G. Siokas (✉) · A. Tsakanikas
Laboratory of Industrial and Energy Economics, National Technical University of Athens, Athens, Greece
e-mail: geosiok@mail.ntua.gr; atsaka@central.ntua.gr

P. Fitsilis (ed.), *Building on Smart Cities Skills and Competences*, Internet of Things,
https://doi.org/10.1007/978-3-030-97818-1_5

numerous socioeconomic, legal, financial and technological challenges (Monzon, 2015). Based on these ascertainments, the major challenge cities face today has to do with the need for strategies to be in line with the exploitational development of technology, and the constant shifting of stakeholders' expectations and desires (Kummitha & Crutzen, 2017). Thus, a city needs to function as a pole to attract and activate the participation of its stakeholders. Through numerous mechanisms, a municipality can give value to the role of the stakeholders. Therefore, smart cities can partner with stakeholders to attain social change and sustainability and tackle the main urban challenges and possible solutions (Lam & Yang, 2020).

Under this concept, a municipality forms its goals based on (i) improving services to stakeholders, (ii) making more efficient use of available resources, (iii) protecting the environment, (iv) strengthening local economic development and (v) highlighting the local cultural heritage (Neirotti et al., 2014). Towards these goals, smart initiatives can accelerate the process of achieving the desired goals. Therefore, strategic planning is key for successfully implementing smart initiatives, especially, when the initiatives cover aspects such as the e-government, e-services, city planning, education, finance and resource management (Neirotti et al., 2014; Chourabi et al., 2012; Siokas et al., 2021). Additionally, the successful digital transformation of an urban environment presupposes its appropriate preparation by the public authorities. Therefore, having a tailor-made strategy can trigger factors, which enable the community to transform smoothly to smart urban environments (Siokas et al., 2021).

1.2 The Role of Smart Specialization in Smart City Strategies in Greece

As has been stated before, both smart strategies (Siokas et al., 2021) and smart specialization have a conceptual and policy meaning (Hassink & Gong, 2019). According to the literature, there is a need for converging the two otherwise separate policy concepts: (i) smart city and (ii) smart specialization strategies (Caragliu & del Bo, 2015). This statement is based on an econometric empirical analysis from the literature in which policy is formulated in the EU between smart specialization strategies and successful smart city initiatives (Caragliu & del Bo, 2015). Smart specialization focuses on skills which differ from the traditional ones. These skills are associated with deeper learning, analytic reasoning, problem-solving and teamwork. Subsequently, a higher degree of urban "smartness" in the digital transformation can be achieved via smart specialization strategies (Caragliu & del Bo, 2015). But before the smart specialization can have an effect, there is a need for smart strategies both for specialization and cities. This way, urban municipalities can strengthen their capacity of urban systems and citizens and effectively confront challenges and risks. Mainly, there is a need for institutional, organizational, societal development in a Greek city which targets to overcome barriers and close gaps created during the economic recession and austerity (Marava et al., 2019). Under this concept, this chapter tries to identify the importance of a smart strategy that focuses on the improvement of the social and institutional dimension of an urban environment.

1.3 Scope and Purpose of the Chapter

Taking into consideration the level of complexity of planning and executing an urban strategy, this chapter aims to identify patterns in the planning and implementing phase of a smart strategy in a Greek municipality. At the same time, the research tries to identify the Greek municipalities' potentials towards their smart transformation. In more detail, we will try to classify all participating municipalities in levels depending on the completion and execution of their smart initiatives. Therefore, a set of characteristics defying the Greek municipalities is fostered, helping both themselves and the general government curve the ideal or potentially tailor-made strategies for the different urban environments. Therefore, this chapter is divided into three sections. The first section analyses the theoretical background and the hypotheses formulating the empirical analysis of the chapter. Following, the second section presents all the relevant information about the data and the methodology implied in the following analysis. Finally, the chapter closes with the analytical presentation of the results, the discussion and the conclusion on the empirical analysis.

2 Theoretical Developments of the Research Model

2.1 Developing Smart Urban Ecosystems

Nowadays, multidimensional definitions are developing a relevant framework for smart cities (Fernandez-Anez et al., 2018; Caragliu et al., 2011), but it cannot adequately address all the unique needs of an urban environment for a balanced and sustainable development (Musa, 2018). The most common elements, mechanisms and processes of a smart city fall under three main dimensions: (i) technology, (ii) people and society and (iii) institutions and policies (Al-Hader et al., 2009; Heaton & Parlikad, 2019; Zanella et al., 2014). With different initiatives regarding these three dimensions, a city aims to improve the quality of life, security and privacy and to meet the needs of its inhabitants. These different aspects can be attained through the digital integration of infrastructure, the improvement of the efficiency of services and the attraction of citizens and knowledge-intensive organizations.

The city's advantages stem from the ability to integrate multiple technological solutions, securely, to manage its assets, which include, among others, information systems of local departments, schools, libraries, transportation systems, hospitals, production stations energy, law enforcement and other services. The city can, also, function as a mechanism to produce knowledge, which can help address inefficiency and optimize the urban ecosystem (Musa, 2018). Interactive platforms can offer to cities holistic solutions and services, which have an advanced graphical user interface (e.g., toolbars, reports, web interface, maps), a digital control system (e.g., common platforms, automated controls) and access to databases (e.g., big data, data warehouse, exchange platforms) (Zanella et al., 2014; Al-Hader & Rodzi, 2009). Through the proper use of the platforms, the governance, the democratic process and the interactions of the actors are strengthened. Public authorities can understand the interactions between economic, social, physical and other systems and make decisions to determine optimal solutions. People need to familiarize themselves with the new technologies and intergrade them into their lives. In this way, the quality of energy supply and renewable resources, the security and protection of citizens and the satisfaction of their daily needs are ensured. In other words, technological solutions can describe life in a city and help policymakers make better decisions and improve planning capacity and implementation efficiency (Gaur et al., 2015).

2.2 The Role of Smart Specialization in Smart City Strategies

The digital transformation of cities requires the investment in smart specialization. Based on the literature, the conceptual meaning of smart specialization is descripted as "*the capacity of an economic system (e.g., a city) to generate new specialities through the discovery of new domains of opportunity and the local concentration and agglomeration of resources and competences in these domains*" (Foray, 2014). This notion advocates the consistent correlation between the technological abilities and competences of municipalities and their investments in knowledge and human capital (Capello & Kroll, 2016). In addition, there is a continuous effort in investing and transforming existing productive structures with the help of various diversification processes of local resources and capabilities and new, but related explorative, research activities (Hassink & Gong, 2019; Foray, 2014; Foray, 2016). Additionally, smart specialization explains and reduces the productivity gap between different regions (e.g., cities) (Foray et al., 2009). Based on this, the aftermath of the smart specialization depicts on successful smart city initiatives and vice versa (Caragliu & del Bo, 2015). Therefore, the different generated processes based on the rationale of smart specialization reflect the capacity of an urban economy to generate new activities, while aiming at its transformation (Foray, 2016). Contrariwise, there is the need for a smart specialization strategy to tackle potential issues and difficulties (Foray, 2016). Two main issues faced nowadays are (i) the national-level fragmentation of public research systems and (ii) the duplication of knowledge bases (Foray, 2014).

2.3 The Urban Environment as a System of Systems

Smart cities are made up of different basic and fragmented systems, each forming different networks relating to infrastructures, information and applications that provide products and services to citizens and private and public organizations (Suzuki & Finkelstein, 2018). These may include geographic information systems, traffic, healthcare, infrastructure, security, transport, water supply, waste and environmental and energy monitoring (Awad et al., 2017). Although city systems provide control over critical resources, they are difficult to adapt, maintain and expand. This is due to the cost, complexity and features of systems, which public authorities face compatibility problems between the different technologies (Suzuki & Finkelstein, 2018; Gann et al., 2011). This situation results in making a significant effort for the management and communication of multiple independent systems (Awad et al., 2017; Atzori et al., 2010). To facilitate the analysis of the systems and the different characteristics that exist in a city, we present a hierarchy on the five layers of application in a smart urban ecosystem, as follows (Nam & Pardo, 2011; Lim et al., 2018; Berger, 2017; Singh et al., 2020; Negre & Rosenthal-Sabroux, 2014; Giffinger et al., 2007; Edmonton City, 2018; Shichiyakh et al., 2016; Σαριδάκης, 2019):

- **First layer**: The level concerning the urban ecosystem, the city, has a horizontal effect on all other levels and corresponds to (i) technology, (ii) human and (iii) institutions.
- **Second layer**: The aspects of a strategy which a municipality plans and implements.
- **Third layer**: The different sectors of a strategy for implementation: (i) government and governance; (ii) education, economy and financial environment; (iii) human resources, citizens, hospitality and healthcare; (iv) infrastructure (building, home) and technology; (v) exploitation of technology (security, safety, mobility, logistics and transportation); and (vi) natural resources, energy, environment and farming.
- **Fourth layer**: The type of applications in a strategy (e.g., environment, devices, policies) has an application across all five layers.
- **Fifth layer**: The type of partners (e.g., available resources, government, private and public sector, citizens and other stakeholders) has an application across all five layers.

Due to the wide range of these layers, for this study we will focus on the second level, taking into consideration all the other levels when forming the methodology and discussing the results. It is assumed that the second layer together with the third layer have the most important impact in smart specialization. Therefore, we will try to analyse the smart strategy of the second layer based on the spectrum of smart specialization. But before continuing on, we will analyse a catalytic parameter, which is the different stages of developing a strategic map.

2.4 Stages of Developing a Strategic Map

The fundamental goal of a municipality for implementing a digital strategy, a combination of strategies for e-government and smart city, is completing its digital transformation (Siokas et al., 2021). Digital transformation can be configured in different ways, at different speeds and with different priorities (Huawei, 2015). A catalyst for the digital transformation is the presence of an integrated strategy emphasizing the implementation of digital actions based on the city's needs (Siokas et al., 2021). Regardless of a municipality's path to digital transformation, by providing targeted advice and research findings, policymakers can be assisted in designing an efficient navigation path to digital transformation. It requires a coordinated effort by all the actors involved in an urban ecosystem. According to the literature, a successful strategy is divided into three levels, the macro-, mezzo- and micro-level (Letaifa, 2015). In more detail, the macro-level includes the phases of strategic planning and utilization of resources and partners. Following, the mezzo-level refers to the phases of ownership of the participants with the projects and implementation of a clear roadmap. Finally, the micro-level focuses on the phase of digital transformation for the implementation of new high value-added services (Letaifa, 2015).

Primarily, careful, and gradual initiatives are required while maintaining a degree of flexibility. This allows uncertainties to be reduced and additional investments to be made. At the same time, it requires the development of different methods of active participation and research of users, as well as special conditions for the formation of effective cooperation between the municipality, the organizations and the residents (London, 2021a, 2021b). It is stressed that designs do not always have to be about digital technologies but may be related to different aspects which contribute to the development of a digital-friendly culture (i.e., bridging the digital divide). An interesting approach is shown by the city of Leuven in Belgium. It has implemented the "Leuven 2030" strategy, which is a six-step strategy and aims to transform ideas into actions for the city with the active participation of citizens (Leuven, 2021). The steps are (i) sharing the content or ideas; (ii) adopting the ideas and actions from the community; (iii) shaping, mixing or adapting the content and ideas; (iv) funding or supporting resources for initiatives the implementation of the actions; (v) production, creation or execution of the selected initiatives; and (vi) formation of a co-ownership regime on the content, elements or initiatives. Through this plan, the municipality was able to activate its citizens to engage with the public and to focus on actions that society needs and that the citizens themselves demand. Given the wide range of citizen engagement mechanisms, the tools needed to achieve stakeholder engagement and activation throughout the decision-making and policy cycle were designed and developed.

At a European level, the development of smart cities for the EU is part of the "Digital Agenda 2020", created by the "European Innovation Partnership (EIP)", and aims at disseminating knowledge and innovative solutions in European cities in terms of energy saving, quality improvement of urban air, mitigation of traffic problems and reduction of greenhouse gas emissions. These are supported from different organizations (e.g., municipalities, universities and research institutions, enterprises, NGOs, etc.) of Europe through their high participation in consortia and collaborations in the fields of ICT, energy and transport. According to the published texts, the need for common infrastructures is highlighted, with an emphasis on ICT tools and methods that collect, store, acquire and manage digital data. This allows the development of a measurable city management mechanism (European Commission, 2014; European Commission, 2020). Based on these facts, we will finalize our theoretical approach by briefly presenting facts regarding the smart city strategies by green public authorities.

2.5 Smart City Strategies in Greece

Smart cities in Greece show a need to use strategies as a tool to support their digital transformation and needs (Siokas et al., 2021). Many cities in Greece invest in digital technologies and infrastructure because there is a prerequisite to effectively develop digital tools and provide publicly digital information (Siokas et al., 2021). In many cases (e.g., city of Heraklion), infrastructure can enhance the overview of the urban environment and support sustainability and prosperity (Kalaitzakis et al., 2019). Taking into consideration Covid-19, other Greek cities (e.g., city of Trikala) are continuing implementing technological innovations (Anthopoulos et al., 2021). Under this scope, Greek municipalities are presented with great opportunities to make feasible projects and apply them to their ecosystem (Kanellos & Siokas, 2021). Towards this target, sometimes cities are faced with lack of funding and participation from the local and national public bodies and government (Alexopoulos et al., 2018).

Apart from the technological investments, a major challenge for cities (i.e., city of Korydallos) is creating a collaborative approach based on collaborations with main stakeholders (i.e., business, research, policy and citizen groups) and achieving an alignment of local, regional and European policy levels (Alexopoulos et al., 2018). This has driven Greek cities to gradually strengthen their collaboration with local decision-making bodies (i.e., municipality's staff and local community) and establish trust between them, the community and the local leaders (Anthopoulos et al., 2021).

Consequently, a smart strategy and a series of smart initiatives may lead to the city's digital transformation and the holistic integration of smart technologies in the daily routines of the city's stakeholders. A combination of the needs and the capabilities of the stakeholders, which potentially may lead to their smart specialization, may improve the social and economic development of the local community and foster enhanced planning strategies, social services and decision-making techniques (Siokas et al., 2021).

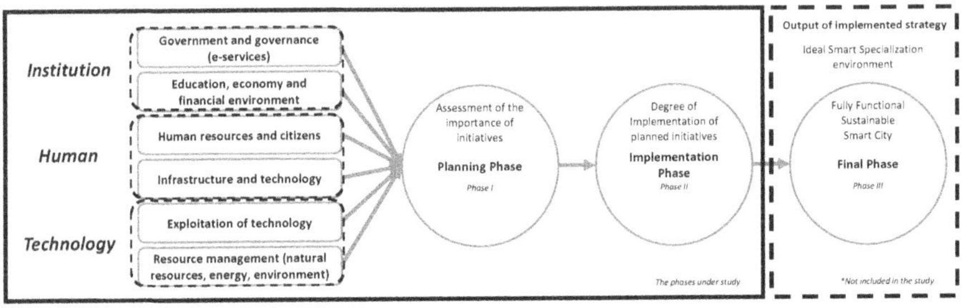

Fig. 1 The research model for our analysis

2.6 *Research Model*

Based on the importance of the strategy, the paper examines the characteristics of the implemented strategies which influence the planned strategies and the importance of digital projects in their immediate design and implementation. Having taken all the above under consideration, we could identify two phases leading to the desired outcome. First, the planning phase (phase I), in which the public authorities assess the importance of the smart city initiatives, and second, the implementation phase (phase II), which shows the degree of implementation of the planned smart initiatives in the previous phase.

Even though most cities have embedded various forms of smart projects in their strategies, there is an absence of a coherent framework to efficiently utilize their strategies. In this direction, it was deemed necessary to study the characteristics of the mechanism of digital transformation and to capture the process of transformation. By examining their strategies, we are trying to map the crucial factors that lead to successful mechanisms in developing smart strategies. Therefore, our research question, accompanying us to the conclusions, is: "*Which dimension impacts planning and implementing a smart strategy and what factors affect the formation of a smart city-oriented strategy among the different types of municipalities?*"

Considering the purpose of this chapter, we constructed a base model, studying the differences between phases I and II in different Greek municipalities. The features studied include city planning; quality of life; transport system; financial, innovation and education system; natural resource management; e-governance; and e-public services. The research model is presented in Fig. 1.

3 Research Methodology

The methodology used in this paper combines various theoretical approaches and exploits different econometrical tools. The relevant steps are presented in Fig. 2. First, we constructed a questionnaire for collecting all the essential data for

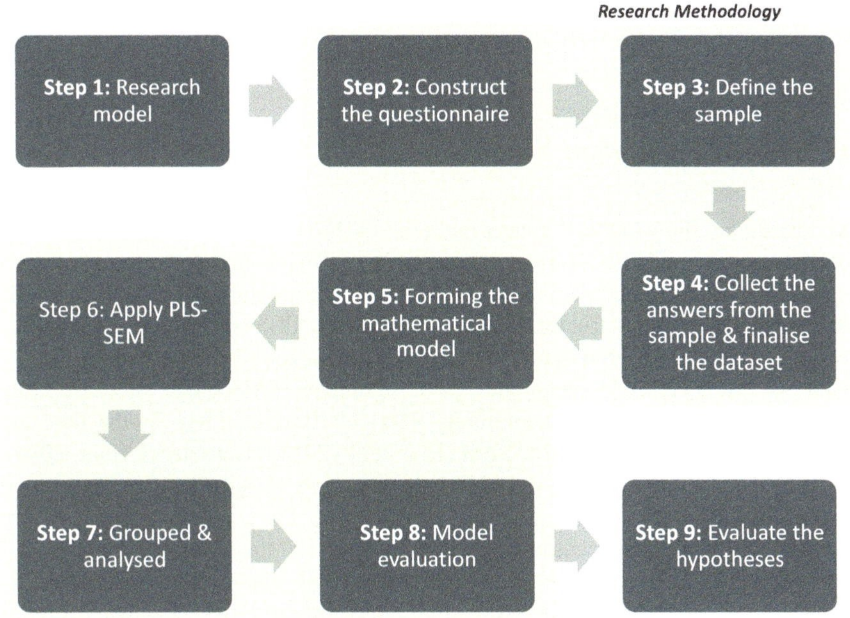

Fig. 2 The steps of the methodology

our research model while targeting the whole sample, which is all the Greek municipalities. Second, we collected the answers from the sample and finalized the dataset (Siokas & Tsakanikas, 2022). Following, we formulated the mathematical model representing our initial empirical model. Then, the model and the results are grouped and analysed according to the Greek distinction made by the Ministry of Interior. Finally, the output and the models can be evaluated and lead to the confirmation of the hypotheses.

3.1 Data Specifications

The analysis of the conceptual framework is based on a structured questionnaire, which was based on three main concepts. The first concept refers to the difficulties and the challenges public authorities face while trying to implement a strategy. The second refers to the level of diverse collaborations and partnerships which are necessary to develop a strategy. The final concept refers to the level of integration and use of digital technologies by the municipality and the local stakeholders in a city. The final sample consists of 252 municipalities across all 13 administrative regions of Greece (NUTS II level). The final number of participants corresponds to over 70% of the Greek municipalities and population, respectively. Regarding the data source's validity, each city's respondent, via her/his role, had explicit knowledge of the municipality's relevant strategy.

3.2 Research Method

We investigate the variables' conceptual relationship using the advanced statistical technique, structural equation modelling (SEM).[1] This technique has two predominantly favoured types, covariance-based-structural equation modelling (CB-SEM) and partial least square-structural equation modelling (PLS-SEM), which are widely recognized by the research community as highly resourceful (Bagozzi & Yi, 2012). Although the two SEM techniques have significant differences, they are used respectively by the research community. Particularly, CB-SEM was the first and the newly developed type is PLS-SEM (Hair et al., 2011a; Henseler & Chin, 2010). For the needs of our study, we chose PLS-SEM as a more suitable technique, based on a series of criteria.

Firstly, PLS-SEM is preferred for the prediction and theory development rather than the theory testing and confirmation of structural relationships (vs. CB-SEM) and has the unique ability for estimating causal relationships between variables (Hair et al., 2011a). Secondly, it focuses on maximizing the variance of the dependent variables requiring the least residual of distributions by the independent ones, instead of reproducing the empirical covariance matrix. Thirdly, it is a widely accepted method for most research predictions in the fields of economics and strategic planning. Fourthly, it is suited for a small sample and models with single-item constructs (Hair et al., 2012; Tenenhaus et al., 2005).

Finally, the municipalities will be grouped and analysed according to the Greek distinction made by the Ministry of Interior[2]. The categories for the municipalities are based (i) on their population, (ii) their geomorphological characteristics, (iii) the main characteristics of economic activity within their boundaries, (iv) their degree of urbanization, (v) their inclusion or not in wider metropolitan urban complexes and (vi) their position in the administrative division of the country. Therefore, we have four main categories:

- Municipalities of metropolitan centres (labelled: *metropolitan municipality*): This category includes all the municipalities of the region Central, North,

[1] PLS-SEM technic is operating on the logic of having multiple regression models calculated simultaneously. At the same time, it conducts confirmatory factor analysis, allowing different types of variables to be calculated and tested (Hair et al., 2011a).

[2] The official site for the categorization is http://www.opengov.gr/ypes/?p=5962 (the official language of the site is in Greek).

South and West Athens and Piraeus, as well as the municipalities of the region Thessaloniki close to the metropolitan centre.[3]

- Former capitals of regions (labelled: *former city capital*). This category includes all the mainland and insular municipalities which are the capitals of the urban and rural regions before the last changes in the public government.
- Regional municipalities (labelled: *regional municipalities*). This category includes all the mainland municipalities with a population[4] of over 10,000 which are not included in any other category.
- Small municipalities (labelled: *small municipalities*). This category includes all the mainland and insular municipalities with a population of fewer than 10,000 inhabitants.

Hence, it allows us to identify the determinants of policy planning for the process of digital transformation. Additionally, these municipalities will be evaluated for their planning and implementing stage. The evaluation will be based on four categories: (1) follower, (2) challenger, (3) contender and (4) leader. This categorization is based on the UK Smart Cities Index 2017 (Woods et al., 2017).

4 Model Evaluation

4.1 Measurement Model Verification

Before the model can express the corresponding correlations, a set of criteria are examined and verified (Hair et al., 2011a). First, the model's individual item reliability is acceptable for expressing conceptually the construct with loadings between 0.559 and 0.935 ($p < 0.001$) and satisfactory t-test scores (1.96 is the lower threshold value) (Hair et al., 2011a, 2017). Based on the output, the construct reliability and validity of the reflective latent variables are the values of Cronbach's alpha (CA) range from 0.654 to 0.963, the Dijkstra-Henseler rho (ρ_A) from 0.721 to 1.266, the composite reliability (CR) from 0.769 to 0.960 (lower threshold value 0.700) and the average variance extracted (AVE) from 0.401 to 0.797. The lower threshold value for the first three criteria is 0.700 and the last one 0.500. If a latent variable showed a value lower than the threshold, we cross-checked the other value. Since all the other criteria were satisfied, we kept the construct. This allowed us to keep the conceptual meaning of the latent variable intact (Hair et al., 2011a, 2017).

We, also, tested the discriminant validity of the reflective latent constructs using the heterotrait-monotrait (HTMT) ratio technique (Hair et al., 2011b; Hair Jr. et al., 2017). For achieving a higher conceptual distinction between two constructs, the

[3] The municipalities are (i) Thessaloniki, (ii) Ampelokipoi-Menemeni, (iii) Kalamaria, (iv) Kordelio-Evoske, (v) Neosmos, (vi) Neos and (vii) Pylaia-Chortiatis.

[4] Based on the actual population of the latest census (2011) of the Hellenic Statistical Authority.

values of the constructs need to be closer to 0 with an upper limit of 0.80 (Hair et al., 2011a). The results comply with the criteria HTMT <0.850 (Henseler et al., 2015), showing an acceptable conceptual distinction among them and the measurement model's discriminant validity. Therefore, all criteria are above the desired threshold values, ensuring construct reliability and convergent validity (Hair et al., 2017; Latan, 2018; Nicolas et al., 2020; Bagozzi & Yi, 1988).

4.2 Structural Model Verification

The evaluation of the measurement model succeeds the corresponding verification structural model (Hair et al., 2017). Consequently, the cross-validated redundancy and predictive relevance[5] of the main latent variables "*Degree of implementation of planned initiatives*" (0.465) and "*Assessment of the importance of initiatives*" (0.454) (Hair et al., 2011a) are positive and higher than zero. Therefore, the model can be used as a predictor for the latent constructs (Hair et al., 2011a). Additionally, the values of R square and the R square adjusted[6] are close to 1.000 (Hair et al., 2011a). Supplementary, the outer and inner VIF is lower than 3.3, indicating no serious multicollinearity problems (Diamantopoulos & Siguaw, 2006). The effect size (f^2) for the main latent variables "*Assessment of the importance of initiatives*" and "*Degree of implementation of planned initiatives*" (0.354) is acceptable (Hair et al., 2011a). All the values mentioned are acceptable and correspond to the main sample and all subsamples, indicating the conceptual models' predictive ability in relevant samples (Hair et al., 2017). The structural model of the main sample is presented in Fig. 3.

5 Results

5.1 Strategy Shaping Factors and Maturity

It is understood that the existence of a smart city strategy corresponds to the existence of e-government, and its absence rings a bell of ambiguity. As shown in Fig. 4, most municipalities do not have an adequate digital strategy for the expected digital transformation. Only 14% have an e-government strategy and 15% have a complete digital strategy. The paradox of the answers is that there is a very small set of municipalities that state that they have a strategy for smart cities but not for e-government. Moreover, 71% and 81.3% of the municipalities stated that they do not

[5] These criteria were based on the algorithm of blindfolding (Stone-Geisser's Q^2).

[6] R square values above 0.75 for endogenous latent variables in the structural model can be described as substantial.

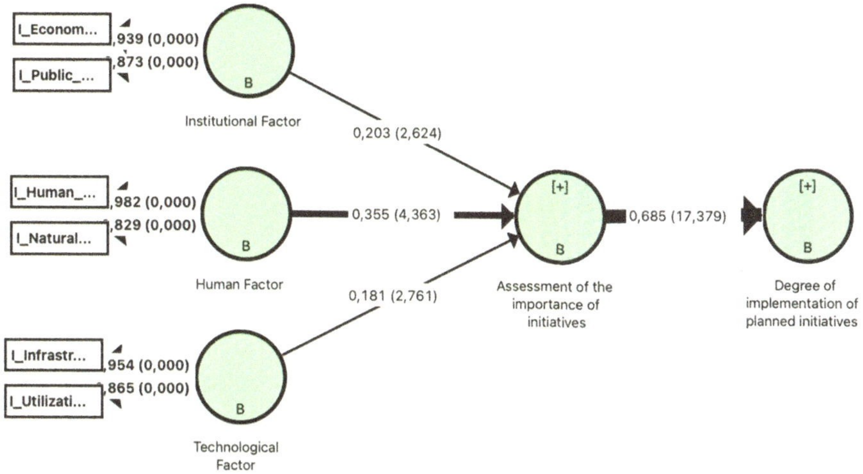

Fig. 3 The structural model with factor loadings and t-values (in parenthesis) of the inner and outer models

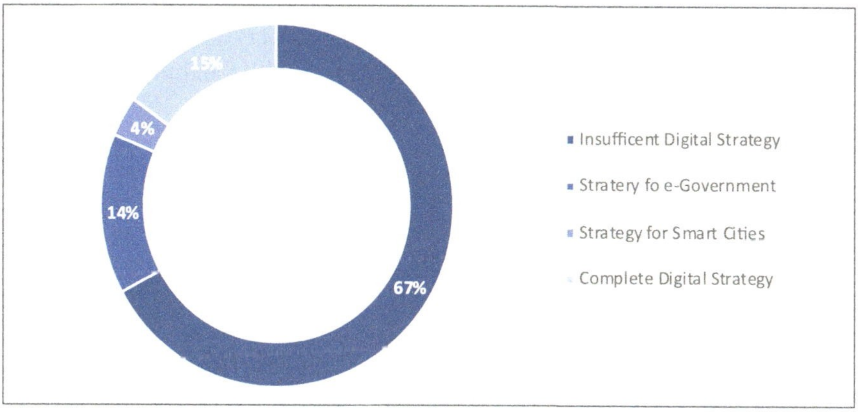

Fig. 4 The level of the digital strategy of the Greek municipalities ($N = 252$) (Source: Author)

have a digital strategy for e-government and smart cities, respectively. Especially, municipalities in islands or regional areas show an absence of a digital strategy at over 80% in contrast with the capital of each region.

There is a sense of familiarity by the public authorities for the concept of "smart city" (M.O. = 3.41). According to the results, there is a significant correlation of understanding the term and the category of city [$F(3, 250) = 8055$, $p = 0.000$]. Through the post hoc comparisons using the Tamhane test (LS[7] = 3257, $p = 0.022$),

[7] The abbreviation stands for Levene statistics.

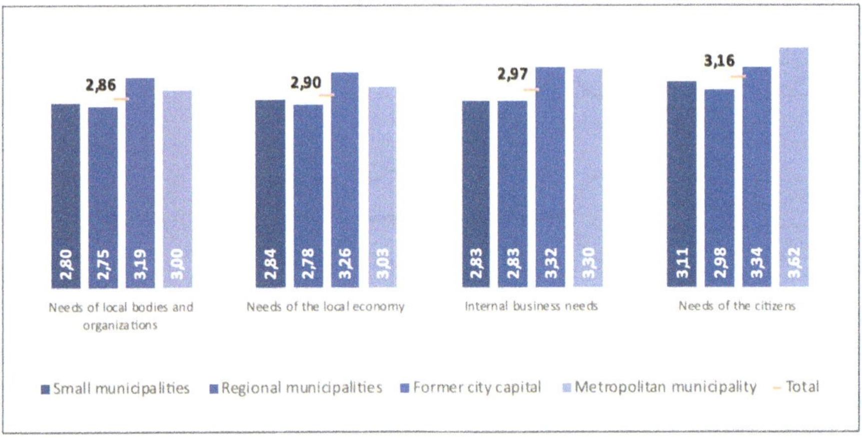

Fig. 5 The degree to which the municipalities consider various factors for the formation of a digital strategy per category of municipality ($N = 250$) (Source: Author)

the average score of the small municipalities (M.O. = 2,94, SD[8] = 0.982) was significantly different from the regional municipalities (M.O. = 3.42, SD = 1.108), the former capitals of regions (M.O. = 3.71, SD = 1.001) and the metropolitan municipality (M.O. = 3.88, SD = 0.954).

Respectively, the degree to which the municipalities consider the factors that concern various needs of stakeholders is moderate (Fig. 5). However, what prevails are the needs of the citizens and the internal business needs of the local government. There was a significant impact on the type of municipality and the functions that influence the strategy. Specifically, the difference is in the functions related to the internal business needs of the municipality [$F(3, 234) = 2644$, $p = 0.050$] and the citizens [$F(3, 236) = 4167$, $p = 0.034$]. Through the post hoc comparisons using the LSD test (LS = 0.228, $p = 0.877$) for the internal business needs, the average score of the regional municipalities (M.O. = 2.83, SD = 1.185) is significantly different ($p < 0.05$) versus the former city capitals (M.O. = 3.32, SD = 1.077) and the metropolitan municipality (M.O. = 3.30, SD = 1.175). Respectively, from the LSD control (LS = 0.876, $p = 0.454$) for the needs of the citizens, the average score of the municipalities of metropolitan centres (M.O. = 3.62, SD = 1.063) is significantly different ($p < 0.05$) from the corresponding regional municipalities (M.O. = 2.98, SD = 1.228) and small municipalities (M.O. = 3.11, SD = 1.279).

[8] The abbreviation stands for standard deviation.

5.2 *Planning and Implementing a Digital Strategy*

This section presents the characteristics of the variables related to the strategic map of Greek municipalities for a digital strategy. In the beginning, we conduct a descriptive analysis of the central model's variables, and, then, the structure of the respective latent variables is presented. Specifically for the first part, via the evaluation process, a cluster of cities with similar characteristics was performed. Regarding the planning and implementation of smart actions, the overall scenery emphasizes the inability of most municipalities to implement actions based on their strategic planning (Fig. 6). Almost half of the municipalities (46.6%) are characterized as "Challenger", and about one-fifth of the municipalities show a competitive position. The former city capitals show good performance, where 32.4% are characterized as "Contender" while 61% of the municipalities of metropolitan municipalities are accumulated in the penultimate level. According to the results of the statistical analysis, $\chi 2$ (9, N = 247) = 20.370, p = 0.016, a statistically significant difference is identified. The former capitals show the best performance overall and try their actions to be in line with the corresponding level of smart strategy they have established. The following is a synoptic presentation of each thematic area.

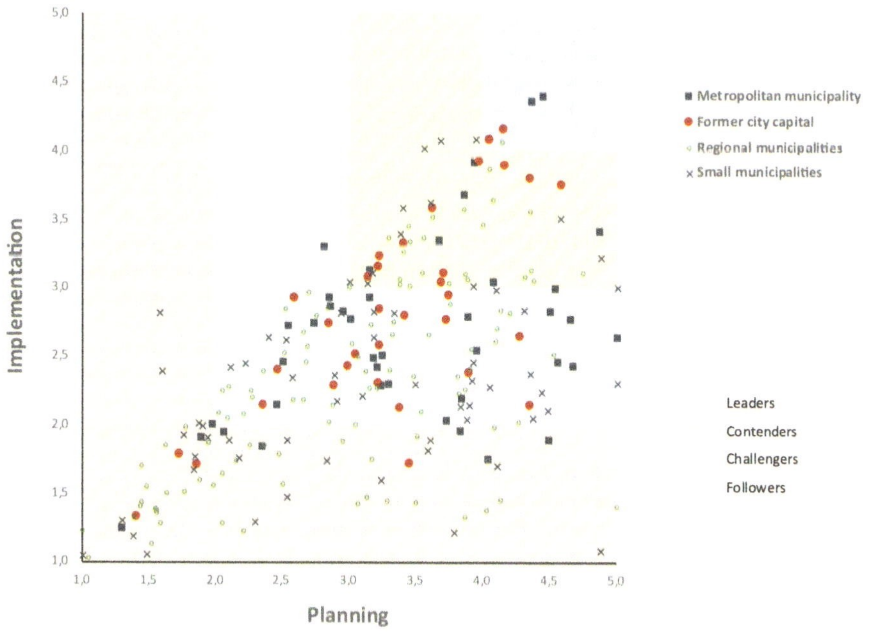

Fig. 6 Ranking of the different municipalities based on the intensity of planning and implementation of smart actions in their entire strategy (Source: Author)

Based on the intensity of planning and implementing smart actions of e-government and e-services, most municipalities appear in the category "Challengers" with 43.6%, and only 4% is in the leading position. It has been observed that many of the metropolitan municipalities (85.0%) and capitals of former prefectures (73.5%) have accumulated in the categories "Challengers" and "Contenders", while most of the regional mainland and island municipalities (78.6%) and small municipalities (84.4%) are concentrated in the last two categories. In addition, there is a statistically significant relationship (>99%) between the intensity of implementation of smart actions of the government and e-services based on the category of municipalities, x^2 (9, $N = 250$) = 24.389, $p = 0.004$. The municipalities with a central role seem to have integrated digital actions in their strategy, and, at the same time, they show better performance in their implementation, compared to other regional municipalities.

For actions related to urban planning, transportation system and quality of life, the picture is different from that of the previous category. The majority of municipalities (47.0%) are accumulated at the lowest level as "Followers", and while they appear to have integrated digital actions in their strategy, the level of implementation is low. More specifically, more than half of the regional and small municipalities (55.0% and 50.8%, respectively) are characterized as "Followers", and in the other two categories, 48.8% of the metropolitan municipalities and 41.2% are characterized as "Challengers". It appears that the category lacks the implementation of the relevant actions. Supplementary, there is a statistically significant relationship (>95%) between the intensity of implementation of smart actions and the category of municipalities, x^2 (9, $N = 247$) = 18.625, $p = 0.029$. Municipalities with a central role appear to be in a better position compared to other regional municipalities.

The image formed by the actions of management of natural resources and energy is similar regardless of the category of the municipality. In other words, the majority of metropolitan municipalities (48.7%), former city capitals (35.3%), regional municipalities (55.6%) and small municipalities (65.1%) are characterized as "Followers". The absence of a statistically significant relationship between the groups is confirmed by the control results, x^2 (9, $N = 244$) = 10.782, $p = 0.291$. Therefore, most municipalities (54.1%) show a low degree of implementation of actions related to energy and natural resource management.

Vis-à-vis the strategy on the environment, about one-third of the municipalities are accumulated in the category "Challengers" (38.4%), and about 6% are characterized as "Leaders". Specifically, 43.6% of the metropolitan municipalities are characterized as "Challenger", and the regional municipalities and small municipalities are in the category "Follower" by 44.0% and 41.3%, respectively. Only the group of former city capitals show a differentiation with 35.3% being in the category "Contenders". The differentiation per category of municipality shows a statistically significant difference at the level of 95%, x^2 (9, $N = 245$) = 19.238, $p = 0.023$. Therefore, they appear to have implemented more actions related to the environment compared to the other municipalities of the country.

For the actions on the economy, education and innovation, the largest part of the sample (54.5%) is characterized as "Follower". It is observed that the regional municipalities and the small municipalities show a high accumulation in this category (64.5% and 52.4%, respectively). The same pattern applies for the metropolitan municipalities and former city capitals in the last two categories (39.5% and 44.7% against 44.1% and 35.5%, respectively). Considering the results of the significance of the differences between the groups, x^2 (9, $N = 242$) = 21,774, $p = 0.010$, the larger municipalities or municipalities with central roles appear to invest more in relevant issues.

In the category relevant to the infrastructure, there is a critical mass of municipalities characterized as "Challenger" (40.1%). While there is an uneven distribution of municipalities, almost 20% is considered a "Contenders" municipality. Along with the visual observation, there is a statistically significant difference (level of 95%), x^2 (9, $N = 247$) = 17,734, $p = 0.038$. It is understood that at the level of infrastructure, the central actions of a state and the private bodies, which implement the investments, play an important role. Specifically, several municipalities show a high degree of implementation of corresponding actions compared to other categories that have been studied so far.

5.3 *The Roadmap to a Smart City*

The results for evaluating the different direct and indirect correlations are shown in Table 1. In general, the creation of a strategy is affected from planning to implementation. According to the results of the model, multiple human and social ($\beta = 0.355$, $t = 4.363$, $p = 0.000$), institutional ($\beta = 0.203$, $t = 2.624$, $p = 0.008$) and technological ($\beta = 0.181$, $t = 2.761$, $p = 0.005$) factors impact the stage of planning and prioritizing the actions for a smart city strategy. Regarding the former city capital, the planning stage is affected by the human and social ($\beta = 0.587$, $t = 3.996$, $p = 0.000$) and institutional ($\beta = 0.323$, $t = 2.339$, $p = 0.021$) factors. A different picture is drawn for the metropolitan municipalities. The most significant factor for planning is technology ($\beta = 0.346$, $t = 2.431$, $p = 0.015$). For the regional municipalities, the human and social ($\beta = 0.310$, $t = 2.908$, $p = 0.004$) and technological ($\beta = 0.391$, $t = 4.116$, $p = 0.000$) factors have a significant impact. For the fourth category, the small municipalities show no impact at a level of significance of 95%. Instead, the corresponding 90%, the human and social ($\beta = 0.359$, $t = 1.909$, $p = 0.064$) and institutional ($\beta = 0.349$, $t = 1.850$, $p = 0.056$) factors show a significant impact.

Table 1 The total effects of the different factors for each group category

Total effect (p-value)	Complete	Former city capital	Metropolitan municipality	Regional municipality	Small municipality
Direct effect → degree of implementation of planned initiatives					
Assessment of the importance of initiatives	0.685(0.000)	0.899(0.000)	0.829(0.00[illegible])	0.610(0.000)	0.612(0.000)
Direct effect → assessment of the importance of initiatives					
Human factor	0.355(0.000)	0.587(0.000)	0.189(0.96[illegible])	0.310(0.003)	0.359(0.062)
Institutional factor	0.203(0.008)	0.323(0.021)	0.359(0.91[illegible])	−0.023(0.835)	0.349(0.059)
Technological factor	0.181(0.005)	0.061(0.569)	0.346(0.01[illegible])	0.391(0.000)	−0.050(0.737)
Indirect effect → degree of implementation of planned initiatives via *assessment of the importance of initiatives*					
Technological factor	0.124(0.008)	0.055(0.571)	0.288(0.6[illegible])	0.238(0.000)	−0.031(0.740)
Human factor	0.243(0.000)	0.528(0.000)	0.157(0.9[illegible])	0.189(0.008)	0.220(0.083)
Institutional factor	0.139(0.011)	0.291(0.021)	0.297(0.9[illegible])	−0.014(0.836)	0.214(0.073)

6 Discussion

The uniqueness of this analysis is the indirect effects that each factor has on planning and implementing for each category. According to Table 1, the former city capitals show a significance for the human and institutional factors, and the regional municipalities show a significance for the technological and human factors. Whereas the metropolitan municipalities do not show any signs, the smaller municipalities show a significance for the human and institutional factors at 90%.

So far, most applications focus on the environment and mobility, while governance and living issues are deteriorating. In other areas such as people and the economy, they are treated with due importance, but their actions and reporting are small. Most of the time, businesses and governments cannot invest in everything at the same time due to insufficient resources. Each municipality shows a unique set of characteristics, which are justified by their general characteristics. In smaller municipalities, where there are lower levels of interactions between different organizations, there is a need for them to participate in different collaborative projects with other municipalities to help them gain access to knowledge and experience and learn from others. In former city capitals, policies and central funding can help them implement projects and get started on their digital transformation. On the contrary, the regional municipalities do not have strong institutional organizations in their urban environment and focus more on the human and technological factors. Finally, the metropolitan municipalities have a strong connection with the technology. They, already, have concentrated on the major institutional organizations and human capital, and they can focus on the technological aspects and implementations.

Finally, international experience provides a guarantee of success. Common infrastructures and directly related interdisciplinary innovative actions can be recognized as pillars of local community development and the emergence of collaborative innovation poles. Regarding Greece, cities need to focus on the eight priority areas of the National Strategic Framework for Research, Technological Development, and Innovation: (1) Agri-food; (2) Life Sciences, Health & Medicine; (3) ICT; (4) Energy; (5) Environment & Sustainable Development; (6) Transport & Supply Chain; (7) Materials & Constructions; and (8) Culture, Tourism & Creative Industries. At the same time, there is a need to mobilize human resources per region with experience and know-how in cutting-edge sectors and form synergies with innovative private and public sector actors. For example in Europe, the involvement of academic research teams in synergies with public authorities for the design, development and expansion of smart cities has shown great potential.

7 Conclusion

A municipality needs to form a department or office responsible for the digital transformation of the municipality. Of course, depending on the size of the municipality and the characteristics of the urban ecosystem, it can, also, form an office of technology and innovation. The reason for the creation of the department lies in the need for strong coordination and constant monitoring of the usable resources and the progress of the transformation. The department should be responsible for shaping and developing the leadership capacity of the relevant organizations, the organizations' collaborations and the innovation. Additionally, the municipalities should look for partnering with other organizations and participating in national and European projects regarding their digital transformation. This will help them find the right solutions for their urban environment, gain from the experience of other cities and fund projects with short- and long-term effects.

In conclusion, this study can be the stepping stone for further studying the effect and implications of the key dimensions in a smart city. One important factor is the indirect effect a policy may have at the stage of implementation. The results could, also, help formulate targeted workshops or case studies that could provide targeted insights on each category of municipality. This helps overcome potential challenges and apply strategies according to the initial planning. Additionally, stakeholders (e.g., workers, students, elderly and citizens) familiarize themselves with new technologies, thus supporting indirectly the need for smart specialization in the urban environment. At the same time, the development of a roadmap with a vision, accompanied by long-term and short-term goals, transforms the municipality into a technological, social and cultural mechanism of political and digital actions. Therefore, a roadmap properly adapted to the specifics of each municipality can be a mechanism of strategic development in a multidimensional and rapidly growing urban environment while ensuring the appropriate conditions for the economic and innovative flourishing of the urban system.

References

Alexopoulos, C., Charalabidis, Y., Kolokotronis, D. E., & Vogiatzis, N. (2018). A taxonomy for analysing smart cities developments in Greece. *ACM International Conference Proceeding Series*, 537–549. https://doi.org/10.1145/3209415.3209471

Al-Hader, M., & Rodzi, A. (2009). The smart city infrastructure development & monitoring. *Theoretical and Empirical Researches in Urban Management, 4*, 87–94.

Al-Hader, M., Rodzi, A., Sharif, A. R., & Ahmad, N. (2009). Smart city components architecture. In *CSSim 2009—1st International conference on computational intelligence, modelling, and simulation* (pp. 93–97). doi: https://doi.org/10.1109/CSSim.2009.34

Anthopoulos, L., Ziozias, C., & Siokis, A. (2021). Shaping a digital transformation strategy for smart cities under the COVID-19 pandemic: Evidence from Greece. *The web conference 2021—Companion of the world wide web conference, WWW 2021* (Vol. 2, pp. 657–660). doi: https://doi.org/10.1145/3442442.3453470.

Atzori, L., Iera, A., & Morabito, G. (2010). The internet of things: A survey. *Computer Networks, 54*, 2787–2805. https://doi.org/10.1016/j.comnet.2010.05.010

Awad, J., Hyder, A., & Irfan, A. (2017). Development of smart cities from fiction to reality in member states of the gulf cooperation council. In *Smart city networks* (pp. 43–63).

Bagozzi, R. P., & Yi, Y. (1988). On the evaluation of structure equation models. *Journal of the Academy of Marketing Science, 16*, 74–94. https://doi.org/10.1007/BF02723327

Bagozzi, R. P., & Yi, Y. (2012). Specification, evaluation, and interpretation of structural equation models. *Journal of the Academy of Marketing Science, 40*, 8–34. https://doi.org/10.1007/s11747-011-0278-x

Berger, R. (2017). *Smart city, smart strategy* (pp. 1–15).

Capello, R., & Kroll, H. (2016). From theory to practice in smart specialization strategy: Emerging limits and possible future trajectories. *European Planning Studies, 24*, 1393–1406. https://doi.org/10.1080/09654313.2016.1156058

Caragliu, A., del Bo, C., & Nijkamp, P. (2011). Smart cities in Europe. *Journal of Urban Technology, 18*, 65–82. https://doi.org/10.1080/10630732.2011.601117

Caragliu, A., & del Bo, C. F. (2015). Smart specialization strategies and smart cities: An evidence-based assessment of European Union policies. In *The rise of the city*. Edward Elgar Publishing.

Chourabi, H., Nam, T., Walker, S., et al. (2012). Understanding smart cities: An integrative framework. *Proceedings of the Annual Hawaii International Conference on System Sciences*, 2289–2297. https://doi.org/10.1109/HICSS.2012.615

Diamantopoulos, A., & Siguaw, J. A. (2006). Formative versus reflective indicators in organizational measure development: A comparison and empirical illustration. *British Journal of Management, 17*, 263–282. https://doi.org/10.1111/j.1467-8551.2006.00500.x

Edmonton City. (2018) *Smart city strategy 2017—City of Edmonton*. Edmonton.

European Commission. (2014). *European innovation partnerships—Innovation union—European Commission*. Retrieved February 15, 2021, from https://ec.europa.eu/info/research-and-innovation/strategy/goals-research-and-innovation-policy/open-innovation-resources/european-innovation-partnerships-eips_en

European Commission. (2020). *Digital agenda for Europe | Fact sheets on the European Union | European Parliament*. Retrieved February 15, 2021, from https://www.europarl.europa.eu/factsheets/en/sheet/64/digital-agenda-for-europe

Fernandez-Anez, V., Fernández-Güell, J. M., & Giffinger, R. (2018). Smart City implementation and discourses: An integrated conceptual model. The case of Vienna. *Cities, 78*, 4–16. https://doi.org/10.1016/j.cities.2017.12.004

Foray, D. (2014). *Smart specialisation: Opportunities and challenges for regional innovation policy*. Routledge.

Foray, D. (2016). On the policy space of smart specialization strategies. *European Planning Studies, 24*, 1428–1437. https://doi.org/10.1080/09654313.2016.1176126

Foray, D., David, P. A., & Hall, B. (2009). Smart specialisation–the concept. *Knowledge economists policy brief, 9*, 100.

Gann, D. M., Dodgson, M., & Bhardwaj, D. (2011). Physical–digital integration in city infrastructure. *IBM Journal of Research and Development, 55*, 8:1–8:10. https://doi.org/10.1147/JRD.2010.2095750

Gaur, A., Scotney, B., Parr, G., & McClean, S. (2015). Smart city architecture and its applications based on IoT. *Procedia Computer Science, 52*, 1089–1094. https://doi.org/10.1016/j.procs.2015.05.122

Giffinger, R., Fertner, C., Kramar, H., & Meijers, E. (2007). *City-ranking of European medium-sized cities* (pp. 1–12). Centre of Regional Science.

Hair, J. F., Hult, G. T. M., Ringle, C. M., et al. (2017). Mirror, mirror on the wall: A comparative evaluation of composite-based structural equation modeling methods. *Journal of the Academy of Marketing Science, 45*, 616–632. https://doi.org/10.1007/s11747-017-0517-x

Hair, J. F., Ringle, C. M., & Sarstedt, M. (2011a). PLS-SEM: Indeed a silver bullet. *Journal of Marketing Theory and Practice, 19*, 139–152. https://doi.org/10.2753/MTP1069-6679190202

Hair, J. F., Celsi, M., Money, A., et al. (2011b). *Essentials of business research methods*. ME Sharpe.

Hair, J. F., Ringle, C. M., & Sarstedt, M. (2012). Partial least squares: The better approach to structural equation modeling? *Long Range Planning, 45*, 312–319. https://doi.org/10.1016/j.lrp.2012.09.011

Hair, J. F., Jr., Matthews, L. M., Matthews, R. L., & Sarstedt, M. (2017). PLS-SEM or CB-SEM: Updated guidelines on which method to use. *International Journal of Multivariate Data Analysis, 1*, 107. https://doi.org/10.1504/ijmda.2017.10008574

Hassink, R., & Gong, H. (2019). Six critical questions about smart specialization. *European Planning Studies, 27*, 2049–2065. https://doi.org/10.1080/09654313.2019.1650898

Heaton, J., & Parlikad, A. K. (2019). A conceptual framework for the alignment of infrastructure assets to citizen requirements within a smart cities framework. *Cities, 90*, 32–41. https://doi.org/10.1016/j.cities.2019.01.041

Henseler, J., & Chin, W. W. (2010). A comparison of approaches for the analysis of interaction effects between latent variables using partial least squares path modeling. *Structural Equation Modeling, 17*, 82–109. https://doi.org/10.1080/10705510903439003

Henseler, J., Ringle, C. M., & Sarstedt, M. (2015). A new criterion for assessing discriminant validity in variance-based structural equation modeling. *Journal of the Academy of Marketing Science, 43*, 115–135. https://doi.org/10.1007/s11747-014-0403-8

Huawei. (2015). Global Connectivity Index 2015.

Kalaitzakis, M., Bouloukakis, M., Charalampidis, P., et al. (2019). Building a Smart City ecosystem for third party innovation in the City of Heraklion. In *Mediterranean cities and Island communities* (pp. 19–56). Springer.

Kanellos, N., & Siokas, G. (2021). Smart municipalities in Europe and Greece: A comparative SWOT analysis. In D. P. Sakas, D. K. Nasiopoulos, & Y. Taratuhina (Eds.), *Business intelligence and modelling* (pp. 149–155). Springer International Publishing.

Kummitha, R. K. R. (2018). Entrepreneurial urbanism and technological panacea: Why Smart City planning needs to go beyond corporate visioning? *Technological Forecasting and Social Change, 137*, 330–339. https://doi.org/10.1016/j.techfore.2018.07.010

Kummitha, R. K. R., & Crutzen, N. (2017). How do we understand smart cities? An evolutionary perspective. *Cities, 67*, 43–52.

Lam, P. T., & Yang, W. (2020). Factors influencing the consideration of Public-Private Partnerships (PPP) for smart city projects: Evidence from Hong Kong. *Cities, 99*, 102606.

Latan, H. (2018). Chapter 4 PLS path modeling in hospitality and tourism research: The golden age and days of future past.

Letaifa, S. B. (2015). How to strategize smart cities: Revealing the SMART model. *Journal of Business Research, 68*, 1414–1419. https://doi.org/10.1016/j.jbusres.2015.01.024

Leuven. (2021). *Home | Leuven2030*. https://www.leuven2030.be/. Accessed 15 Feb 2021.

Lim, C., Kim, K. J., & Maglio, P. P. (2018). Smart cities with big data: Reference models, challenges, and considerations. *Cities, 82*, 86–99. https://doi.org/10.1016/j.cities.2018.04.011

London, S. (2021a). *Smart London Camp 2018—How it happened | by Smart London | Medium*. https://smartlondon.medium.com/smart-london-camp-2018-how-it-happened-ce5ffe84fc80. Accessed 15 Feb 2021.

London, S. (2021b). *Six trends in smart city strategic planning from the U.S. & Canada | by smart London | medium.*. https://smartlondon.medium.com/six-trends-in-smart-city-strategic-planning-from-the-u-s-c553c2f88e02. Accessed 15 Feb 2021.

Marava, N., Alexopoulos, A., & Stratigea, A. (2019). Tracking paths to smart governance: The case of Korydallos municipality—Greece. In *Mediterranean cities and island communities* (pp. 81–112). Springer.

Monzon, A. (2015). Smart cities concept and challenges. In *2015 International conference on smart cities and green ICT systems (SMARTGREENS)* (pp. 17–31). doi: https://doi.org/10.1007/978-3-642-33489-4_4

Musa, S. (2018). Smart cities—A road map for development. *IEEE Potentials, 37*, 19–23. https://doi.org/10.1109/MPOT.2016.2566099

Nam, T., & Pardo, T. A. (2011). Conceptualizing smart city with dimensions of technology, people, and institutions. *ACM International Conference Proceeding Series*, 282–291. https://doi.org/10.1145/2037556.2037602

Negre, E., & Rosenthal-Sabroux, C. (2014). Recommendations to improve the smartness of a City. In R. P. Dameri & C. Rosenthal-Sabroux (Eds.), *Smart city* (pp. 101–115). Springer International Publishing.

Neirotti, P., de Marco, A., Cagliano, A. C., et al. (2014). Current trends in smart city initiatives: Some stylised facts. *Cities, 38*, 25–36. https://doi.org/10.1016/j.cities.2013.12.010

Nicolas, C., Kim, J., & Chi, S. (2020). Quantifying the dynamic effects of smart city development enablers using structural equation modeling. *Sustainable Cities and Society, 53*, 101916. https://doi.org/10.1016/j.scs.2019.101916

Satterthwaite, D. (2008). Cities' contribution to global warming: Notes on the allocation of greenhouse gas emissions. *Environment and Urbanization, 20*, 539–549.

Shichiyakh, R. A., Klyuchnikov, D. A., Balashova, S. P., et al. (2016). Smart city as the basic construct of the socio-economic development of territories. *International Journal of Economics and Financial Issues, 6*, 157–162.

Singh, S., Sharma, P. K., Yoon, B., et al. (2020). Convergence of blockchain and artificial intelligence in IoT network for the sustainable smart city. *Sustainable Cities and Society, 63*. https://doi.org/10.1016/j.scs.2020.102364

Siokas, G., & Tsakanikas, A. (2022). Questionnaire dataset: The Greek smart cities-municipalities dataset. *Data in Brief, 40*, 107716. https://doi.org/10.1016/j.dib.2021.107716

Siokas, G., Tsakanikas, A., & Siokas, E. (2021). Implementing smart city strategies in Greece: Appetite for success. *Cities, 108*, 102938. https://doi.org/10.1016/j.cities.2020.102938

Suzuki, L., & Finkelstein, A. (2018). *Data as infrastructure for smart cities*. Institution of Engineering and Technology.

Tenenhaus, M., Vinzi, V. E., Chatelin, Y. M., & Lauro, C. (2005). PLS path modeling. *Computational Statistics and Data Analysis, 48*, 159–205. https://doi.org/10.1016/j.csda.2004.03.005

Woods, E., Labastida, R. R., Chow, T., et al. (2017). UK smart cities index 2017: Assessment of strategy and execution for the UK's leading smart cities.

Yigitcanlar, T., Kamruzzaman, M., Buys, L., et al. (2018). Understanding 'smart cities': Intertwining development drivers with desired outcomes in a multidimensional framework. *Cities, 81*, 145–160. https://doi.org/10.1016/j.cities.2018.04.003

Zanella, A., Bui, N., Castellani, A., et al. (2014). Internet of things for smart cities. *IEEE Internet of Things Journal, 1*, 22–32. https://doi.org/10.1109/JIOT.2014.2306328

Σαριδάκης ΙΧ (2019) Ελληνικό Πρότυπο Διαχείρισης Βιωσιμότητας Πόλεων.

Sustainable Digital Transformation of Urban Landscape Through Disruptive Technologies and Standards

N. Kishor Narang

1 Introduction

Cities nationally and internationally are main drivers of economic activity, growth and, in the current context, recovery, but this output depends on a comprehensive infrastructure to deliver physical and social resources—the fuel of a city's "economic engine." The economic performance of a city is inextricably linked to its physical and communications infrastructures, and the delivery of resources through these infrastructures. Rapid urbanization over the past two decades has led to the mushrooming of megacities (accepted as those with a population more than 10 million) around the world. The sheer size and scale of these cities place huge pressure on infrastructure development, public services provision, and environmental sustainability. If we add economic, social, and ethnic stratification, as well as health, safety, and security risks to the list of challenges, the task facing the leader of any megacity seems overwhelming and is certainly one that cannot be solved by technology alone. As traditional resource delivery systems approach the limits of their capability, there is an urgent need for innovative delivery systems to effectively manage and control resource use in cities. Ensuring that the world's cities offer citizens a rich and rewarding lifestyle requires that cities exploit technology to enrich people's lives, deliver services, and ensure sustainable growth.

The breadth and scope of this task touches on many areas and requires a holistic approach that not only looks at core technical issues but also needs to consider the management, process, and strategies associated with smart cities. As always, standards play a key role in facilitating the adoption of new technologies, as well as efficient management and governance of the cities, and are critical to the growth

N. K. Narang (✉)
Narnix Technolabs Pvt. Ltd., New Delhi, India
e-mail: kishor@narnix.com

P. Fitsilis (ed.), *Building on Smart Cities Skills and Competences*, Internet of Things,
https://doi.org/10.1007/978-3-030-97818-1_6

of smart cities worldwide. One of the most crucial imperatives to make cities truly smart, sustainable, and resilient is the availability of adequate skilled manpower to successfully implement the vision through efficient management and governance and deploy the required technology solutions and infrastructures. Smart cities development is a complex and cross-domain subject and needs a combination of diverse skill sets and competencies based on the class of stakeholders one belongs to.

2 Urban Landscape

A vast body like a city consists of a large number of heterogeneous information resources. These include sensors, exchanges between or information from citizens, the various workflows and processes, events that occur, etc. that can together complement the integrated management of smart cities. The relationships of these diverse information resources are complicated and could be complementary, reinforced, or redundant. The data gathered can further be processed and modeled, correlated with historic data and other activities performed on it before it can be made insightful and can be presented to offer analysis, decision support, or forecasts. There is also a recursive cycle to the data in a smart city. Information that is generated is information that is consumed which in turn adds to the information generated which becomes new information to be used again.

Since smart city is a complex system of systems, involving many different domains, infrastructures, organizations, and activities, it is imperative that all these need to be integrated and work together effectively and homogeneously for that city to become smart in the real sense. Knowing that transformation of existing cities into smart and sustainable cities is a daunting and complex task, to purposely transform them, experts with different knowledge, expertise, competencies, and skills shall need to examine their complex needs from different points of view, and all those views and models must be systemically aligned. This shall considerably increase the probability that desired transformation will be valid, successful, and scalable because various aspects and components will be designed to fit together.

All smart city programs and projects pursue many common goals including sustainable development, better efficiency, resilience, safety, and wider support for citizen's engagement and participation. However, each individual city tends to follow its own approach in smart city programs and projects. It is not surprising that the numerous technology activists are very vocal on various smart cities' forums even though cities cannot be reduced to just "big data" and "IoT."

3 Disruptive Technologies Landscape

The pace of advancements in technologies is increasing at an unprecedented rate. This rapid change is largely fueled by the democratization of innovation as cutting-edge technologies like cloud computing are providing modern companies access to new disruptive technologies. They have a lot to offer to an organization to transform its traditional business processes, advancing efficiency and leading to competitive advantages at the same time.

A little crystal ball gazing—the three mega trends:

- AI everywhere—deep learning, deep reinforced learning, artificial general intelligence, autonomous vehicles, cognitive computing, commercial UAVs (drones), conversational user interface, enterprise taxonomy, ontology management, machine learning, smart dust, smart robots, and smart workplace
- Transparently immersive experience—4D printing, augmented reality, virtual reality, brain-computer interface, connected homes, human augmentation, nanotube electronics, and volumetric displays
- Digital platforms—5G, digital twin, edge computing, blockchain, IoT platform, neuromorphic hardware, quantum computing, serverless PaaS, software-defined security, etc.

Disruption is everywhere and the future is uncertain—no one knows what the world will look like even a decade from now. As we head into the future, we are surrounded by disruptive innovation. Whether it's artificial intelligence, driverless cars, space exploration, or quantum computing, it can be hard for even the most enthusiastic technophiles to stay up to date with all the rapid advances taking place and coming down the line. These innovations are changing the world as we know it, e.g., how the combination of AI, IoT, and blockchain is creating both disruption and opportunity in the enterprise world, not just for consumer facing businesses but for all enterprises. Over the coming year, what will be the most important developments in disruptive technology?

As we look to the years and decades ahead, tech disruption will be driven as much by the methods and systems as it is by the devices; we associate with tech disruption. The pace of innovation is incredibly fast, with new things getting discovered daily. The future trends in technology are very diverse, very intertwined, and very promising. There are several developments that have and will continue to shape business strategies. From automation to sustainability, organizations are adapting to a whole new wave of consumer preferences.

4 Standards Landscape

Standards development organizations (SDOs) are busy mapping the imperatives for standardization in the smart cities and smart infrastructure domain including IEC, ISO, ITU, IEEE, and IETF along with 3gpp, oneM2M, and other regional and national SDOs like ETSI, CEN, CENELEC, NIST, ANSI, BSI, DKE, DIN, JSA, and BIS. Furthermore, many standardization bodies and industry fora from the ICT and infrastructure industry consider smart city as a priority issue. All the global SDOs, industry consortia, and fora have been addressing the development of reference frameworks, architectures, and standards in this domain. Given the scale, moving forward cannot be successfully, efficiently, and swiftly accomplished without standards. The role of standards to help steer and shape this journey is vital. Standards provide a foundation to support innovation. Standards capture tacit best practices and standards set regulatory compliance requirements.

Innovation and technology development are accelerating. Strategic plans and roadmap are needed to help ensure that the market is suitably served with best practices that are pertinent to the goals and context of this very large market. The standards support our need to balance agility, openness, and security in a fast-moving environment. The standards provide us with a reliable platform from which we can innovate, differentiate, and scale up our technology development. They help us control essential security and integrate the right level of interoperability. Standards help ensure cybersecurity in ICT and IoT systems. The respective ecosystems of smart infrastructure, smart manufacturing, health, education, banking, administration, governance, etc. will require an unprecedented integration of systems across domains, hierarchic boundaries, and lifecycle phases. System standards will be needed for the automation and digitalization of our systems and solutions.

The world has never been as competitive as today, yet cooperation is a must to deliver solutions for increasingly complex systems. No technical committee and no standards organization can single-handedly develop all the standards that are needed. We all need to work together. However, standards and even SDOs are not at the forefront of cities', utilities, or users' minds. There are misconceptions on what standards are for, and the case for use of standards has not been made. Liberalization and markets have a lot of great virtues, but they cannot create their own conditions of existences: they must be designed! (Fig. 1).

Furthermore, standards need to be supported by enabling technical regulations and policies to ensure that the developed standards are provided a conducive policy and regulatory framework to be seamlessly adopted by all the stakeholders (Narang, 2020).

4.1 Categorizing Standardization Activities

A useful way to categorize these different types of standardization activities is to group them by level of abstraction into strategic, process, and technical:

- Level 1: Strategic. These are smart city standards that aim to provide guidance to city leadership and other bodies on the "process of developing a clear and

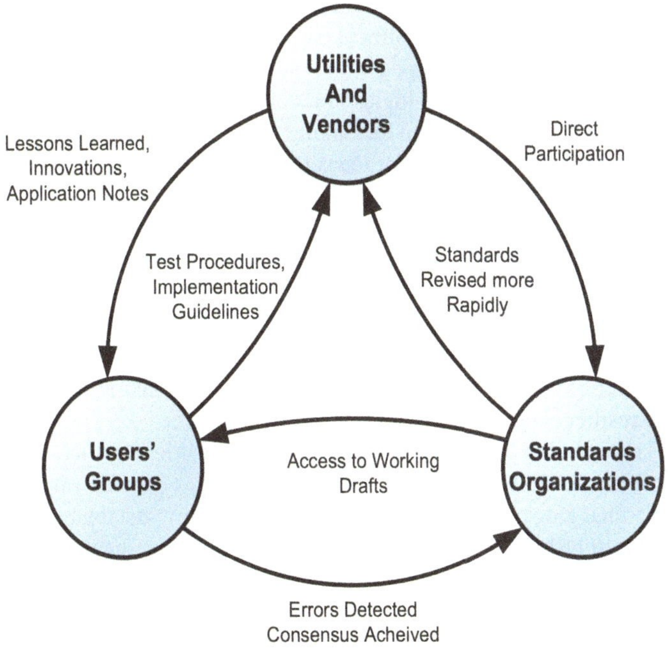

Fig. 1 Standardization ecosystem (credit—IEEE SA)

effective overall smart city strategy." They include guidance in identifying priorities, how to develop a roadmap for implementation, and how to effectively monitor and evaluate progress along the roadmap.

- Level 2: Process. Standards in this category are focused on procuring and managing smart city projects—particularly those that cross both organizations and sectors. Essentially these offer best practices and associated guidelines.
- Level 3: Technical. This level covers the myriad technical specifications that are needed to implement smart city products and services so that they meet the overall objectives.

5 Digital Transformation

The new paradigm of smart grid, smart home, smart building, smart manufacturing, and smart city already complicated by the "Internet of Things" and Internet of "Everything" made further complex by the 5G, artificial intelligence, machine learning, blockchain, quantum computing, and ever-evolving cyber-threat landscape demand a fresh perspective in critical/civic infrastructure design.

The recent evolution of disruptive technologies and digitalization compounded by the Covid-19, changing geopolitical situations, and increasing cyberattacks from not-so-friendly nations brings a whole new set of challenges for the security and security evaluation methodologies for complex nature and architectures of critical infrastructures of any nation leveraging the IT and communication networks evolving to meet these rising needs of the society. On one hand, we have the highly protected networks for the "Critical Information Infrastructures"; on the other hand, these very "highly protected networks" need to give access to the consumers for consumer engagement and participation in these smart (digital) infrastructures to meet the true drivers of setting them up. These large smart networks are actually highly complex "Systems of Systems" and "Networks of Networks," and thus create fresh challenges in architecting approaches that need to develop comprehensive and granular and yet flexible and scalable standards to ensure making the infrastructure sustainable, resilient, and future proof.

All the nations are going through an urban transition; rather, some might say transformation. Without doubt, modern technologies provide opportunities to deliver game-changing outcomes, which will deliver a more sustainable and resilient society, and must be built intelligently into the fabric of that transformation process. In many ways the opportunity is to reinvent the model for urban living, a model that ensures responsible resource consumption, and one that ensures prosperity, equality, societal cohesion, and happiness. All the ecosystems, be it smart cities, smart grid, smart buildings, or smart factories, now find themselves making three classes of transformations:

- Improvement of infrastructure—to make it resilient, sustainable, etc.
- Addition of the digital layer—which is the essence of the smart paradigm
- Business process transformation—necessary to capitalize on the investments in smart technology

The genesis of digital transformation—in digital transformation in any paradigm, domain, or ecosystem—sustainability is the true destination, resilience is the core characteristic, smart is merely the accelerator, and standards are the chromosomes of digital infrastructure. Hence, digital transformation is not a technology. It's a complex paradigm with domain-specific implications. We are living in an ephemeral world.

6 Imperatives

11th UN Sustainable Development Goal—"Make cities and human settlements Inclusive, Safe, Resilient & Sustainable." One of the Sustainable Development Goals (SDG) "Sustainable Cities and Communities" (SDG11) is about an uber-complex domain (also called "the urban landscape") that urgently requires its considerable improvement for many good reasons including but not limited to already stated by the UN. The well-being of residents is critical in sustainable cities, which means guaranteed access to quality education, safe health centers, easy-to-access public transportation, garbage collection services, safety, and good air quality, among other modern living necessities. The case for sustainable urban development can be made in terms of interlinked economic, social, and environmental benefits. Economically the benefits include agglomeration economies, lower infrastructure costs, and reduced congestion cost while reducing carbon emissions and other environmental pressure.

6.1 Citizens' Needs

Cities face a whole set of challenges in providing for the needs of its residents—the citizens. This is because sustainable development of any nation depends on the development of sustainable cities, which can only be achieved through the wide-reaching rollout of integrated, scalable, smart/sustainable city/community solutions. Sustainable, smart cities and communities will contribute to sustainable development and resilience through soundly based decision-making, and the adoption of both a long- and a short-term strategy.

6.2 Dealing with Change

If providing for these many needs were not enough, cities also need to deal with change. There are short-term shocks, such as pandemics, natural disasters, and terrorist attacks. Cities need to always continue transforming themselves to align with the various ever-changing and evolving needs and expectations of their respective citizens. There are changes that take place over a few years, such as the changes in people's lifestyle due to technology innovation and gradual modification of social norms. People are increasingly moving to online shopping, undermining the viability of town centers and even of retail parks and shopping malls. The increasing use of electric vehicles is putting strain on the existing electricity infrastructure. The increasing use of online entertainment is driving the need to ensure ubiquitous, high-speed wired and wireless broadband everywhere.

Then, there are the longer-term challenges that a city needs to prepare for. There is climate change, which may increase the risk of natural disasters, may make areas of the city no longer viable for use, and where increasing temperatures may impact on people's quality of life. There is demographic change, not just the challenge of caring for an aging population but the need to consider the move to smaller and more flexible family groups, potentially requiring major change in housing provision. Industry and the world of work is also changing, not simply in terms of old industries dying and very different ones replacing them but also in terms of the number of workers required and the changing skill sets needed in existing industries.

6.3 Covid-19 Pandemic

The Covid-19 pandemic, a humanitarian challenge, has caused widespread disruption in the global business community and society, at large. The current global challenge of Covid-19 pandemic has surpassed the usual provincial, radical, conceptual, spiritual, social, and pedagogical boundaries. The Covid-19 crisis has upended urban life, as we know it. Cities were under lockdown, and the once bustling streets of Paris, New York, London, Rome, Bombay, and more sat virtually empty for many months. Technology has been critical to the way cities and society have coped with the crisis. Online delivery companies have been essential for getting food and supplies to residents, while their restaurant delivery counterparts have helped keep restaurants up and running during the lockdown. Urban informatics has helped track the virus and identify infection hot spots. As cities begin to reopen, digital technologies are being leveraged to better test and trace the virus as well as to ready urban infrastructure, like airports, public transportation, office buildings, and businesses, to open back up safely.

The current health crisis which has gripped the world can be seen as an inflection point between digital transformation and businesses. It has also impressed upon various stakeholders to invest more robustly in digital technologies. It is also a challenge to the security planners who must guard against security threats and ensure business continuity. Tackling the challenges of complexity and interdependence of solutions is not easy. Technology can offer new and effective solutions, but these often require new management processes and changing business models. The collection and analysis of data can aid decision-making, but attention needs to be paid to issues of privacy and security.

Again, the different needs tend to be interdependent and require holistic approaches. Ensuring good health is not just the role of the health service providers; the city also needs to ensure clean air and support healthy forms of transport such as walking and cycling. Attracting companies into the city to provide employment is not just a matter of, for instance, providing financial incentives. Companies want to move to cities where there is a good quality of life, affordable and attractive housing, good educational provision, and easy transportation. With these many complex and interconnected challenges facing cities, they need a clear set of frameworks to help them review all of them in a holistic manner and make sensible decisions about the best set of actions to achieve their goals and reference architectures to guide them in building the organizational structures and processes that will lay the right foundation for the future. A smart city needs smart governance, smart businesses, and smart citizens. A smart city is one that can effectively leverage technology, infrastructure, public policy, and citizen engagement to create an urban environment that fosters economic growth and productivity, innovation, social mobility, inclusiveness, and sustainability.

6.4 Digital Transformation of Smart Cities

This is a massive and collective digital transformation (DX) of many similar and diverse socio-technological systems to be carried out in a systemic and industrialized manner. One of the essential characteristics of a matured industrialized domain is its reference architecture (as a template for solution architectures which realizes a predefined set of requirements). An important driving factor for reference architecture is to improve the effectiveness of creating products and product lines and managing synergy; providing guidance, e.g., architecture principles and best practices; providing an architecture baseline and an architecture blueprint; and capturing and sharing (architectural) patterns.

An ideal smart cities reference architecture (SCRA) as a basis for systemic DX (digital transformation) of cities into smart cities will bring the following capabilities to sustainable cities and communities (International Electrotechnical Commission, 2018a; International Electrotechnical Commission, 2018b):

- Ability to use the best worldwide knowledge, practices, and solutions
- Ability to deploy ready-to-use digital solutions which can be calibrated to unique needs of each city and each community
- Ability to collaborate in the evolution of domain solutions and thus develop local economy
- Ability to coordinate the evolution of domain solutions
- Ability to evolve with its own pace and priorities, and under civil-led governance
- Ability to consider various economic, cultural, legal, and other specific needs of citizens, businesses, and city administrations (Fig. 2)

This is an explicit confirmation that smart cities are not only about information and communication technologies (ICT), and all various views must be considered and reviewed. ICT is mandatory but not sufficient to achieve smart cities. A few

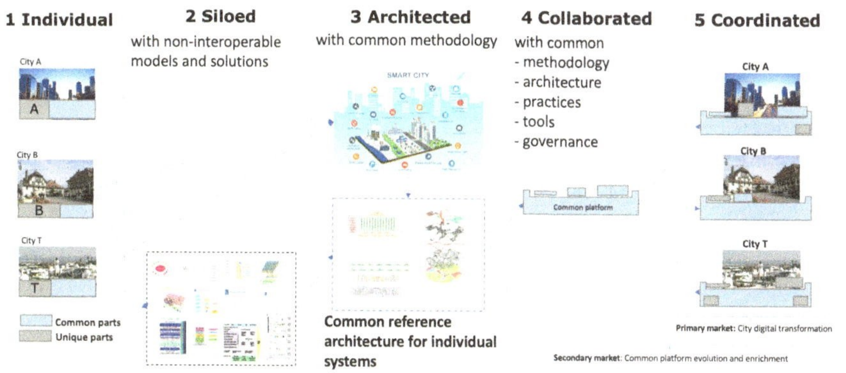

Fig. 2 Levels of maturity of digital transformation for smart cities

other crucial views and imperatives would include but not limited to sustainability, resilience, governance, vibrancy, and economic prosperity.

7 Concerns and Challenges

ICT has been recognized as a true enabler of the smartness in every aspect of the smart city paradigm. But there is a need of consensus among city administration, consulting companies, service companies, and technology companies on what ICT components are necessary and how cities should approach this agenda. Smart technologies and city-scale ICT are part of a new and emerging market where many of the products—both hardware and software—in a multi-vendor environment and across sectors and services are still being developed. But this, almost nascent smart technologies, market suffers from several barriers—interoperability, technical, and institutional—that need to be overcome if the market is to grow and mature.

In most of the initiatives to make our homes, buildings, cities, and/or our planet earth green or sustainable, we are extensively leveraging the ICT (information and communication technologies) solutions to monitor/control, and hence manage the various aspects of O&M (operation and management) of any infrastructure and services. While, with intervention of ICT tools, we can achieve major optimization in the energy consumption and environment contamination including but not limiting to GHG emission, we also need to keep in check the carbon footprint of the ICT infrastructure itself. In the gold rush of getting our buildings and cities certified as green or sustainable, we are adding a plethora of SCADA and automation systems in every aspect of the utilities and infrastructures. In any smart building or smart city, every service and utility is being automated and being reenforced with ICT backbone to monitor and control its operation in a most optimized manner. While the attending benefits of ICT backbone for any service/utility are quite commendable, there is little focus to optimize the design of the ICT infrastructure itself.

The technological trends in "smart homes," "smart buildings," "smart grid," "smart water," "smart transport," and "smart cities" are being considered and pursued in isolation from each other, by the respective stakeholders. This is although they form a very tightly interwoven and homogenous confluence of similar technologies being applied in different domains for a common cause of making our planet earth "smart, secure & sustainable." There is no common framework and architecture defined for the various physical infrastructures to be deployed in the proposed smart cities to work in an integrated, harmonized, and optimized manner. Thus, data sharing among the multiple stakeholders of a smart city shall be a major challenge. In fact, there is a recursive cycle to the data in a smart city. Information that is generated is information that is consumed, and then cycle repeats quite a many time.

Several barriers currently exist to widespread deployment of effective and powerful smart city solutions. One key barrier is that many current smart city information and communications technology (ICT) deployments are based on custom systems that are not interoperable, portable across cities, extensible, or cost-effective. Another is that architectural design efforts currently underway (in, e.g., ISO/IEC JTC1, IEC, IEEE, ITU, and consortia) have not yet converged, creating uncertainty among stakeholders. There is a lack of consensus on both a common language/taxonomy and smart city architectural principles. The result is that groups are likely to generate standards that are divergent, perhaps even contradictory, which does not serve the global smart city community well. A third barrier is the insufficient interoperability and scalability of underlying Internet of Things (IoT), and cyber-physical system (CPS) technologies that provide the foundation for many smart cities' applications. Additional barriers include lack of skilled resources, leadership, prioritization, capability, and experience. Thus, it is imperative to address these problems holistically and comprehensively to improve interfaces, avoid unnecessary overlaps, and deliver high-quality services to all city residents.

8 A Systemic Approach

The multiplicity of technologies and their convergence in many new and emerging markets, however particularly those involving large-scale infrastructure, demand a top-down approach to standardization, starting at the system or system architecture rather than at the product level. A systems-level approach in design and standardization is likely to not only enable newer and better services but also allow far greater synergies and cost-effective deployments, reducing the lifecycle (total) cost of ownership of any infrastructure, be it the smart grid, a home, a building, or even a city, with attendant environmental benefits, including carbon reductions. Therefore, the systems work can help define and strengthen the systems approach throughout the technical community to ensure that highly complex market sectors can be properly addressed and supported. System standards, having implications for the conformity assessment systems and processes, are also increasingly required in sectors with cross-domain implications.

Smart cities are mainly self-evolving cities with thousands of individual needs and capabilities to be handled systemically. Such systemic problem can be comprehensively solved only via a systems approach. In the case of smart cities, the systems approach should be a holistic and iterative discovery process that helps with first defining the right problem in complex situations and then with finding elegant, well-designed, and working solutions. It should be used not only in design of technical solutions but also architecting systems with many various additional aspects (human, economic, social, etc.). Use of the systems approach enables to establish and formalize explicitly:

- A set of system elements
- Relationships between elements (including a structure of the system)
- Relationships of elements with the environment.
- Relationships between relationships (as a higher level of organization)

- How relationships govern system behavior
- How system behavior is manifested in the form of qualities (characteristics), capabilities, and capacity of the system

Whatever architectures and frameworks we design that provide overseeing guidelines to the stakeholders of respective components and layers of the overall smart infrastructure paradigm, it is imperative to work on sufficiently fine granularity of each component and layer for standardization, as well as harmonization, and ensuring the interoperability among various similar components addressing different applications at semantic as well as syntactic levels. Further, the standards being adopted for the smart home or smart building deployments must be harmonized with standards in all other relevant ecosystems and integrated smart infrastructure paradigms. There is a need to create and suggest frameworks to achieve the interoperability among all the devices and layers at every interface in the networks, be it a smart home network, a smart building network, a smart city/community network, or the smart grid network that shall enable the stakeholders to prepare a set of detailed standards-based specifications to cater to specific/defined/fixed use cases followed by development of a compliance testing frameworks.

9 Functional Architecture

The existing siloed manner of operation—a review of ICT infrastructures of utilities currently in practice globally—shows that current applications live in silos as shown below in the siloed architecture (Fig. 3).

9.1 The Paradigm Shift: From Siloed to Unified: From Vertical to Horizontal

All sectors in the infrastructure framework are influenced by the unified ICT backbone paradigm. However, a common infrastructure pool enables the creation of an interconnected and truly homogenous system with seamless communication between services. Coordination, collaboration, and harmonization can be better implemented by the effective use of standards based on open, common, and shareable information and communication technologies. The disconnect among technological trends being pursued by the stakeholders of the now homogenous smart infrastructure needs to be bridged without any further delay to maintain the lifecycle cost/TCO (total cost of ownership) of these individual components within viable economic thresholds (Fig. 4).

Cities that are serious about getting smart know that they cannot rely on traditional ways of doing things. Vertical rollouts, where each infrastructure, utility, or IoT use case is propped up by a dedicated network, use case-specific data

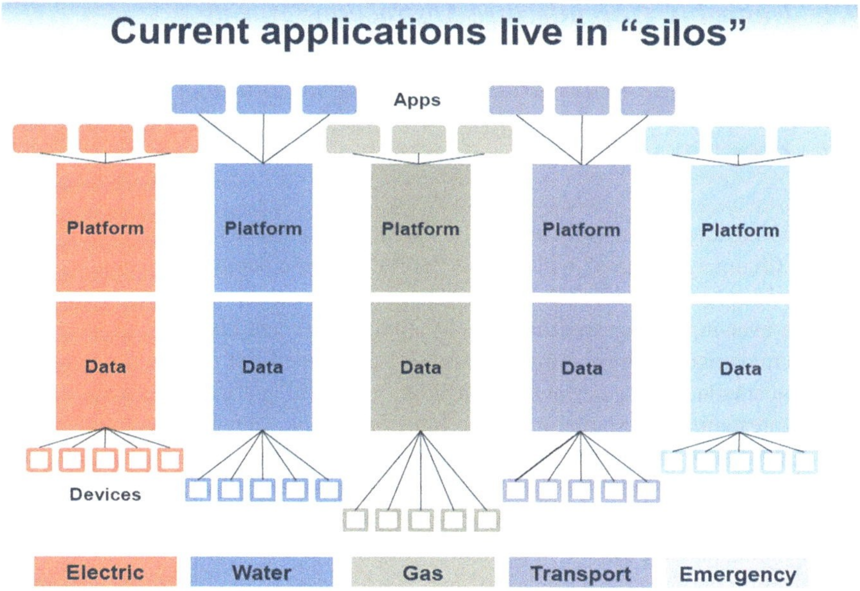

Fig. 3 Siloed architecture of city infrastructure

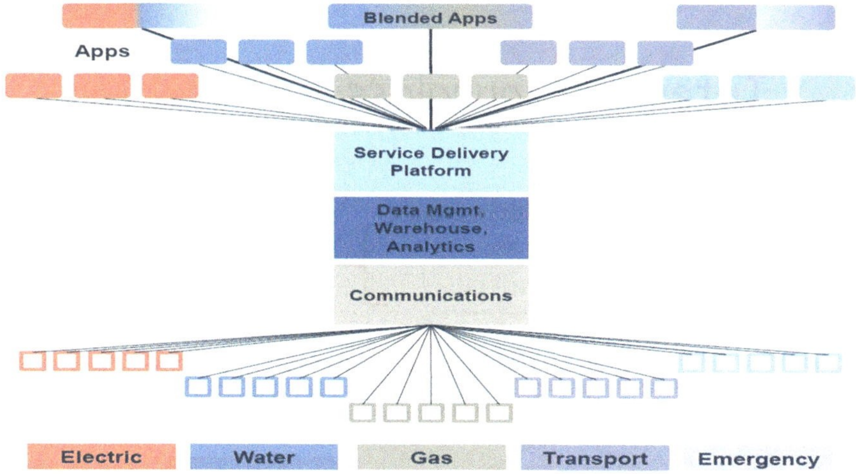

Fig. 4 A converged common ICT infrastructure pool

exchange mechanisms, and single-use devices do not scale. As city planners strive for greater cross-departmental synergies, it is essential that networks and devices, as well as data, can be used for more than one purpose. Even better if various functions, such as device management, security, and communication management, can be shared by multiple use cases and applications (Bhawan & Marg, 2020).

9.2 *Unified ICT Architecture Abstraction: Moving from Hourglass Architecture to Classic Saucer Champagne Glass Architecture (Figs. 5 and 6)*

9.3 *Classic Saucer Champagne Glass Architecture Model*

The evolved comprehensively unified ICT architecture can be modeled as a "Classic Saucer Champagne Glass" with a wide flat bottom base depicting the multitude of field devices, sensors, etc. The saucer-shaped bowl on the top depicts being filled with an ever-increasing spectrum of city applications and citizens' services. The long stem depicts all the common layers, viz., the unified last mile communication, common standardized gateways (application or vertical agnostic), common service layer representing the common service functions in the gateways, as well as in the cloud and the smart city middleware and city data reservoir in the cloud (Fig. 7).

It is the "Long Stem" of the "Champagne Glass Model" instead of the Short and Narrow Neck in the "Hourglass Model" that brings the comprehensive harmonization, standardization, and interoperability in the architecture leading to optimization in operational efficiency and lifecycle cost of the ICT infrastructure in any smart city. This architecture model helps dramatically bring down the CAPEX (capital expenditure), OPEX (operational expenditure), and carbon footprint of the digital infrastructure in any city to about "one fifth" of the "business as usual"

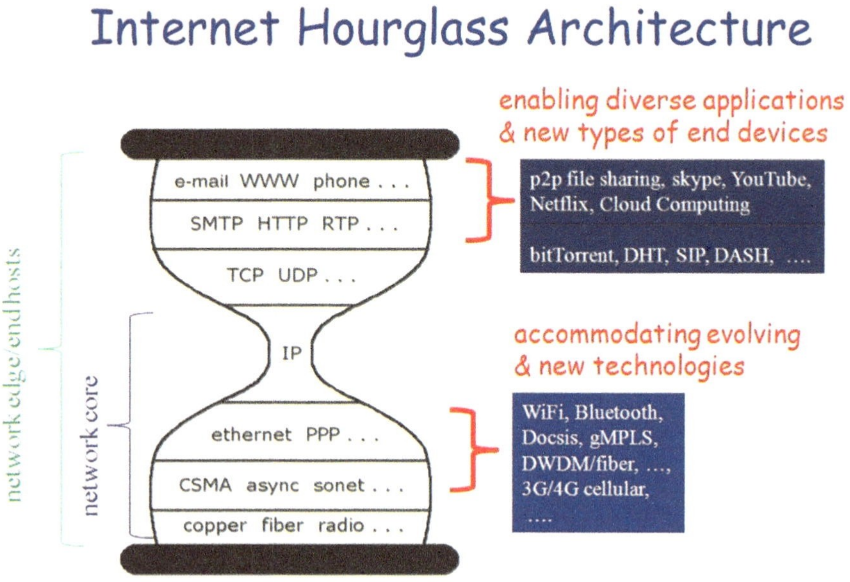

Fig. 5 Internet protocol hourglass model (credit—IETF)

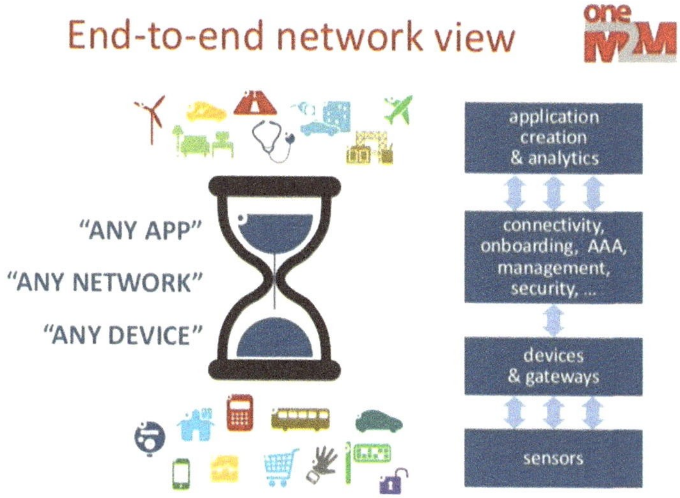

Fig. 6 oneM2M hourglass model (credit—oneM2M)

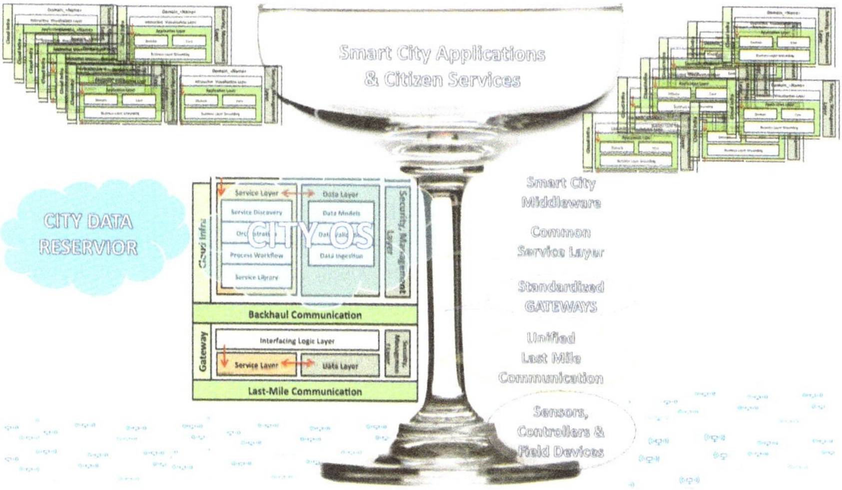

Fig. 7 Classic saucer champagne glass architecture model (Bureau of Indian Standards, 2017)

scenario. Further, since it provides a city with a well-architected unified digital (ICT) infrastructure, it becomes much easier to make it (the digital infrastructure) comprehensively cybersecure and resilient.

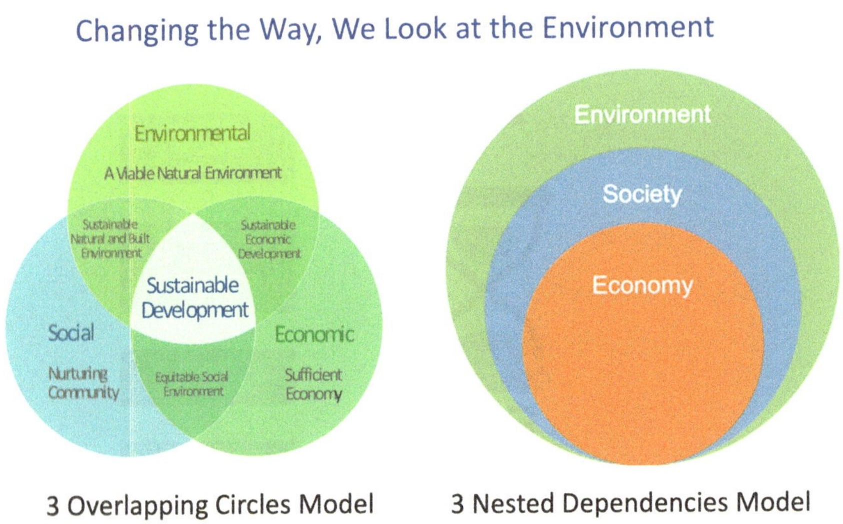

Fig. 8 Shifting perspectives in sustainability modeling

10 An Eco-Centric Approach to Sustainability

Sustainable development is largely about people, their well-being, and equity in their relationships with each other, in a context where nature-society imbalances can threaten economic and social stability (Fig. 8).

In the 3 Nested Dependencies model, the three sectors are co-dependent, while the 3 Overlapping Circles imply the economy can exist without the environment. The 3 Nested Dependencies model reminds us that there is no planet B: without the environment, the society and the economy cannot exist. The 3 Nested Dependencies remind us that we must live within our means or face the very real threat of ecological (and subsequently societal and economic) collapse.

In short, the 3 Nested Dependencies approach is eco-centric. It acknowledges the inherent value of the environment and prioritizes the health of our planet over economic gains. Rather than the three sectors competing, as might be the case in the 3 Overlapping Circles model, the 3 Nested Dependencies remind us that without clean water, fresh air, and healthy ecosystems, the society and the economy cease to function.

10.1 The Future: A Circular Vision for Smart Cities

Cities can transform from black holes sucking in food, energy, and other resources to engines of a regenerative food system and bioeconomy. The transition to a circular economy will see production which regenerates rather than harms the natural systems upon which it relies. Through circular economy strategies, countries and cities can take actions in "food waste, eco-design, organic fertilizers, guarantees for consumer goods, and innovation and investments." Circular economy principles must also need to be gradually integrated in industrial best practices, green public procurement, the use of cohesion policy funds, and through new initiatives in the construction and water sectors (Kristoffersen et al., 2020).

10.2 Circular Cities: Reimagining Urban Centers

Circular cities hold the key for transformational change, unlocking a new economic model that will transition away from fossil fuels, provide a reliable platform for waste to become a resource, and advance more equitable, inclusive value chains for our communities. Important topics of the circular economy like resources and raw materials or trash and waste are closely linked with the strategies of smart cities. Every city that sees itself as a smart city must have the goal of transforming the local economy to a circular economy.

Circular city encourages the use of systems thinking to provide economic, social, and environmental benefits for its citizens while also looking to improve the quality of life. Cities can follow seven principles in its transition toward a circular economy. These principles can be extended to define a vision and an action roadmap on circularity in cities:

- Closed-loop—all materials enter an infinite cycle (technical or biological).
- Reduced emissions—all energy comes from renewable sources.
- Value generation—resources are used to generate (financial or other) value.
- Modular design—modular and flexible design of products and production chains increases the adaptability of systems.
- Innovative business models—new business models for production, distribution, and consumption enable the shift from the possession of goods to (use of) services.
- Region-oriented reverse logistics—logistics systems shift to a more region-oriented service with reverse logistics capabilities.
- Natural systems upgradation—human activities positively contribute to ecosystems, ecosystem services, and the reconstruction of "natural capital."

10.3 Circular City Strategies

To facilitate the transition to a circular economy, smart city strategies may encompass:

- Ecological public procurement policies

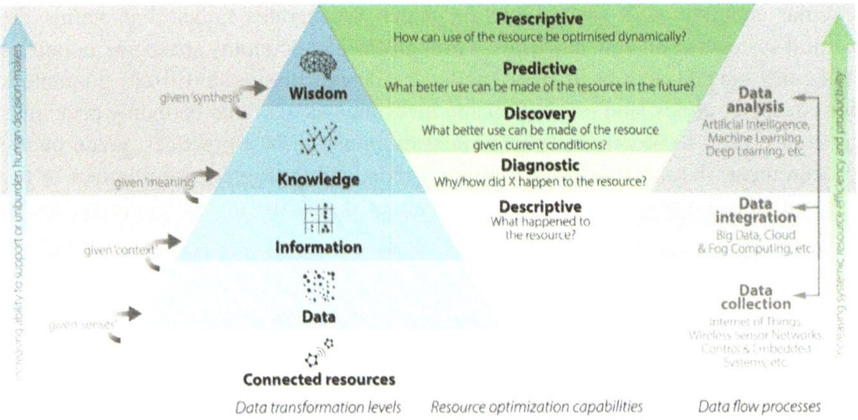

Fig. 9 Digital circular economy (Kristoffersen et al., 2020)

- Local production, repair, and reuse initiatives
- Eco-design principles in the built environment
- Bio-intensive urban farming
- Energy generation from biomass
- Innovation in water and waste management systems
- Infrastructure solutions for e-mobility and low-energy districts (Fig. 9)

10.4 A Crucial Imperative

We must include *CIRCULARITY* as one of the baseline *TARGETS* in the UN SDG 11 to ensure enhanced focus on circularity recognizing its increasing role in making cities comprehensively sustainable.

11 Capacity Building, Skills, and Competencies

With the emergence of new technologies and ICT domains like artificial intelligence, big data, robotics, cloud computing, and IoT, the importance of standardization goes beyond interoperability required for completing the global digital single market. Given the fast pace of change in our world, and its possible implications for our societies and workforce, all nations' policymaking aims to reap the maximum benefits from digital transformation. In some instances, the availability of standards can become a precondition for implementing policy or legislation. The safety and security of "smart" "products, automated devices and IoT," together with the reliability and validity of artificial intelligence, data, and privacy protection, are all

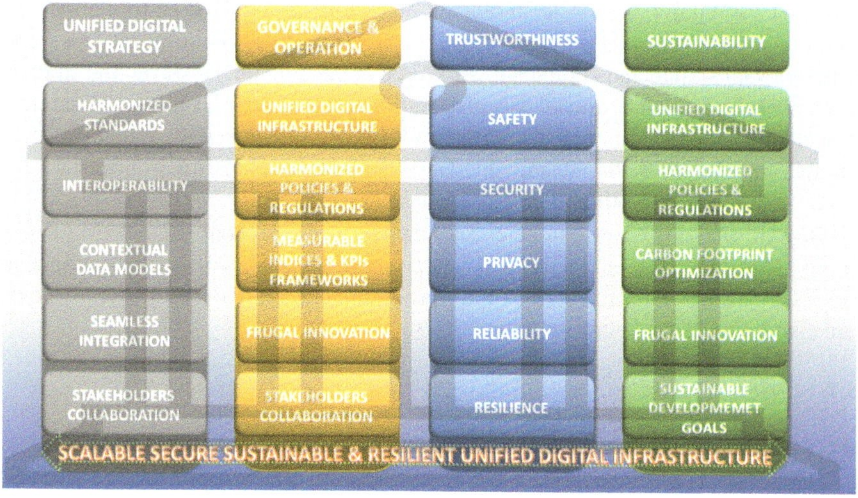

Fig. 10 Strategy framework for smart, sustainable, and resilient cities

challenges that may require standards to be developed and used for regulatory or public policy purposes.

However, standards per se do not make the cities smart and/or sustainable by themselves. They need to be supported by well-harmonized policies and regulatory frameworks, proactive city administrations who work in a completely collaborative manner with all other stakeholders be it the citizens or the different infrastructure and services providers, technology providers, local and global civic agencies, and district, state, and central (national) administrations. One of the most crucial imperatives to make cities truly smart, sustainable, and resilient is the availability of adequate skilled manpower to successfully implement the vision and deploy the required technology solutions and infrastructures.

As enumerated in the above sections, smart cities development is a complex and cross-domain subject and needs a combination of diverse skill sets and competences based on the class of stakeholders one belongs to. For the top-level officials and professionals, it needs understanding of town planning, architecture, project management, systems engineering, administration, governance, etc. Sustainable and/or digital transformation of the cities needs further in-depth understanding of the respective domains and relevant technologies, strategies, standards, etc. (Fig. 10)

This strategy framework for scalable, secure, sustainable, and resilient cities provides a systemic approach to developing and deploying essential enabling building blocks and the digital infrastructure across cities, districts, and states rather even at the national level to bring homogeneity in technology and business ecosystems by creating economy of scale. This framework also enumerates the core constituent domains that are integral to the development of smart, sustainable, and resilient cities. Normally, while considering the smart cities development, the focus is essentially on the technologies and more so on ICT (information and communication technologies) and digital transformation, and building an ecosystem with relevant skilled and competent workforce in the domain. However, the strategy framework brings other crucial aspects and domains to the focus like sustainability, frugal innovation, harmonized policies and regulations, coherent standards, measurable indices and KPIs, Sustainable Development Goals, trustworthiness, safety, security, privacy, reliability, Resilience, etc. (Narang, 2021).

Beyond the skills and competences mapped to the strategy framework, a few more subjects that need focus to develop comprehensive skilled and competent professionals and workforces for future-proof cities development and deployments are elaborated in the below sections.

11.1 Design Paradigm

As technologies are evolving and changing very fast, the task of designing a new product, system, or solution is becoming very much difficult and challenging. Apart from the complexity of design and engineering, the problem is also to understand user needs and preferences, many of them untold or implicit needs, and then translating those to product/system development specifications and features (Fig. 11).

Design is the practice of intentional creation to enhance the world. It is a field of doing and making, creating great products and services that fit human needs, which delight and inform. Design is exciting because it calls upon the arts and humanities, the social, physical, and biological sciences, engineering, and business. One of the critical components of the product realization process is the engineering design, which deserves a special attention in the engineering education to better prepare engineers to meet the demands of the industry.

The evolution of a design paradigm in engineering curriculum is based on integration of several design experiences into theoretical courses and the design-specific courses. This approach is imperative to enhance the employability of engineering students in the industry by virtue of filling the wide gap between the theoretical aspects of learning in the education system and the practical applications of the concepts learnt in college, which are essential for employment in the industry. It is imperative to address the vital gap in the fundamentals taught and their applications in product/system realization thru the understanding of the design paradigms. It leverages all the theoretical learnings and introduces the students to the vast applications of the concepts learnt thru the exposure to an approach to the integration of comprehensive design experience with the "Engineering Curriculum."

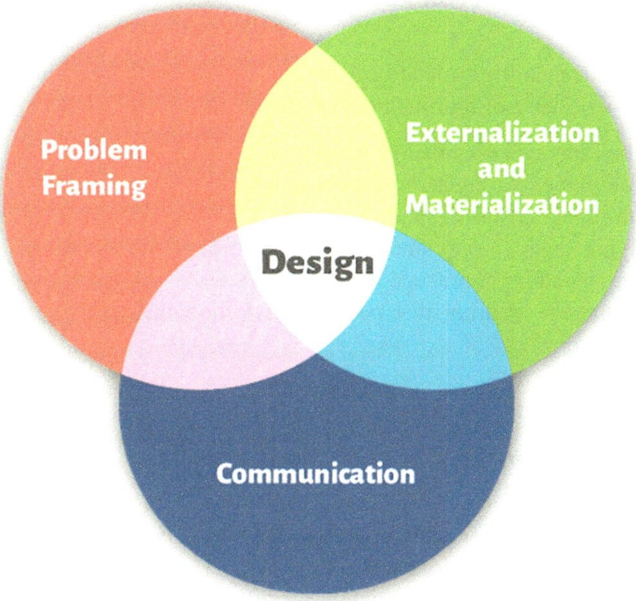

Fig. 11 Design paradigm

11.2 90–90 Formulae in Design

According to this, you spent 90% of the time in completing the 90% of design. And then you end up spending another 90% of the time in completing rest of the 10%. Such a situation occurs simply because you did not invest (enough time and effort) in developing a structured design.

If design is to live up to its promise, it must create new, enduring curricula for design education that merge science and technology, art and business, and, indeed, all the knowledge of a university. In earlier years, designers were trained in form, function, materials, and aesthetics. Today, culture and emotions are central, plus knowledge of societal issues, techniques for subtle persuasion, and the intricacies of complex, interdependent systems. Since design is a field of doers and makers, in the practical world, successful products and services require generalists who can cut horizontally across many of the deep, vertical specialties. Generalists cannot succeed without close collaboration with specialists, while the knowledge of a specialty is too limited to create an effective service or product for people without the aid of design generalists.

11.3 Core Characteristics of Design

Design has the capability to lead because it cuts across all disciplines. Design is transformative because of four major characteristics:

- Design thinking: ensuring that the correct problem is being solved
- Systems thinking: cutting across and encompassing all disciplines
- Critical thinking: ensuring an intellectually disciplined process of conceptualizing, applying, analyzing, synthesizing, and/or evaluating information
- Integrative: blending of practice and theory
- Human-centered: assuring that people and technology work harmoniously as collaborative players

11.4 Design Thinking

Design thinking is the new paradigm for approaching solving of complex problems. Designing products and solutions fall in this category—be it design of an electronic product, or architecting and constructing a bridge, or designing a new motorway to ease traffic congestion around a city (Dam & Siang, 2020). Design thinking is a human-centered approach to innovation that draws from the designer's toolkit to integrate the needs of people, the possibilities of technology, and the requirements for business success. Design attitude is a critical factor in integrating design thinking for nurturing INNOVATION, etc.

11.5 Systems Thinking

Systems thinking is an approach to integration that is based on the belief that the component parts of a system will act differently when isolated from the system's environment or other parts of the system. Systems thinking is particularly useful in addressing complex or wicked problem situations. These problems cannot be solved by any one actor, any more than a complex system can be fully understood from only one perspective. Moreover, because complex adaptive systems are continually evolving, systems thinking is oriented toward organizational and social learning—and adaptive management (Goodmans, n.d.).

11.6 System Design

System design is at the heart of engineering effort in improving human lives. Every system is built to address user needs and preferences. But user needs are seldom clearly delineated. It is left to the genius of the system designers to understand the unstated or implicit needs and to translate those into product specifications and features. The features of systems come at a cost; designers must weigh what optional features to include and what not. The challenge is compounded by the fact that technology is evolving very fast. System designers have a daunting task to keep pace with the rapidly changing technology in realizing their designs.

11.7 Systems Engineering

An important subject to focus for enhancing system design capability in the students is systems engineering. Understanding systems, systems of systems, and systems approach is crucial to design any complex products, systems, and solutions in the following.

11.8 System

A group of interacting, interrelated, or interdependent elements forming a purposeful WHOLE of a complexity that requires specific structures and work methods to support applications and services relevant to the stakeholders.

11.9 Systems Approach

A holistic, iterative, discovery process that helps first defining the right problem in complex situations and then in finding elegant, well-designed, and working solutions. It incorporates not only engineering but also logical human and social aspects. The systems approach helps put a structure to our thinking and design methodology:

- Identify and understand the relationships between the potential problems and opportunities in a real-world situation.
- Gain a thorough understanding of the problem and describe a selected problem or opportunity in the context of its wider system and its environment.
- Synthesize viable system solutions to a selected problem or opportunity situation.
- Analyze and trade off between alternative solutions for a given time/cost/quality version of the problem.
- Measure and provide evidence of correct implementation and integration.
- Deploy, sustain, and apply a solution to help solve the problem (or exploit the opportunity).

- All the above are considered within a lifecycle framework which may need concurrent, recursive, and iterative applications of some or all the systems approach.

We shall need to apply the systems approach to develop a comprehensive suite of solutions and tool chain to help realize the true vision of smart sustainable and resilient cities (International Electrotechnical Commission, n.d.). The systems approach shall entail:

- Systemic and structured study and analysis of the domain, its sub-domains, the market, and technology trends
- Identifying all the stakeholders and understanding their respective needs, concerns, expectations, and objectives
- Capturing a comprehensive inventory of use cases and applications and their relationships and interdependencies
- Extracting the detailed list of requirements to develop a reference architecture/framework with all the components and their interplay, etc.
- Mapping the different technologies, solutions, tools, etc. to meet all the requirements
- Undertaking field trials/pilots
- Iterating and fine-tuning the solutions, tools, etc. to meet the stakeholders' expectations and needs

Without such a systemic approach, we may NOT be able to develop a comprehensive solutions bouquet to establish a well-coordinated and coherent system of systems foundation for any city or community.

11.10 Sustainable Engineering

Sustainability today is a global imperative, and we need to meet the new environmental and social regulations and yet offer quality and a competitive price. Our industry must believe in and implement the principle of sustainable development, which makes us responsible and accountable to meeting the needs of the present generation without compromising the ability of future generations to meet their needs. Environment and social considerations have influenced the business environment of the global electronics sector, bringing to the fore some new regulations to be followed by the members of all the ecosystems. These include:

- Restriction on hazardous substances (RoHS)
- Waste electrical and electronic equipment (WEEE)
- Ozone-depleting substances (regulation and control) rule (ODS), 2000 – Montreal Protocol

12 Digital Pedagogy

Digital transformation of the education practices is not new, with multiple attempts and initiatives over the past many years. However, the Covid-19 pandemic has brought an unprecedented urgency to this. It has added fresh perspectives and new nuances to the evolving complex and yet intricate digital pedagogy paradigm (JISC, 2021).

12.1 Envisioning Digital Pedagogy

Digital pedagogy is not just using digital technologies for teaching and learning but rather for using digital tools to enhance both outcomes and impact of learning. Digital transformation in education can enhance the traditional face-to-face learning environment. It can enable teachers to innovate pedagogical models for creating better connect with students. It unites disruptive technologies and the learning sciences (education, psychology, neuroscience, linguistics, sociology, and anthropology) to promote the development of adaptive learning environments (a digital learning environment that adapts teaching and learning approaches and materials to the capabilities and needs of individual learners) and other digital tools that are flexible, inclusive, personalized, engaging, and effective. We need to devise new pedagogies, implementing innovative digital systems, developing new areas of knowledge, and informing policymakers and skilling/education stakeholders. We need to propose concrete options that will allow education stakeholders to make the potential of digital pedagogy real at the system level—that is, at the scale that will allow it to support the teaching profession broadly and impact positively on the learning experience of each, and every, learner. In other words, what we need is a degree of specificity about digital pedagogy that allows everyone to assess, invest, plan, deliver, and test.

Digital pedagogy is equally relevant to the upskilling of employed workforces in different domains including the smart cities ecosystem. It shall enable an environment where all the professionals could upskill themselves in their chosen domains/subjects at their own pace. *For effective and lasting proliferation of digital pedagogy, it needs to be woven into the national skilling and education policy implementation strategy, framework, and roadmap of every nation.*

Digital pedagogy shall leverage all the contemporary and forthcoming disruptive technologies including but NOT limited to big data, AI (artificial intelligence), VR (virtual reality), AR (augmented reality), IoT (Internet of Things), virtual learning companions, etc. to make the teaching and learning experiences immersive and comprehensive. Teachers and trainers need to be central agents in the digital pedagogy. The digital tools, and the data-driven insights that these tools provide, will empower teachers/trainers to decide how best to marshal the various resources at their disposal. This participatory design methodology will ensure that the messiness of real classrooms is considered and that the tools deliver the support that educators need—not the support that technologists or designers think they need. It can help teachers create more sophisticated learning environments than would otherwise be possible.

12.2 Taking the Digital Pedagogy to the Next Level to Help Us Respond to the Biggest Unsolved Issues in Education and Skilling

12.2.1 Tackling Achievement Gaps

Currently, we are failing to meet the needs of all learners. The gap between those who achieve the most and those who achieve the least is a challenge that teachers, educators, administrators, and government officials face every day, in every country. The reasons behind the achievement gaps in different countries vary, but the fact remains that not all learners are achieving their potential at school, colleges, universities, and/or profession training academies. Technology has provided a level playing field for learners from all sections of society. Digital pedagogy can offer a new set of tools for addressing this challenge. Moreover, these digital learning systems will scale broadly as the reduction in their cost makes them increasingly affordable to training and skilling institutions.

Digital learning systems/tools could also help teachers/trainers find and share the best teaching/training resources. Intelligent support for teachers/trainers could also help address the issue of teacher/trainer retention where we see many skilled professionals leaving the profession due to "burnout." However, education and skilling systems will need to be nimble to take advantage of the rich real-time systems-level analysis that will be continuously available.

Finally, there are three powerful forces that must be combined if we are to deliver on the promise of technology to catapult learning dramatically forward. One is pedagogy, or the science of how we teach and learn; the second is technology itself; and the final component is system change, or our understanding of how to deliver change so that it has a positive impact on each, and every, learner. The future ability of digital learning systems to tackle real-life challenges in education and skilling depends on how we attend to each of these three dimensions—that is: (i) we need intelligent technologies that embody what we know about great teaching and learning in (ii) enticing consumer grade products, which (iii) are then used effectively in real-life settings that combine the best of human and machine.

13 Conclusion

This chapter enumerates the shifting paradigms and the required skills and competencies for sustainable digital transformation of urban landscape. In this chapter attempted to first understand the current scenarios relevant to smart cities including but not limited to global status and trends in cities' transformation into the smart cities, the disruptive technologies landscape, standards landscape, the paradigm of digital transformation, and its impact on smart cities' transformation. Then, we reviewed the imperatives for the smart cities' ecosystem stakeholders beginning with United Nations Sustainable Development Goal 11, citizens' needs, dealing with change, short-term and long-term like Covid-19 pandemic, climate change, cyber resilience, and complex and interdependent systems in the cities. Then, we delve into the concerns and challenges for the sustainable digital transformation of the urban landscape like fragmented ICT ecosystem and diverse siloed technology solutions, the increasing carbon footprint of digital infrastructure in the smart cities' deployments, and absence of harmonized system standards from global SDOs leading to lack of interoperability and data sharing capability among different city systems. This leads to elaborating the importance of adopting the systems approach when dealing with large-scale infrastructure that demand a top-down approach to standardization to not only enable newer and better services but also to allow far greater synergies and cost-effective deployments, reducing the lifecycle (total) cost of ownership of any infrastructure.

The following section explores the contemporary and future function architectures of the digital infrastructure in a city. This is followed by a detailed review of new imperatives in sustainability focusing on the eco-centric 3 Nested Dependencies model, circular economy strategies for the smart cities.

The next section is dedicated to understanding capacity building imperatives and identifying the skills and competencies that would be crucial for the sustainable digital transformation of the urban landscape in the future in context of the learning from earlier sections in the chapter. The skills and competencies identified beyond the obvious and well-understood ICT domain-specific skills include some nonconventional learnings like design thinking, systems thinking, critical thinking, system design, sustainable engineering, etc. Last but not the least, we focus on the new paradigm of digital pedagogy that has become an integral part of the education, skilling, and training paradigm in the times of comprehensive digital transformation of every aspect of the business, industry, and society, at large.

The key takeaway from the chapter for the proponents of the sustainable digital transformation of the urban landscape could be summarized as follows:

We humans need to radically change our relationship, not just with the planet but with the objects with which we fill our lives.

We need to change how we think about technology and innovation. Rather than allowing technological advancement to steer our narratives, innovation and technology should help us build bridges between the worlds we inhabit now and the ones we imagine for tomorrow.

The pandemic showed how spending on scientific innovation and technologies makes a huge difference. But it also showed how the failure to develop a global response has very worrying results—we cannot let this happen with climate change. We need a global approach for the comprehensive sustainable digital transformation of the urban landscape.

References

Bhawan, M. D., & Marg, J. N. (2020). *White Paper on smart cities in India: Framework for ICT infrastructure*. Retrieved from https://trai.gov.in/sites/default/files/White_Paper_22092020pdf_0.pdf

Bureau of Indian Standards. (2017). *Pre-standardization study report on unified secure & resilient ICT framework for smart infrastructure*. Bureau of Indian Standards. Retrieved from http://www.bis.org.in/other/USR_ICT_FSI_V_1_0.pdf

Dam, R. F., & Siang, T. Y. (2020). *Design thinking*. Retrieved from https://www.interaction-design.org/literature/article/5-stages-in-the-design-thinking-process

Goodmans, M. (n.d.). *System thinking*. Retrieved from https://thesystemsthinker.com/systems-thinking-what-why-when-where-and-how/

International Electrotechnical Commission. (2018a). *IEC SRD 63188 ED1: Smart cities reference architecture methodology*. Retrieved from https://www.iec.ch/dyn/www/f?p=103:23:0::::FSP_ORG_ID:13073

International Electrotechnical Commission. (2018b). *IEC 63205 ED1: Smart cities reference architecture*. Retrieved from https://www.iec.ch/dyn/www/f?p=103:23:0::::FSP_ORG_ID:13073

International Electrotechnical Commission. (n.d.). *A systems approach*. Retrieved from https://www.iec.ch/standards-development/systems-approach

JISC. (2021). *Digital pedagogy toolkit*. Retrieved from https://www.jisc.ac.uk/full-guide/digital-pedagogy-toolkit

Kristoffersen, E., Blomsma, F., Mikalef, P., & Li, J. (2020). The smart circular economy: A digital-enabled circular strategies framework for manufacturing companies. *Journal of Business Research, 120*, 241–261.

Narang, N. K. (2020). Mentor's musings on IoT 2.0: IoT coming of age. *IEEE Internet of Things Magazine*. Retrieved from https://ieeexplore.ieee.org/stamp/stamp.jsp?tp=&arnumber=9125421

Narang, N. K. (2021). *IoT in smart cities: Homogeneous communication architecture imperative for the heterogeneous applications environment*. Retrieved from https://iot.ieee.org/newsletter/september-2021/iot-in-smart-cities-homogeneous-communication-architecture-imperative-for-the-heterogeneous-applications-environment.

Smart Cities: Emerging Risks and Mitigation Strategies

Konstantinos Kirytopoulos, Theofanis Christopoulos, and Emmanuel Dermitzakis

1 Introduction

The urbanization phenomenon is encountered in every corner of the earth and directly affects almost 50% of its population, with the forecasts of experts showing even higher percentages in the years to come (Shen et al., 2016). The population concentration and growth in urban areas lead to the need for continuous improvement in the management of resources, goods, services and infrastructure. In order to meet these needs of the growing urban life and achieve better decision-making, it is necessary for cities to transform into smart cities. This transformation will positively impact the quality of life of their citizens, support economic development and promote environmental sustainability (Silva et al., 2018). This transition has already been attempted by a number of cities, and experience has shown that it directly or indirectly affects all urban activities while engaging all stakeholders, city members and institutions (state, companies, universities and citizens) (Shamsuzzoha et al., 2021).

For the long-term sustainability of these multidimensional cities, special attention to risks is required by those responsible for the development and operation of smart cities. Particular attention has already been devoted both in research and in the implementation of new technologies, in order to address relevant risks. Typical examples are the models that have been developed for assessing the personal information risks managed within a smart city (Yan et al., 2020), or cyber-security risks related to digital assets (Sheehan et al., 2021). Still though, we are far from saying that risks have been addressed entirely. Risk management aims to prepare those responsible for possible incidents in order to avoid improvising responses

K. Kirytopoulos (✉) · T. Christopoulos · E. Dermitzakis
School of Mechanical Engineering, National Technical University of Athens, Athens, Greece
e-mail: kkir@mail.ntua.gr

P. Fitsilis (ed.), *Building on Smart Cities Skills and Competences*, Internet of Things,
https://doi.org/10.1007/978-3-030-97818-1_7

when they occur (Pym, 1987). Identified risks and their potential treatment strategies are key elements in the design and management of smart cities. Nevertheless, a detailed risk taxonomy for the risks of smart cities is still missing from the literature (Ullah et al., 2021). The aim of this chapter is to alert smart cities' designers and other stakeholders on the potential risks that can occur, as well as present some high-level strategies to overcome such risks.

2 The Smart System as a Multisystem Construct

In order to achieve the transformation of a city into a smart city, the integration of new technologies is required, so that the digital and the physical world can merge. Therefore, sophisticated technologies compose the basis on which the philosophy of a smart city is built. Fundamental technologies that build a smart city are Information and Communications Technologies, with their main applications being:

- Internet of Things (IoT): a network which includes a plethora of technologies (e.g. sensor nodes, software solutions, information technologies), aiming at the generation, transfer and exploitation of data (Nižetić et al., 2020).
- Big data: massive volumes of data produced from multiple sensors (Rathore et al., 2015).
- Cloud computing: The National Institute of Standards and Technology's defines cloud computing as "*a model for enabling ubiquitous, convenient, on-demand network access to a shared pool of configurable computing resources (e.g., networks, servers, storage, applications, and services) that can be rapidly provisioned and released with minimal management effort or service provider interaction*" (Mell & Grance, 2011).

All the above-mentioned technologies resonate in every urban activity. The main identified dimensions in the literature are smart economy, smart mobility, smart environment, smart people, smart living and smart governance (Giffinger et al., 2007). The "smartness" of these categories is more than just a fancy term. It describes the tendency to improve the economic, social and environmental conditions within cities in a people-oriented way (Silva et al., 2018).

As these urban activities overlap, so do the actuated technologies. A typical example is big data generated in the context of the smart city. More specifically, the debate in the scientific community raises the question whether it is more efficient to collect and manage data under a single roof (centralization of data) (Economic and Social Council of United Nations, 2016) or whether separating data by city dimension is an effective solution to avoid failures and vulnerabilities (decentralization) (Tariq et al., 2020). Another example are the sensors within smart cities, which are the core of smart cities and produce a large amount of data (Ahad et al., 2020). The data from the same sensor can be useful to stakeholders from different fields of activity and can contribute significantly to their decision-making.

Finally, the overlapping functional areas are favoured by the existence of the Internet of Things, which not only allows the extraction of data from existing infrastructure but also supports their fully autonomous operation through the use of artificial intelligence (Mainzer, 2020).

These interrelations and overlaps of activities and technologies have an impact on the smart city's stakeholders. The main stakeholders are the government and local authorities, industries, universities and citizens (Fernandez-Anez, 2016). Each one has their own role, their own contribution and their own requirements in relation to the smart city. Therefore, it is almost impossible to carry out changes and developments in the city's sectors, either in terms of structure or technologies, without affecting their sustainability, since the interests of the stakeholders are also often conflicting (Shamsuzzoha et al., 2021). However, the need to synthesize and manage these technologies in smart cities with a citizen-centred approach cannot be overlooked (Anthopoulos et al., 2007).

The multilayered application, integration and interconnection of new and complex technologies in an intertemporal establishment like a city comes along with risks, both threats and opportunities, while their management is more complicated than the elements that they compose it (Ullah et al., 2021).

Risk management is a systematic process and consists of the following subprocesses: identification, analysis, evaluation, treatment and monitoring and review of risks (International Organization for Standardization, 2018a). Identifying those risks has a complexity proportional to that of the technological systems being installed as well as the number of interconnections, but it is the first step towards their management.

3 Methodology

3.1 Systematic Literature Review

In order to identify the risks of smart cities and their possible mitigation strategies, a systematic literature review (SLR) was undertaken. SLR contributes to the research by setting objective criteria for the selection of the literature to be included, in order to minimize as much as possible the bias and subjective judgement of the researchers (Nightingale, 2009). The literature review process that was followed is summarized in Fig. 1.

The Preferred Reporting Items for Systematic Reviews and Meta-Analyses (PRISMA) guideline is used to present the results of the literature review. This guideline is a statistical approach to the results to promote transparency and full inclusion of the results of the literature review (Page et al., 2021).

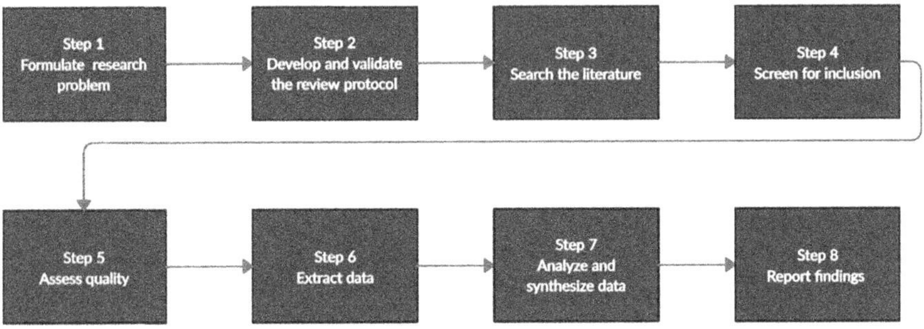

Fig. 1 Steps for SLR (Xiao & Watson, 2019)

3.2 *Selecting Articles*

For the search of the journals, the Scopus electronic database was used. Scopus database has a plethora of publishers, whose number exceeds 5000, while the number of peer-reviewed journals amounts to 34,500 (Gupta et al., 2019). Moreover, Scopus covers a wide range of scientific fields, for example, Computer Sciences, Social Sciences and Information Science (Mat Ludin et al., 2017).

In order to select the articles for analysis, certain selection criteria were applied, as presented in Table 1. Following the search method of article "title/abstract/keywords" (Derakhshanfar et al., 2019), 2378 results were identified. Next, only journals were selected (excluding conference proceedings, book series, books and other types of literature), to ensure the quality of the included publications, which is guaranteed through the peer review that journals go through (Prater et al., 2017). To include the most up-to-date literature, the search was limited to the last decade. Afterwards, only relevant subject areas were kept, excluding Mathematics, Environmental Science, Physics and Astronomy, Materials Science, Earth and Planetary Sciences, Chemistry, Medicine, Biochemistry, Genetics and Molecular Biology, Chemical Engineering, Agricultural and Biological Sciences, Pharmacology, Toxicology and Pharmaceutics, Neuroscience, Nursing and Immunology and Microbiology. The limitation of the subject areas resulted in 263 journal papers, followed by the restriction of the language to English, coming to 251. The document type was limited to articles, excluding conference papers, reviews, editorials, notes and undefined types, concluding to 231 papers. Finally, based on the research goals, 43 papers were finally selected, reviewing their titles, abstracts and then their content.

Table 1 Stages of setting criteria for the SLR

Search stage	Keyword string	Number of results
1	TITLE-ABS-KEY ((smart OR sustainable OR digital) AND (city OR cities OR town) AND (risk OR risks OR uncertaint*) AND (management OR identification OR mitigation OR response) AND ("sustainable development" OR "risk management" OR "Smart City" OR "sustainability" OR "risk mitigation" OR "risk assessment"))	2378
2	Limiting the search to journals	1308
3	Limiting the search to up to date journals	1079
4	Excluding irrelevant subject areas	263
5	Limiting the search to English journals	251
6	Limiting the search to articles	231
7	Reviewing titles, abstracts and content of the articles	43

3.3 Risk Clustering

To ensure homogeneity of language and lack of repetition and to avoid misinterpretation within the presentation of identified risks, intervention on the description of certain risks is necessary (Le et al., 2019). For this cause, the description of many risks was fine-tuned, while risks with slightly different names but with the same meaning were unified. More specifically, in many cases risks did not follow the typical risk metalanguage, and there was a mix up of causes, risks and impacts. For example, "overestimating the positive impacts of technology" (Ambrosino et al., 2015) was described in one study as a cause for other risks, while "delay in actual deployment of new technologies" (Lee et al., 2013) was described as a cause from strategic and legal risks. Also, as shown in Table 2, the same risk could be described by different authors in different terms.

4 Existing and Emerging Risks in the Development and Operation of Smart Cities

Risk as defined in ISO 31000:2018 is the "*effect of uncertainty on objectives*" (International Organization for Standardization, 2018b), while emerging risks are "*either new risks or familiar risks that become apparent in new or unfamiliar conditions*" (International Risk Governance Council, 2015). In the case of smart cities, this uncertainty is compounded by the city's exposure to new technologies on which cities are founded. This exposure to new technologies is the factor that increases the uncertainty within a smart city compared to a "non-smart city". More specific, these risks have their source in precisely this interwovenness and interconnection of the technologies used, their breakthrough nature and rapid pace

Table 2 Examples of different terminology used for the same risk

No.	Alternative terms used in the literature	Proposed risk's terminology	References
1	Natural disasters	Natural disasters	Yan et al. (2020), Ganin et al. (2019), Hayat (2016)
	Natural hazards		Yang et al. (2018)
2	Unstable power grid	Unstable power supply	Vincent et al. (2020)
	Unstable electric network		Jiménez-Bravo et al. (2018)
	Unstable power generation		Vincent et al. (2020)
	Premature energy depletion of sensors		Soyata et al. (2019)
	Power outage		Vitunskaite et al. (2019)
	Lack of energy supply of sensors		Ayala-Ruiz et al. (2019)
3	Absence of participation by civil society	Lack of participation from citizens	Mainzer (2020)
	Lack of participation from citizens		Gupta and Hall (2020)
	Disinterested citizens		Gupta and Hall (2020)
4	Institutional conditions can be deeply embedded in governance systems (structural inertia)	Organizational incapacity to manage change	Vu and Hartley (2018)
	Lack of capabilities to manage change		Vu and Hartley (2018)
	Organizational incapacity to adopt smart applications		Soyata et al. (2019)

of development. In addition, the smart systems that consist of these technologies are creating greater risks than the risk of each separate component (Axelrod, 2013).

All of the above leads us to the conclusion that, the importance of risk identification is particularly high, as the failure to identify certain risks implies failure to analyse them and subsequent exposure to that risk in the development and operation phases. The number of risks identified from literature review after properly naming and grouping them is 65, and they are presented in Table 3, in a Risk Breakdown Structure (RBS), accompanied by the percentage of the number of appearances of the risks in relation to the total number of risks. The RBS is used to present the identified risks. The risks in the RBS are categorized in groups with a hierarchical structure, allowing the reader to concentrate on the subjects that concern him.

5 Risk Management Skills and Mitigation Strategies for Safe Smart Cities

5.1 Risk Management Skills

To ensure the sustainability of smart cities, it is essential for both developers and managers of the cities to be equipped with risk management skills and knowledge to identify, analyse and treat risks. Only in that way they will be able to enhance opportunities and mitigate threats. Achieving effective risk management requires the risk manager to be equipped with certain competencies and skills. Risk management in smart cities is a complex and multidimensional process, and this requires at least the following:

- Technical skills: Technical skills refer to the ability to implement the processes of risk management (Marx & de Swardt, 2019). Standards for risk management, such as ISO 31000, set guidelines for integrated risk management, but the effectiveness of implementing standards varies from manager to manager.
- Smart city concept understanding: The risk manager should be able to see the big picture in the smart cities' concept. Knowledge of the key objectives of smart cities, the stakeholders and the factors that create uncertainty are necessary to be known, in order to manage the emerging risks.
- Project management skills: Tasks such as creating timelines, long-term planning and setting budget are included in project management procedures. These tasks will be used either directly or indirectly in the risk management procedure too.
- Soft skills: Building the right team, effective cooperation with all its members, effective transmission of information, wise judgement and communicating the risk management procedures to external stakeholders are just some of the soft skills that a risk manager may need (Carvalho & Rabechini Junior, 2015). This means that technical skills alone are not sufficient for successful risk management.

Table 3 The RBS of smart cities

Description of categories/risks		Frequency count	Percentage in relation to total risks
1.	**Economic**		
1.1.	Inappropriate cost planning	3	1.49%
1.2.	Economic distress	2	1.00%
1.3.	Low investment returns	2	1.00%
1.4.	Lack of funding	2	1.00%
1.5.	Financial losses during operation	1	0.50%
2.	**Social**		
2.1.	Lack of technology and information awareness among citizens and external stakeholders	7	3.48%
2.2.	Lack of participation from citizens	3	1.49%
2.3.	Lack of acceptance from society	2	1.00%
2.4.	Social inequality	1	0.50%
3.	**Organizational**		
3.1.	**Partnership risks**		
3.1.1.	Lack of technical know-how and expertise from contractors	3	1.49%
3.1.2.	Conflict of interest of multiple stakeholders	3	1.49%
3.1.3.	Unreliable partners due to vulnerability to cyberattacks	2	1.00%
3.1.4.	Underestimation of critical issues dealing with the interaction activities between providers or suppliers	1	0.50%
3.1.5.	Possible obstacles for technology's application to related industries	1	0.50%
3.1.6.	Trust issues with government officials	1	0.50%
3.1.7.	Lack of fixed tenure of companies that plan, release funds, implement, manage and evaluate the smart cities projects' CEOs	1	0.50%
3.2.	**Human resources**		
3.2.1.	Lack of personnel	2	1.00%
3.2.2.	Lack of staff training	6	2.99%

(continued)

Table 3 (continued)

Description of categories/risks		Frequency count	Percentage in relation to total risks
3.3.	**Operational risks**		
3.3.1.	Lack of standard management	3	1.49%
3.3.2.	Organizational incapacity to manage change	3	1.49%
3.3.3.	Lack of coordination across city's agencies	6	2.99%
3.3.4.	Lack of unified taxonomy from city governments to smart infrastructure systems	2	1.00%
3.3.5.	Great recovery time from disasters and malefactions	1	0.50%
3.3.6.	Limited consideration of interdependency issues between infrastructure systems	1	0.50%
3.4.	**Implementation risks**		
3.4.1.	Lack of project planning	3	1.49%
3.4.2.	Delays in implementation of projects	3	1.49%
3.4.3.	Lack of interest of constructors	2	1.00%
3.4.4.	Lack of project implementation knowledge	1	0.50%
3.4.5.	Challenges in land acquisition	1	0.50%
3.4.6.	Questionable quality of work	1	0.50%
3.4.7.	Unrealistic sociotechnical projects	1	0.50%
4.	**Environmental**		
4.1.	Natural disasters	4	1.99%
4.2.	Climate change	2	1.00%
5.	**Technological and technical**		
5.1.	**Infrastructure**		
5.1.1.	Insufficient maintenance of infrastructure systems	6	2.99%
5.1.2.	Information systems' errors	4	1.99%
5.1.3.	Failure to integrate technology projects into the social structure	3	1.49%
5.1.4.	Unstable power supply	6	2.99%
5.1.5.	Failure of digitization of existing infrastructure	1	0.50%

(continued)

Table 3 (continued)

Description of categories/risks		Frequency count	Percentage in relation to total risks
5.2.	**Requirements**		
5.2.1.	Failure of infrastructure assets to meet quality requirements	4	1.99%
5.2.2.	Poor service and device research	2	1.00%
5.2.3.	Lack of energy estimation techniques of IoT applications	1	0.50%
6.	**Strategic**		
6.1.	Lack of insight of smart city concept	6	2.99%
6.2.	Lack of clear strategy across municipality	4	1.99%
6.3.	Institutional resistance to change their approaches	2	1.00%
6.4.	Overestimating the positive impacts of technology	1	0.50%
6.5.	Insufficient focus on the consequences of infrastructure asset failure, especially on the community side	1	0.50%
7.	**Political**		
7.1.	Reputational damage	2	1.00%
7.2.	Political pressure	2	1.00%
7.3.	Lack of political will	1	0.50%
7.4.	Political uncertainty	1	0.50%
8.	**Legal**		
8.1.	Limitations of existing laws and regulations	6	2.99%
8.2.	Uncertainty in data's security responsibility	3	1.49%
8.3.	Strict regulations	1	0.50%
8.4.	Lack of strict policy enforcement	1	0.50%
9.	**Security**		
9.1.	**Cyber-risks**		
9.1.1	Cyberattacks	27	13.43%
9.1.2.	Private information and data disclosure risk	27	13.43%
9.1.3.	Installation of supervisory control devices	1	0.50%

(continued)

Table 3 (continued)

Description of categories/risks		Frequency count	Percentage in relation to total risks
9.1.4.	Lacking of personal information protection technologies	1	0.50%
9.1.5.	Unpredictable user behaviour	1	0.50%
9.2.	**Health and safety**		
9.2.1.	Harm of human beings from smart technologies	2	1.00%
9.2.2.	Terrorism	2	1.00%
9.2.3.	Human-induced incidents	1	0.50%
9.3.	**Physical resources risks**		
9.3.1.	Stealing devices	2	1.00%
9.3.2.	Sabotage infrastructures for war efforts	1	0.50%
9.3.3.	Deliberate damage of hardware equipment	1	0.50%
Total		201	100.00%

Competence in statistics: The risk manager will be required to use a significant number of mathematical models, simulations and statistics to analyse risks. For example, quantitative analysis follows risk identification and is used to express the probability of occurrence and consequences of identified risks in mathematical form (Baker et al., 1998). This analysis allows the comparison of risks in order to derive a priority for dealing with them, as the budget for this purpose is not limitless.

5.2 Risk Mitigation Strategies

It is necessary to develop appropriate risk response strategies to address the risks that threaten the existence of smart cities. This purpose is served by risk response strategies which address the causes, likelihood and consequences of risks, before or after their occurrence. PMI suggests as risk response strategies: avoidance, transfer, mitigation and acceptance. Definitions of each one are (Project Management Institute, 2017):

- Avoidance: "eliminate the threat or protect the project from its impact".
- Transfer: "shifting ownership of a threat to a third party to manage the risk and to bear the impact if the threat occurs".
- Mitigation: "reduce the probability of occurrence and/or impact of a threat".
- Acceptance: "acknowledges the existence of a threat, but no proactive action is taken".

First of all, cyberthreats could be assessed by training the personnel responsible for data management, for cyberattacks (Sheehan et al., 2021). Such a measure would help in avoiding potential errors that would create breaches in the smart systems for hacker attacks. Moreover, equipping city's personnel with cyber-attack assessment skills and knowledge would create one more layer of safety from such risks. Finally, since every sector of city is operating in smart technologies, it is not enough to train IT staff, but equipping all staff with good practices in operating technology systems is essential (Kitchin & Dodge, 2019).

Transfer strategy is served by cyber insurance companies, for example, in cyber-security issues, the number of which is increasing not only because of the growing need for their services but also because of legal considerations (Sheehan et al., 2021). By exploiting such excesses, the city is relieved of the cost of a cyberattack.

In case of data storage, encrypting data stored in clouds could prevent their retrievement, even if the attacker succeeded accessing in the cloud (Krämer et al., 2019). Also keeping backups for the important data would eliminate the losses of their potential delete by hackers. Finally, as mentioned before, decentralization of data management by city's sector would avoid exposing all city's data to the attackers. Each one strategy would reduce the impact of a breach of the databases.

There is however another evolving tool for avoiding the vulnerabilities of databases called blockchain. Blockchain is a decentralized storage technology which was initially developed for cryptocurrency transactions and then adopted for other applications, as in smart cities too. The key features of blockchain that make it suitable for replacing databases, as they are known today, are decentralization, resistance to cyberattacks, transparency and scalability (Bhushan et al., 2020; Cui et al., 2018).

Another example, which this time would address the possibility of the risk occurring, is the introduction of standards when creating smart systems in terms of security, encryption, verification and other factors (Sengan et al., 2020). By setting standards, no technology will fall short of safety measures, and the probability of data breach would be reduced. This fact is of crucial importance as in interdependent technological systems, their overall security is equal to the security provided by the weakest component (Kitchin & Dodge, 2019).

Standards can be applied not only to the technologies to be included in smart cities but also to the companies involved. Companies in smart cities are an extension of cities, as they generate, manage and move data to and from city services. Therefore, no matter how many measures the smart city takes for potential risks, it will remain vulnerable to the security flaws of the partner companies. For this reason, it is proposed in the literature that companies that want to participate in the smart city environment should commit themselves to following the standards that have already been developed and which deal with data ownership issues and security procedures for data protection (Vitunskaite et al., 2019).

Artificial intelligence (AI) is a crucial tool for managing data and countering cyberattacks at the same time. By utilizing the machine learning capability of AI applications, these applications are able to identify patterns for optimal data management (Bellam, 2018). The same pattern recognition capability can be also used to identify cyber-attack patterns, while its self-learning capacity offers the ability to anticipate new cyber-risks (Srivastava et al., 2017). For the autonomous assessing of cyberthreats by AI, algorithms have already been developed and analysed in the literature, such as the neural network model (Krundyshev, 2020).

To extract data from the physical environment of the city and convert it into digital data, the deployment of a plethora of sensor nodes within the smart city is essential. The number of those makes it impossible to check their functionality and reliability in hardware and software level by physical testing. For this cause, dynamic trust measurement models have been developed and tested (Gong et al., 2018). Such measurement models consist of algorithms for the production and evaluation of the signatures of the nodes and for their comparison with trusted nodes (Gong et al., 2018).

Another measure to counter cyberattacks is to keep IoT devices and systems up to date (Andrade et al., 2020). The methods and means for cyberattacks are also evolving rapidly. Therefore, neglecting to upgrade the software used in the smart city environment will create vulnerabilities, as the systems will be outdated against the advances in attack methods.

Risks such as cyberattacks, private information and data disclosure or information systems' errors are often dealt by using other technological systems and automations. This fact raises new issues, as in studies the technologies are often the problem that generates a risk and not the means to solve it (Soyata et al., 2019). Consequently, the situation as it stands at present gives the impression of a vicious cycle. For this reason, but also because new technologies are being integrated and their complexity increases in each smart city, it is necessary to constantly reassess and identify new risks. Risk management is a process that follows the whole lifecycle of a smart city.

6 Conclusions

The number of smart cities is growing rapidly, a trend driven by changing conditions and needs within cities. Along with smart cities, the number of stakeholders who are required to participate in, adapt to and take decisions is growing. In decision-making both in the process of designing smart cities and in their operation, a risk management plan is necessary to ensure the sustainability of the endeavour. Risk management is becoming a complex process, similar in complexity to the interaction of the technologies that build smart cities.

This chapter aims to familiarize developers, managers and other stakeholders of smart cities with the risks to which smart cities may be exposed. Lack of risk awareness will threaten the existence of the smart city in the future, as these risks emerge. Identifying them is therefore the first step in addressing them.

In order to identify the risks that affect the design and operation of a smart city, an SLR was conducted, while the results were presented as PRISMA guidelines suggest. From the SLR the following categories of risks have been identified,

(1) economic, (2) social, (3) organizational, (4) environmental, (5) technological and technical, (6) strategic, (7) political, (8) legal and (9) security, with their subcategories presented in detail in Table 3.

For the effective implementation of risk management, risk managers need to be qualified with certain skills and competences. Such skills are technical skills, smart cities' concept understanding, project management skills, soft skills and mathematical skills. In addition to their personal skills, risk managers can use strategies that have already been developed to mitigate risks. Particular emphasis is given in the literature to address security risks, as the direct and indirect protection of citizens is a priority for any smart city. Following the avoidance, transfer and mitigation strategies and by utilizing new technologies, a number of methods are presented.

One potential limitation of this study is the bias of the researchers on the naming of risks, the grouping of common risks and finally their categorization in the RBS. To address this, detailed reviews were carried out by all researchers, and lengthy discussions were held. As in every RBS, the researchers may differentiate the final result; however, it is the authors' belief that the information (i.e. risks appearing in the RBS) is complete and accurate.

The generated RBS for smart cities' risks could be an advisor in identifying the risks that smart city managers will be asked to undertake. Moreover, high-level strategies are presented, as addressing methods of the identified risks. The equipment of those actively involved in smart cities with risk management skills and knowledge for risk mitigation strategies is an essential step to ensure the sustainability of smart cities from the top level of their management.

References

Ahad, M., Paiva, S., Tripathi, G., & Feroz, N. (2020). Enabling technologies and sustainable smart cities. *Sustainable Cities and Society, 61*. https://doi.org/10.1016/j.scs.2020.102301

Ambrosino, G., Finn, B., Gini, S., & Mussone, L. (2015). A method to assess and plan applications of ITS technology in public transport services with reference to some possible case studies. *Case Studies on Transport Policy, 3*(4), 421–430. https://doi.org/10.1016/j.cstp.2015.08.005

Andrade, R., Yoo, S., Tello-Oquendo, L., & Ortiz-Garces, I. (2020). A comprehensive study of the IoT cybersecurity in smart cities. *IEEE Access, 8*, 228,922–228,941. https://doi.org/10.1109/ACCESS.2020.3046442

Anthopoulos, L., Siozos, P., & Tsoukalas, I. (2007). Applying participatory design and collaboration in digital public services for discovering and re-designing e-government services. *Government Information Quarterly, 24*(2), 353–376.

Axelrod, C. W. (2013). Managing the risks of cyber-physical systems. *2013 IEEE Long Island Systems, Applications and Technology Conference (LISAT)*. doi:https://doi.org/10.1109/lisat.2013.6578215.

Ayala-Ruiz, D., Castillo Atoche, A., Ruiz-Ibarra, E., Osorio de la Rosa, E., & Vázquez Castillo, J. (2019). A self-powered PMFC-based wireless sensor node for Smart City applications. *Wireless Communications and Mobile Computing, 2019*. https://doi.org/10.1155/2019/8986302

Baker, S., Ponniah, D., & Smith, S. (1998). Techniques for the analysis of risks in major projects. *Journal of the Operational Research Society, 49*(6), 567–572.

Bellam, S. (2018). Robotics vs. machine learning vs. artificial intelligence: Identifying the right tools for the right problems. *The Credit and Financial Management Review, 27*(2), 1–10.
Bhushan, B., Khamparia, A., Sagayam, K., Sharma, S., Ahad, M., & Debnath, N. (2020). Blockchain for smart cities: A review of architectures, integration trends and future research directions. *Sustainable Cities and Society, 61*. https://doi.org/10.1016/j.scs.2020.102360
Carvalho, M., & Rabechini Junior, R. (2015). Impact of risk management on project performance: The importance of soft skills. *International Journal of Production Research, 53*(2), 321–340.
Cui, L., Xie, G., Qu, Y., Gao, L., & Yang, Y. (2018). Security and privacy in smart cities: Challenges and opportunities. *IEEE Access, 6*, 46,134–46,145.
Derakhshanfar, H., Ochoa, J., Kirytopoulos, K., Mayer, W., & Tam, V. W. Y. (2019). Construction delay risk taxonomy, associations and regional contexts: A systematic review and meta-analysis. *Engineering, Construction and Architectural Management, 26*(10), 2364–2388.
Economic and Social Council of United Nations. (2016). *Smart cities and infrastructure. Nineteenth session of The Commission on Science and Technology for Development*, Geneva, 9–13 May. Retrieved May 24, 2021, from https://unctad.org/system/files/official-document/ecn162016d2_en.pdf
Fernandez-Anez, V. (2016). Stakeholders approach to smart cities: A survey on Smart City definitions. *Smart Cities*, 157–167.
Ganin, A., Mersky, A., Jin, A., Kitsak, M., Keisler, J., & Linkov, I. (2019). Resilience in intelligent transportation systems (ITS). *Transportation Research Part C: Emerging Technologies, 100*, 318–329.
Giffinger, R., Fertner, C., Kramar, H., Kalasek, R., Pichler-Milanovic, N., & Meijers, E. (2007). *Smart cities—Ranking of European medium-sized cities*. Centre of Regional Science. Retrieved June 8, 2021, from http://www.smart-cities.eu
Gong, B., Wang, Y., Liu, X., Qi, F., & Sun, Z. (2018). A trusted attestation mechanism for the sensing nodes of internet of things based on dynamic trusted measurement. *China Communications, 15*(2). https://doi.org/10.1109/CC.2018.8300276
Gupta, K., & Hall, R. (2020). Exploring smart city project implementation risks in the cities of Kakinada and Kanpur. *Journal of Urban Technology, 28*(1–2), 155–173.
Gupta, P., Chauhan, S., & Jaiswal, M. P. (2019). Classification of smart city research—A descriptive literature review and future research agenda. *Information Systems Frontiers*. https://doi.org/10.1007/s10796-019-09911-3
Hayat, P. (2016). Smart cities: A global perspective. *India Quarterly: A Journal of International Affairs, 72*(2), 177–191.
International Organization for Standardization (2018a). *Risk management (ISO 31000:2018)*. [online] Retrieved June 14, 2021, from https://www.iso.org/obp/ui/#iso:std:iso:31000:ed-2:v1:en
International Organization for Standardization. (2018b). *Risk management—Guidelines (ISO 31000:2018)*. [online] Retrieved July 11, 2021, from https://www.iso.org/obp/ui/#iso:std:iso:31000:ed-2:v1:en
International Risk Governance Council. (2015). *Guidelines for emerging risk governance*.
Jiménez-Bravo, D., De Paz, J., Villarrubia, G., & Bajo, J. (2018). Dealing with demand in electric grids with an adaptive consumption management platform. *Complexity, 2018*. https://doi.org/10.1155/2018/4012740
Kitchin, R., & Dodge, M. (2019). The (in)security of smart cities: Vulnerabilities, risks, mitigation, and prevention. *Journal of Urban Technology, 26*(2), 47–65. https://doi.org/10.1080/10630732.2017.1408002
Krämer, M., Frese, S., & Kuijper, A. (2019). Implementing secure applications in smart city clouds using microservices. *Future Generation Computer Systems, 99*, 308–320.
Krundyshev, V. (2020). Neural network approach to assessing cybersecurity risks in large-scale dynamic networks. In *13th International Conference on Security of Information and Networks*. doi:https://doi.org/10.1145/3433174.3433603.

Le, P. T., Kirytopoulos, K., Chileshe, N., & Rameezdeen, R. (2019). Taxonomy of risks in PPP transportation projects: A systematic literature review. *International Journal of Construction Management*. https://doi.org/10.1080/15623599.2019.1615756
Lee, J. H., Phaal, R., & Lee, S.-H. (2013). An integrated service-device-technology roadmap for smart city development. *Technological Forecasting and Social Change, 80*(2), 286–306. https://doi.org/10.1016/j.techfore.2012.09.020
Mainzer, K. (2020). Technology foresight and sustainable innovation development in the complex dynamical systems view. *Foresight and STI Governance, 14*(4), 10–19. https://doi.org/10.17323/2500-2597.2020.4.10.19
Marx, J., & de Swardt, C. (2019). Towards a competency-based undergraduate qualification in risk management. *Qualitative Research in Financial Markets, 12*(1), 96–117.
Mat Ludin, K. R., Mohamed, Z. M., & Mohd-Saleh, N. (2017). The association between CEO characteristics, internal audit quality and risk-management implementation in the public sector. *Risk Management, 19*(4), 281–300. https://doi.org/10.1057/s41283-017-0022-z
Mell, P. M., & Grance, T. (2011). *The NIST definition of cloud computing, tech. Rep., SP 800–145*. National Institute of Standards & Technology.
Nightingale, A. (2009). A guide to systematic literature reviews. *Surgery (Oxford), 27*(9), 381–384.
Nižetić, S., Šolić, P., López-de-Ipiña González-de-Artaza, D., & Patrono, L. (2020). Internet of things (IoT): Opportunities, issues and challenges towards a smart and sustainable future. *Journal of Cleaner Production, 274*. https://doi.org/10.1016/j.jclepro.2020.122877
Page, M., Moher, D., Bossuyt, P., Boutron, I., Hoffmann, T., Mulrow, C., et al. (2021). PRISMA 2020 explanation and elaboration: Updated guidance and exemplars for reporting systematic reviews. *BMJ*. https://doi.org/10.1136/bmj.n160
Prater, J., Kirytopoulos, K., & Ma, T. (2017). Optimism bias within the project management context: A systematic quantitative literature review. *International Journal of Managing Projects in Business, 10*(2), 370–385.
Project Management Institute. (2017). *A guide to the project management body of knowledge* (6th ed.). Project Management Institute.
Pym, D. V. (1987). Risk Management. *PM Network, 1*(3), 33–36.
Rathore, M., Ahmad, A., & Paul, A. (2015). Big data and internet of things. In *Proceedings of the 2015 international conference on big data applications and services*. doi: https://doi.org/10.1145/2837060.2837067.
Sengan, S., Subramaniyaswamy, V., Nair, S., Indragandhi, V., Manikandan, J., & Ravi, L. (2020). Enhancing cyber–physical systems with hybrid smart city cyber security architecture for secure public data-smart network. *Future Generation Computer Systems, 112*, 724–737.
Shamsuzzoha, A., Nieminen, J., Piya, S., & Rutledge, K. (2021). Smart city for sustainable environment: A comparison of participatory strategies from Helsinki, Singapore and London. *Cities, 114*. https://doi.org/10.1016/j.cities.2021.103194
Sheehan, B., Murphy, F., Kia, A., & Kiely, R. (2021). A quantitative bow-tie cyber risk classification and assessment framework. *Journal of Risk Research*.https://doi.org/10.1080/13669877.2021.1900337
Shen, L., Shuai, C., Jiao, L., Tan, Y., & Song, X. (2016). A global perspective on the sustainable performance of urbanization. *Sustainability, 8*(8), 783.
Silva, B., Khan, M., & Han, K. (2018). Towards sustainable smart cities: A review of trends, architectures, components, and open challenges in smart cities. *Sustainable Cities and Society, 38*, 697–713.
Soyata, T., Habibzadeh, H., Ekenna, C., Nussbaum, B., & Lozano, J. (2019). Smart city in crisis: Technology and policy concerns. *Sustainable Cities and Society*. https://doi.org/10.1016/j.scs.2019.101566
Srivastava, S., Bisht, A., & Narayan, N. (2017). Safety and security in smart cities using artificial intelligence—A review. *2017 7th International Conference on Cloud Computing, Data Science & Engineering - Confluence*. doi:https://doi.org/10.1109/CONFLUENCE.2017.7943136.
Tariq, M., Wai, C., & Muttil, N. (2020). Vulnerability assessment of ubiquitous cities using the analytic hierarchy process. *Future Internet, 12*(12). https://doi.org/10.3390/fi12120235

Ullah, F., Qayyum, S., Thaheem, M., Al-Turjman, F., & Sepasgozar, S. (2021). Risk management in sustainable smart cities governance: A TOE framework. *Technological Forecasting and Social Change, 167*. https://doi.org/10.1016/j.techfore.2021.120743

Vincent, R., Ait-Ahmed, M., Houari, A., & Benkhoris, M. (2020). Residential microgrid energy management considering flexibility services opportunities and forecast uncertainties. *International Journal of Electrical Power & Energy Systems, 120*. https://doi.org/10.1016/j.ijepes.2020.105981

Vitunskaite, M., He, Y., Brandstetter, T., & Janicke, H. (2019). Smart cities and cyber security: Are we there yet? A comparative study on the role of standards, third party risk management and security ownership. *Computers & Security, 83*, 313–331.

Vu, K., & Hartley, K. (2018). Promoting smart cities in developing countries: Policy insights from Vietnam. *Telecommunications Policy, 42*(10), 845–859.

Xiao, Y., & Watson, M. (2019). Guidance on conducting a systematic literature review. *Journal of Planning Education and Research.*https://doi.org/10.1177/0739456X17723971

Yan, X., Fan, Y., Lee, H., & Qiu, R. (2020). Research on personal information risk assessment model in smart cities. *Tehnicki vjesnik - technical. Gazette, 27*(5). https://doi.org/10.17559/TV-20190104101416

Yang, Y., Ng, S., Xu, F., & Skitmore, M. (2018). Towards sustainable and resilient high density cities through better integration of infrastructure networks. *Sustainable Cities and Society, 42*, 407–422.

City Resilience and Smartness: Interrelation and Reciprocity

Christos Ziozias and Leonidas Anthopoulos

1 Introduction

The last decades of the twentieth century were characterized, among other things, by an increasing trend toward urbanism (Alqahtani et al., 2018). More than half of the inhabitants around the world live in cities, and this percentage is expected to exceed 65% by 2050 (Michelucci et al., 2016; Purnomo et al., 2016; United Nations Office for Disaster Risk Reduction (UNDRR), 2017). It is estimated that until 2030, 60% of the inhabitants will gather in cities with a population of over 500,000 (Ragia & Antoniou, 2020). This overconcentration in large urban centers, apart from development opportunities, is responsible for the emergence of new risks (Purnomo et al., 2016; United Nations Office for Disaster Risk Reduction (UNDRR), 2017; Ragia & Antoniou, 2020; Zhu et al., 2020). The city administration along with the political authorities has no other choice but to deal with them directly to secure the well-being of their citizens (ARUP, 2015; Clements-Croome, 2012; Li et al., 2017; Makhoul, 2015).

On their way to evolution, every city chooses a different path, leading to variations like sustainable city, smart city, digital city, etc. (Makhoul, 2015). The development of technology, however, was not enough to ensure a secure future for the urban areas and their inhabitants. Problems and uncertain situations existed and will continue to appear in the future, in the form of risks, crises, or disasters (Anthopoulos et al., 2013; Scholl & Patin, 2012). All these threaten the city's ability to provide the expected quality of life for its inhabitants, especially since these are dynamic such as climate change and not static phenomenon, leading to the *new normal* global situation (Scholl & Patin, 2012). To ensure the continuing operation

C. Ziozias (✉) · L. Anthopoulos
Department of Business Administration, University of Thessaly—School of Management and Economic Science, Larissa, Greece
e-mail: cziozias@uth.gr; lanthopo@uth.gr

P. Fitsilis (ed.), *Building on Smart Cities Skills and Competences*, Internet of Things,
https://doi.org/10.1007/978-3-030-97818-1_8

of their critical infrastructure and services, cities must demonstrate resilience against risk and disasters (Makhoul, 2015). Regarding an individual, an organization, or a system (natural or man-made), resilience is a set of competencies and skills that gives the *ability to survive* during threats, risks, or any unpredictable situation (Scholl & Patin, 2012; Chan & Zhang, 2019; Simone et al., 2021). Like any other living organization, a city must demonstrate resilience under stresses, crises, and disasters, to apply the *bounce-back ability* and secure the quality of life of its people (Cavada et al., 2017).

There has been a global tendency for cities to use ICT to improve their services provided to citizens. Smart infrastructures, solutions, and technology, along with human resources, help cities and communities achieve a better quality of life for their citizens, transforming them into "smart" (Zhu et al., 2020). But on the other hand, the more dependent is a system on innovative technologies, the larger is the risk of new and unknown vulnerabilities and danger (Alqahtani et al., 2018). Although resilience is achieved mostly in smart cities, the features of a smart city are not enough to make it resilient (Oke et al., 2020).

Literature review shows that the research on the smart city or resilient city skills and competencies is in the early stages and there are insufficient findings to compare skills and competencies for a smart or resilient city. This chapter presents the major similarities and differences between a smart and a resilient city, as a tool for officials, responsible for planning each city model, to decide and define the proper policies, strategies, and actions. Through this analysis, certain skills and competencies must be highlighted, which will be necessary to support the management along with the implementation of the strategic vision for the smart or resilient city. It is structured as follows: in Sect. 2 there is an overview of smart city and resilient city definition and aspects. In Sect. 3 models and evaluation indicators for each city are presented. Section 4 highlights the way each city should proceed with its strategic vision. In Sect. 5 there is a presentation of how major dimensions of these two cities interact and which approach should be adopted. Finally, Sect. 6 concludes and summarizes the chapter.

2 Defining the City

The reference to "intelligence" emphasizes a higher level of analysis and design that drives better decisions, conclusions, and strategies (Khatibi et al., 2021a). The overwhelming use of ICT changed the way cities managed their ecosystems, in terms of economy, development, and society, and transformed them into "smart cities" (Backhouse, 2020; Mora et al., 2018; Santinha & de Castro, 2010). Decades ago, scholars started studying the "smart city" and its applications, but there is no common definition of the "smart city" and how to "build" one, due to different perspectives and needs of each city's stakeholder and disagreements on ICT's overall contribution (Cavada et al., 2017; Mora et al., 2018; Falconer & Mitchell, 2012; Lafi Aljohani & Alenazi, 2020). Nowadays "smart" is mostly considered to be a city where ICT and innovative solutions are used to improve the well-being of their citizens, without compromising the core subsystems of the city, like the build, natural and social environment, and information ecosystem (Michelucci et al., 2016; Purnomo et al., 2016; Backhouse, 2020; Fujinawa et al., 2015; Lopez & Castro, 2021; Nel & Nel, 2019; Stubinger & Schneider, 2020; Zhu et al., 2019).

As mentioned above, the main pillars of evolution for a smart city are the ICTs and human skills and competencies (Zhu et al., 2020). In the last decade, researchers have not limited their studies to these two factors and defined other, of equal importance, dimensions of a smart city. Most of them concluded in the following six: smart economy, smart environment, smart governance, smart living, smart mobility, and smart people (Khatibi et al., 2021a; Lopez & Castro, 2021; Anthopoulos et al., 2019). Many more dimensions have emerged in the last years like smart community, smart construction, smart development, smart energy, smart health, smart infrastructure, smart innovation, and others, highlighting all the important topics in the daily management of a smart city (Stubinger & Schneider, 2020; Zhu et al., 2019; Joss et al., 2019).

As the smart city, the resilient city is a term that has not a common definition (Nel & Nel, 2019). The resilience of a city is the overall ability to protect its citizens and continue its functionality while sudden phenomena occur (ARUP, 2015). New risks and dangers emerged over time, so resilience must be built with an innovative approach not only for the current challenges but also for the ones who come (Cavada et al., 2017; Lopez & Castro, 2021). Focusing on resilience, an urban system must face and overcome not only natural but also man-made crises and phenomena, preventing them to evolve into disasters (Chan & Zhang, 2019; Lopez & Castro, 2021; ISO, 2019a). Scholars highlighted some of the dimensions of resilience in urban systems that must be considered when planning the overall resilience management; community/social; economic, infrastructural, institutional, environmental/natural, organizational; and technical resilience (Li et al., 2017; Zhu et al., 2019; Patel & Nosal, 2016).

Many researchers identified the resilience aspect in cities and communities under a process of disaster risk reduction and sustainable development (Patel & Nosal, 2016). A resilient city must be managed in such a way that develops urban sustainability while it prepares itself to overcome not one but multiple hazards simultaneously (ARUP, 2015; Khatibi et al., 2021b). As a holistic approach, the resilient city can be considered to be the one that can absorb, adapt, and recover from multiple external pressures and threats, crises, risks, and disaster situations or mitigate the consequences; it is an interaction and co-operation of the social, technical, and ecological subsystems of the city so that it can maintain its functionality and aim at a stronger version of it (Zhu et al., 2020; ARUP, 2015; Makhoul, 2015; Oke et al., 2020; Nel & Nel, 2019; Bujones et al., 2013).

3 Measuring Performance

Various scholars and international standardization bodies have focused on defining a complete set of indicators for the smart city. Since the "smartness" of a city is approached from different points of view, evaluation is a complex procedure, so a unified model for benchmarking cannot be applied in every city (Backhouse, 2020; Anthopoulos et al., 2019; Khatibi et al., 2021b). A commonly agreed approach is for the city administration to evaluate all aspects or dimensions in a smart city that are considered to be critical. These parameters must cover all the critical dimensions of the smart city mentioned in the previous sector (smart economy, smart environment, smart governance, smart living, smart mobility, and smart people). In 2019 ISO set 19 groups of indicators to be used globally, and cover different dimensions and domains, of a smart city's management, enhancing the effectiveness of a smart city (ISO, 2019b):

- "Economy," "Energy," "Finance," "Governance," and "Safety" that can be used as indicators of policy
- "Environment and climate change," "Wastewater," and "Water" that can be related to environmental issues
- "Education," "Health," "Population and social conditions," and "Sport and culture" as social factors
- "Housing," "Recreation," "Solid waste," "Telecommunication," "Transportation," "Urban/local agriculture and food security," and "Urban planning" on urban management.

Like smart cities, resilient cities must be controlled and evaluated through specific indicators. Most of the scholars agree that resilience in a city must be measured in five critical systems: "political" since it reflects citizens' opinion about their government, "security" that covers the personal feeling and rule of law, "economic" that is related to wealth and resources, "social" that represents the quality of public services, and "environmental" for buildings and natural environment (Bujones et al., 2013). Patel and Nosal (Patel & Nosal, 2016) promoted the *PEOPLES* set of indicators, and this acronym refers to "Population and Demographics, Environmental/Ecosystem, Organized Governmental Services, Physical Infrastructure, Lifestyle, and Community Competence, Economic Development, and Social-Cultural Capital," groups that target a certain aspect of a smart city.

In 2015, the ARUP International Development summarized 12 commonly agreed and critical indicators, with 45–54 sub-indicators and 130–150 variables to measure overall resilience in a city. Since it is a complex topic and affected by many dimensions, the research team tried to cover as many aspects as possible. These 12 indicators are grouped into 4 different categories that refer to citizens, management, places, and knowledge, respectively (ARUP, 2015):

- "Minimal human vulnerability," "Diverse livelihoods and employment," and "Adequate safeguards to human life and health" in *Health and Well-being* category

- "Collective identity and mutual support," "Social stability and security," and "Availability of financial resources and contingency funds" in *Economy and Society* category
- "Reduced physical exposure and vulnerability," "Continuity of critical services," and "Reliable communications and mobility" in the *Urban Systems and Services* category
- "Effective leadership and management," "Empowered stakeholders," and "Integrated development planning" in the *Leadership and Strategy* category.

Finally, ISO chooses 19 groups of indicators to evaluate the overall resilience in a city, like the ones in smart cities: "*Economy, Education, Energy, Environment and climate change, Finance, Governance, Health, Housing, Population, and social conditions, Recreation, Solid Waste, Safety, Sport and culture, Telecommunication, Transportation, Urban/local agriculture and food security, Urban planning, Wastewater*, and *Water*" (ISO, 2019a). The groups are the same as the ones for smart cities, although indicators in each group are not the same – for example, the group "Economy" in smart cities has indicators like Percentage of service contracts providing city services which contain an open data policy, the Survival rate of new businesses per 100,000 population, Percentage of the labor force employed in occupations in the information and communications technology (ICT) sector, etc., while the same group in resilient cities has indicators like Historical disaster losses as a percentage of city product, Average annual disaster loss as a percentage of city product, Percentage of properties with insurance coverage for high-risk hazards, etc. Although the specific indicators are different, the fact that the categories are the same reflects that the smartness and resilience in a city's ecosystem are affected by the same dimensions.

4 City Management

Many cities tend to adopt the best practices that other smart cities have implemented. Officials need to manage a smart city and apply its strategic plan, considering that it is a "multi-sectoral, inter-organizational and intergovernmental" procedure (Michelucci et al., 2016). They must evaluate the city's current status, define the transformation plan, and check the process (Falconer & Mitchell, 2012). A smart city must be planned as holistic planning, which relates to all smart city dimensions and aspects (smart infrastructure, smart people, smart economy, smart government, smart environment, etc.), to reach the best possible outcome (Abdoullaev, 2011).

The majority of applied older strategic plans tried to improve the "smartness" level by enhancing ICT infrastructure (Nel & Nel, 2019). Since most of them target the quality of life for citizens, many cities changed their strategic plan to include the "people" parameter (Lopez & Castro, 2021; Nel & Nel, 2019). Nowadays, a complete strategic plan of a smart city must take into account all smart city dimensions (people, infrastructure, economy, government, mobility, environment) and try to improve each and all of them, although there is no smart city that excels in all these dimensions (Purnomo et al., 2016; Anthopoulos et al., 2019; Abdoullaev, 2011).

According to Santinha and Castro (Santinha & de Castro, 2010), a smart city's administration must promote internal characteristics while establishing external connections. Internal characteristics consider being the high quality of provided services, the sustainable environmental planning, the innovative urban design, the recognition of skilled and talented citizens, and the enhanced technological solutions, among others. At the same time, highlighting comparative advantages will not only improve quality of life but also attract citizens, visitors, and enterprises. On the other hand, the external connection can be built through participation in a greater network of cities, giving access to certain information, solutions, and best practices. This will add value to the city's recognition and branding, securing funding, resources, and technology for sustainable development, through global initiatives. Of course, the first step for this endeavor must be the interaction with other cities within the region or the country.

ISO's research concluded to certain suggestions for a successful implementation of a smart city strategic plan (ISO, 2019b):

- "Respond to challenges such as climate change, rapid population growth, and political and economic instability by fundamentally improving how it engages society."
- "Apply collaborative leadership methods, work across disciplines and city systems."
- "Use data information and modern technologies to deliver better services and quality of life to those in the city (residents, businesses, visitors)."
- "Provide a better life environment where smart policies, practices, and technology are put to the service of citizens."
- "Achieve their sustainability and environmental goals in a more innovative way."
- "Identify the need for and benefits of smart infrastructure."
- "Facilitate innovation and growth."
- "Build a dynamic and innovative economy ready for the challenges of tomorrow."

Deciding the proper strategic plan is not an easy task, since most of the research is about technology in SC than management of SC, and this is the main reason that there are many perspectives regarding SC planning (Michelucci et al., 2016; Mora et al., 2018). Smart cities are living organizations that continuously evolve (Clements-Croome, 2012), and the position that all stakeholders must see SC through that lens is considered to be a correct approach. The complexity in operational, finance, and planning procedures along with conflicted interests of the public and private sector, parties, and communities makes a commonly accepted strategy even more difficult (Falconer & Mitchell, 2012). Smart city planners and managers must take these into account to face the less possible reactions.

Managing and planning for a resilient city is a top priority, and the city's administration develops strategies and actions toward that (AlHinai, 2020; Huck et al., 2020). Organizing its strategic plan, a city must be ready to face events beyond 100-year frequency of happening. Even if the financial cost for this is extremely high, this is justified because the ultimate goal of each administration is to drive a safe and sustainable urban development (Alqahtani et al., 2018; United Nations Office for Disaster Risk Reduction (UNDRR), 2017). Risk reduction and implementation of resilience strategy will save lives; drive to a social, economic, and sustainable urban development; and benefit communities by strengthening their trust in their leaders and governance, enhancing citizen participation and protection of culture, promoting innovation and a safe economic environment, creating new job opportunities and balanced ecosystems, and the interconnection between cities at a national or global level (United Nations Office for Disaster Risk Reduction (UNDRR), 2017).

The main actions to build resilience are divided into two major phases, before and after the disaster occurs, without overlooking the necessary and imminent response to activate emergency plans and operations, during the disaster. In the pre-disaster period, the city must evaluate the level of resilience, through analysis of the city's strong and weak points, along with identification, understanding, evaluation of risks, and preparation against them, to mitigate potential upcoming effects. In the post-disaster period, officials must take actions for recovery and rebuild, along with the preparation and implementation of an action plan for the mitigation of consequences of future threats (United Nations Office for Disaster Risk Reduction (UNDRR), 2017; Anthopoulos et al., 2013; Oke et al., 2020; ISO, 2019a; Bujones et al., 2013; Altay & Green, 2006). The complexity of this procedure is highlighted by the United Nations (United Nations, 2015), by mobilizing and applying many different measures like "*economic, structural, legal, social, health, cultural, educational, environmental, technological, political and institutional measures*," while in 2017 four priorities for successful disaster risk reduction were identified: "*Understanding disaster risk*; *strengthening disaster risk governance to manage disaster risk*; *investing in disaster risk reduction for resilience*, and *enhancing disaster preparedness for effective response and to "Build Back Better" in recovery, rehabilitation, and reconstruction*" (United Nations Office for Disaster Risk Reduction (UNDRR), 2017).

Research from Sendai Framework UNISDR lead to ten guidelines for city officials, planners, and decision-makers, called the "Ten Essentials." It is about a holistic approach toward more resilient cities and communities since it includes actions in all city's critical dimensions. These are (United Nations Office for Disaster Risk Reduction (UNDRR), 2017):

1. "Organize for disaster resilience."
2. "Identify, understand and use current and future risk scenarios."
3. "Strengthen financial capacity for resilience."
4. "Pursue resilient urban development and design."
5. "Safeguard natural buffer to enhance the protective functions offered by natural ecosystems."
6. "Strengthen institutional capacity for resilience."
7. "Understand and strengthen societal capacity for resilience."

8. "Increase infrastructure resilience."
9. "Ensure effective disaster response."
10. "Expedite recovery and build better."

Furthermore, resilience planners must not ignore the important role of stakeholders and adopt a people-centered approach. For a multisectoral topic like this, all the relevant stakeholders must participate and help to define and implement the strategy and policy. In that direction synergies and cooperation with cities from other regions and countries will also help, by transferring knowledge about the best practices that have been successfully implemented (United Nations Office for Disaster Risk Reduction (UNDRR), 2017; United Nations, 2015).

5 Setting Strategy

Smart cities, using ICT and innovation, aim to provide a safe, sustainable, and economically secure environment to their citizens while facing the same challenges as other cities, which can disrupt this normality (Nel & Nel, 2019). There is no other way but to enhance their level of resilience. "Smartness" and resilience are two different aspects, but both enhance sustainable development using common systems, and for this, they are both considered to be parts of the city governance model (Li et al., 2017; Khatibi et al., 2021a). Although both topics have been studied for many decades—resilience since the 1970s and smart city since the 1990s—there is limited literature on their similarities, differences, and connection, probably since there is no solid and globally accepted definition, but this seems to change (Chan & Zhang, 2019; Simone et al., 2021; Nel & Nel, 2019; Zhu et al., 2019; Khatibi et al., 2021b).

A smart city is a complex system of many dimensions that can be utilized to form a resilience status (Nel & Nel, 2019). Assets, people, communities, technology, and infrastructure can be used to overcome shocks, stresses, and disasters (Chan & Zhang, 2019; Zhu et al., 2019). After all, smart cities must be prepared and ready to face natural and man-made challenges, for a continuous provision of their services (Ragia & Antoniou, 2020; Khatibi et al., 2021a; Nel & Nel, 2019). Most scholars that studied this topic tend to agree that smart city technologies improve resilience (Zhu et al., 2020; Chan & Zhang, 2019). Smart technology and applications, along with properly educated and trained communities and individuals, will enhance responsiveness and risk management capacity and help overall management by improving aspects like risk reduction, mitigation, preparedness, response, and recovery (AlHinai, 2020; Kakderi et al., 2021).

Smart and resilient cities use the same major indicators to evaluate performance, like "*Economy, People, Governance, Mobility, Living, Environment, Society, Culture, and Infrastructure*" (Khatibi et al., 2021b), something that highlights the interconnection of these two models. According to Zhu et al. (Zhu et al., 2020), improving resilience in physical, social, and environmental dimensions enhances the overall performance of the smart city, and this is something that must be considered during SC strategy planning. Based on their research, "effects of RC on SC are all positive." Improving the physical dimension of a smart city will enhance resilience, but the effects of the other two dimensions on resilience are not yet completely studied. Their research shows the little impact, if not negative, of social and environmental aspects on the resilience of a city.

According to ISO, a smart city must also be resilient to continuously provide the best of services to its citizens, since:

> *Smart is a city that increases the pace at which it provides social, economic and environmental sustainability outcomes and responds to challenges such as climate change, rapid population growth, and political and economic instability by fundamentally improving how it engages society, applies collaborative leadership methods, works across disciplines and city systems, and uses data information and modern technologies to deliver better services and quality of life to those in the city (residents, businesses, visitors), now and for the foreseeable future, without unfair disadvantage of others or degradation of the natural environment* (ISO, 2019b).

On the other hand, scholars come to agree that there are also negative effects during the implementation of these two models. Many agree that the use of smart technology can cause new threats and risks, especially in environmental issues (Chan & Zhang, 2019). The ongoing production of smart technology and machines will increase wastes and pollution while threatening the sustainable use of underground resources. Furthermore, the intense use of smart devices and applications may result in social inequality that threatens social coherence and eventually urban resilience (Chan & Zhang, 2019). Especially innovative solutions may cause future risks that are unknown at present.

In 2020, Zhu et al. concluded on the most important difference between smart and resilient cities. According to them, this is the purpose: the smart city "*promotes creativity and provides better and convenient life*," while resilient city addresses the "*disaster prevention and mitigation*." The first is an active process approach, while the second is a passive one (Zhu et al., 2020). For that reason, SC studies mainly focus on ICT adoption and how to face environmental and social challenges, while RC studies mainly focus on resilience definition and infrastructures, during various natural or man-made challenges. Due to the intense and long-term research, the smart city topic is considered to be in the application phase. On the other hand, for a relevant new topic like the resilient city, research focuses on the self-awareness phase (Zhu et al., 2020). Regarding facing hazards and social problems, the resilient city is considered to be the only one that can overcome both.

In general, the current research highlights the fact that although many smart strategies and policies tend to improve the resilience level in a city, a smart city may not be resilient by default (Oke et al., 2020). There are cases where smart cities are unable to face certain types of risks, and the COVID-19 pandemic is one of them. New kinds of threats, especially if they are at a national or a global level, seem to be extremely difficult to overcome by a city alone, even a smart one, since applied solutions, applications, and projects seem to be insufficient to reduce their vulnerability against such threats.

As mentioned before both types of cities need to adapt and overcome present and future risks with all available assets like infrastructures, communities, institutions, individuals, etc., transforming their economic, social, and political aspects (Makhoul, 2015; Oke et al., 2020). The city stakeholders must agree on a common strategic plan for the city's future. To simply choose one city model over another will result not only in positive but negative outcomes since there is no specific procedure to evaluate the resilience of a smart city or the smartness of a resilient city (Khatibi et al., 2021b). It's of great importance for decision-makers to understand all dimensions and domains of these two city approaches if they want to succeed in facing all kinds of hazards and disasters while improving the quality of life of their citizens. During planning and development, a smart city must be designed to ensure resilience and a resilient city to implement and improve smart solutions and policies (Clements-Croome, 2012). This is not an easy task since the indicators, which presented in the previous sector, focus on every city model separately and there is no unified index for smart and resilient cities (Khatibi et al., 2021b).

According to many researchers, the solution seems to be the convergence of these two cities—*the smart resilient city*—applying policies and solutions from both (Zhu et al., 2020; Khatibi et al., 2021a). ICTs and innovation, along with human assets, will help improve both governance and resilience (Khatibi et al., 2021b). This approach leads to two different frameworks—the smart resilient city and the resilient smart city—with different starting points and different goals (Khatibi et al., 2021a). The first is when a smart city adopts a resilience policy, strategy, and solutions, and the second is when a resilient city chooses to use smart technology solutions (Khatibi et al., 2021a; Khatibi et al., 2021b). The common goal remains the sustainable development and prosperity of cities and communities. Since the smart and resilient city is a new approach, the few present studies tend to focus more on smart cities that want to enhance resilience and less on resilient cities that want to be smart, probably as a direct effect of the overwhelming global movement and well-known smart city development model. After all, existing studies on the resilient city are not sufficient in covering all its aspects.

Khatibi et al. (Khatibi et al., 2021a) defined the smart resilient city, as the city able:

- "To warn against disruption"
- "To predict the type of disruption"
- "To choose the best method to absorb the disruption"
- "To take the fast, economic and straightforward recovery plan"
- "To select the best technic to bounce back better".

Planning for a smart resilient city is not an easy task, since it is a new trend. Scholars, researchers, and organizations must focus their effort on this topic and provide information and the necessary guidelines to city officials and decision-makers. Further global research on this must be conducted so the strategic plan for a smart resilient city will be commonly accepted and implemented.

6 Conclusions

During the last decades, there was an increasing trend toward urbanism. To provide better services and quality of life for their citizens, cities adopted ICTs and innovation for that. This was the birth of smart cities. On the other hand, that environment is not safe against hazards and risks. City administration must face natural and man-made disasters by displaying resiliency—the ability to protect its citizens and continue its functionality during crises. The transformation to a smart or a resilient city is not a simple procedure, since it covers several different, and in many cases, conflicting systems, domains, and aspects of each city model. The first step for that is the evaluation of the current status and the formation of the strategy and policy. For that, many indicators and frameworks have been studied and proposed by scholars and researchers. The management and implementation of the strategic plan have differences and similarities between the two city models and a common purpose—the enhancing of sustainable development. Many cities might choose one model or another, based on their strategic vision. The interconnections between city dimensions cause positive and negative effects on the city's smartness and resilience on both types of cities.

This chapter highlighted key aspects of both city models on definition, performance, management, and strategy. By studying them, decision-makers will be able not only to define the proper policies, strategies, and actions for each city model but also to discover the necessary skills and competencies that professionals, employees, and city staff responsible for a smart or resilient city should possess.

Acknowledgments This work was supported by the Erasmus+ KA2 under the project "DevOps competencies for Smart Cities" (Project No.: 601015-EPP-1-2018-1-EL-EPPKA2-SSA) (Smart-DevOps Project, n.d.). This paper reflects the views only of the authors, and the Commission cannot be held responsible for any use which may be made of the information contained therein.

References

Abdoullaev, A. (2011). A smart world: A development model for intelligent cities. In *11th IEEE international conference on computer and information technology* (pp. 1–28). IEEE.

AlHinai, Y. S. (2020). Disaster management digitally transformed: Exploring the impact and key determinants from the UK national disaster management experience. *International Journal of Disaster Risk Reduction, 51*, 101851. https://doi.org/10.1016/j.ijdrr.2020.101851

Alqahtani, A. S., Tipper, D., & Kelly-Pitou, K. (2018). Locating microgrids to improve Smart City resilience. In *Proceedings of the resilience week 2018, RWS 2018* (pp. 147–154). Institute of Electrical and Electronics Engineers Inc.. https://doi.org/10.1109/RWEEK.2018.8473464

Altay, N., & Green, W. G. (2006). OR/MS research in disaster operations management. *European Journal of Operational Research, 175*(1), 475–493. https://doi.org/10.1016/j.ejor.2005.05.016

Anthopoulos, L. G., Janssen, M., & Weerakkody, V. (2019). A unified Smart City model (USCM) for smart city conceptualization and benchmarking. *Smart Cities and Smart Spaces*, 247–264. https://doi.org/10.4018/978-1-5225-7030-1.ch011

Anthopoulos, L. G., Pantouvakis, J.-P., & Kostavara, E. (2013). An effective disaster recovery model for construction projects. *Procedia—Social and Behavioral Sciences, 74*, 21–30. https://doi.org/10.1016/j.sbspro.2013.03.026

ARUP. (2015). *City resilience framework. ARUP—the Rockefeller Foundation*. Retrieved January 2021, from https://www.rockefellerfoundation.org/wp-content/uploads/City-Resilience-Framework-2015.pdf.

Backhouse, J. (2020). A taxonomy of measures for smart cities. In *Proceedings of the 13th international conference on theory and practice of electronic governance (ICEGOV 2020)* (pp. 609–619). doi: https://doi.org/10.1145/3428502.3428593.

Bujones, A., Jaskiewicz, K., Linakis, L., & McGirr, M. (2013). *A framework for analyzing resilience in fragile and conflict-affected situations* (pp. 10–32). Economic and political development, Columbia University SIPA.

Cavada, M., Hunt, D., & Rogers, C. (2017). The little book of SMART cities. *Imagination Lancaster*. ISBN: 978-0-70442-949-9.

Chan, J. & Zhang, Y. (2019). Urban resilience in the smart city. In *The 12th conference of the international forum on urbanism: beyond resilience*, June 2019, Jakarta (pp. 24–26).

Clements-Croome, D. (2012). Intelligent sustainable liveable cities. In *Proceedings of the 8th International Conference on Intelligent Environments, IE 2012* (pp. 1–9). doi:https://doi.org/10.1109/IE.2012.65.

Falconer, G. and Mitchell, S. (2012). *Smart city framework: A systematic process for enabling smart + connected communities*, CISCO. Retrieved March 2021, from http://www.cisco.com/web/about/ac79/docs/ps/motm/Smart-City-Framework.pdf

Fujinawa, Y., Noda, Y., & Kouda, R. (2015). The resilient smart city (a proposal). *Journal of Disaster Research, 10*(2), 319–325. https://doi.org/10.20965/jdr.2015.p0319

Huck, A., Monstadt, J., & Driessen, P. (2020). Mainstreaming resilience in urban policymaking? Insights from Christchurch and Rotterdam. *Geoforum, 117*, 194–205. https://doi.org/10.1016/j.geoforum.2020.10.001

ISO. (2019a). *ISO 37123:2019 sustainable cities and communities—indicators for resilient cities*. Retrieved April 2021, from https://www.iso.org/standard/70428.html.

ISO. (2019b). *ISO 37122:2019 sustainable cities and communities—indicators for smart cities*. Retrieved April 2021, from https://www.iso.org/standard/69050.html

Joss, S., Sengers, F., Schraven, D., Caprotti, F., & Dayot, Y. (2019). The smart city as global discourse: Storylines and critical junctures across 27 cities. *Journal of Urban Technology, 26*(1), 3–34. https://doi.org/10.1080/10630732.2018.1558387

Kakderi, C., Oikonomaki, E., & Papadaki, I. (2021). Smart and resilient urban futures for sustainability in the post-covid-19 era: A review of policy responses on urban mobility. *Sustainability (Switzerland), 13*(11). https://doi.org/10.3390/su13116486

Khatibi, H., Wilkinson, S., Baghersad, M., Dianat, H., Ramli, H., Suhatril, M., Javanmardi, A., & Ghaedi, K. (2021a). The resilient—Smart city development: A literature review and novel frameworks exploration. *Built Environment Project and Asset Management.*https://doi.org/10.1108/BEPAM-03-2020-0049

Khatibi, H., Wilkinson, S., Dianat, H., Baghersad, M., Ghaedi, K., & Javanmardi, A. (2021b). Indicators bank for smart and resilient cities: Design of excellence. *Built Environment Project and Asset Management.*https://doi.org/10.1108/BEPAM-07-2020-0122

Lafi Aljohani, S., & Alenazi, M. J. F. (2020). Evaluation of WSN's resilience to challenges in smart cities. *International Journal of Computer and Communication Engineering, 9*(4), 193–206. https://doi.org/10.17706/ijcce.2020.9.4.193-206

Li, K., Chen, Y. & Luna-Reyes, L. F. (2017). City resilience as a framework to understand smart cities: Dimensions & measurement. In *ACM international conference proceeding series. Association for computing machinery* (pp. 568–569). doi: https://doi.org/10.1145/3085228.3085249.

Lopez, L. J. R., & Castro, A. I. G. (2021). Sustainability and resilience in smart city planning: A review. *Sustainability (Switzerland)*, 1–25. https://doi.org/10.3390/su13010181

Makhoul, N. (2015). From sustainable to resilient and smart cities. In *IABSE conference—Structural engineering: Providing solutions to global challenges*, Geneva 2015 (pp. 1901–1906). doi: https://doi.org/10.2749/222137815818359393.

Michelucci, F. V., De Marco, A., & Tanda, A. (2016). Defining the role of the smart-city manager: an analysis of responsibilities and skills. *Journal of Urban Technology, 23*(3), 23–42. https://doi.org/10.1080/10630732.2016.1164439

Mora, L., Deakin, M., & Reid, A. (2018). Smart-city development paths: Insights from the first two decades of research. Green energy and technology. *Springer International Publishing AG*. https://doi.org/10.1007/978-3-319-75774-2_28

Nel, D., & Nel, V. (2019). *Governance for resilient smart cities. In The proceedings of the CIB world building congress 2019: Constructing smart cities*, the Hong Kong Polytechnic University, Hong Kong, 17–21 June 2019 (p. 2112–2220) (online version), ISBN: 978-962-367-821-6.

Oke, A. E., Aghimien, D. O., Akinradewo, O. I., & Aigbavboa, C. O. (2020). Improving resilience of cities through smart city drivers. *Construction Economics and Building, 20*(2), 45–64. https://doi.org/10.5130/AJCEB.v20i2.6647

Patel, R. & Nosal, L. (2016). *Defining the Resilient City*. United Nations University Centre for policy research, working paper 6 (pp. 1–21). ISBN: 978-92-808-9034-1.

Purnomo, F., Meyliana, & Prabowo, H. (2016). Smart city indicators: A systematic literature review. *Journal of Telecommunication, Electronic and Computer Engineering, 8*(3), 161–164.

Ragia, L., & Antoniou, V. (2020). Making smart cities resilient to climate change by mitigating natural hazard impacts. *ISPRS International Journal of Geo-Information, MDPI AG*. https://doi.org/10.3390/ijgi9030153

Santinha, G., & de Castro, E. A. (2010). Creating more intelligent cities: The role of ICT in promoting territorial governance. *Journal of Urban Technology, 17*(2), 77–98. https://doi.org/10.1080/10630732.2010.515088

Scholl, H. J., & Patin, B. J. (2012). Resilient information infrastructures: Mobilizing adaptive capacities under extreme events. In *Proceedings of the eighteenth Americas conference on information systems, Seattle, WA, USA* (pp. 4451–4458).

Simone, C., Iandolo, F., Fulco, I., & Loia, F. (2021). Rome was not built in a day. Resilience and the eternal city: Insights for urban management. *Cities, 110*, 103070. https://doi.org/10.1016/j.cities.2020.103070

SmartDevOps Project (n.d.). Retrieved October 2021, from https://smartdevops.eu

Stubinger, J., & Schneider, L. (2020). Understanding smart city—A data-driven literature review. *Sustainability (Switzerland), 12*(20), 1–23. https://doi.org/10.3390/su12208460

United Nations (2015). *Sendai framework for disaster risk reduction 2015–2030. United Nations*. Retrieved February 2021, from https://www.undrr.org/implementing-sendai-framework/what-sendai-framework

United Nations Office for Disaster Risk Reduction (UNDRR). (2017). *How to make cities more resilient: A handbook for local government leaders. A contribution to the global campaign 2010–2020: Making cities resilient—"My City is Getting Ready!"*. United Nations.

Zhu, S., Li, D., & Feng, H. (2019). Is a smart city resilient? Evidence from China. *Sustainable Cities and Society, 50*. https://doi.org/10.1016/j.scs.2019.101636

Zhu, S., Li, D., Feng, H., Gu, T., Hewage, K., & Sadiq, R. (2020). Smart city and resilient city: Differences and connections. *Wiley Interdisciplinary Reviews: Data Mining and Knowledge Discovery, 10*(6), 1–19. https://doi.org/10.1002/widm.1388

Smart City Projects Evaluation: A Bibliometric Approach

Vassilis C. Gerogiannis and Stella Manika

1 Introduction

Nowadays, many initiatives are aimed at evaluating the success of smart city projects. For example, taxonomy surveys of smart city projects are identified by deconstructing project key characteristics, such as project scope, business model, and purpose axis, based on the city logistics approach (Perboli et al., 2014). There are also research works focusing on how constantly evolving smart city projects contribute to the development and resilience of urban areas, businesses' robustness, and innovativeness (Scuotto et al., 2016). At the same time, European organizations and cities' networks emphasize research on the assessment of the smart city projects results and achievements (Bosch et al., 2017). However, there seems to be a lack of generally adopted and standardized metrics/methodologies for evaluating, ranking, and managing smart city projects (Monzon, 2015).

Furthermore, although the benefits of smart cities are widely recognized, very often critical questions arise as to the actual satisfaction (or dissatisfaction) of citizens with the progress and the results of a smart city's development and evolution (Gooch et al., 2015). It is a fact that, at their core, smart city projects are oriented toward the satisfaction of the citizenry as regards the use of information and communication technologies, through the automation of government and public sector processes, the more direct delivery of public services, the dissemination/transparency of public information, as well as the facilitation of the

V. C. Gerogiannis (✉)
Department of Digital Systems, School of Technology, University of Thessaly, Larissa, Greece
e-mail: vgerogian@uth.gr

S. Manika
Department of Planning and Regional Development, School of Engineering, University of Thessaly, Larissa, Greece
e-mail: smanika@uth.gr

P. Fitsilis (ed.), *Building on Smart Cities Skills and Competences*, Internet of Things,
https://doi.org/10.1007/978-3-030-97818-1_9

citizens' mobility. Nevertheless, it is considered important, on the one hand, for the more efficient operation of smart city projects, and, on the other hand (as well as to a greater extent), for the improvement of the quality of life of citizens, to evaluate also smart city projects through the perspective of the citizen's satisfaction (Macke et al., 2018). Because citizens are intended to benefit from smart city projects almost daily, it is reasonable and appropriate to critically evaluate top-down smart city's development endeavors which ignore the satisfaction/dissatisfaction of the targeted population (Ji et al., 2021).

The current chapter attempts to evaluate smart city projects by applying systematically a bibliostatistic (or also called bibliometric) analysis which is conducted with the aim to highlight in the relevant literature the prevailing trends concerning the criteria set used for the evaluation of smart projects. The results of analyzing the literature demonstrate the limited role of the human dimension in terms of evaluating the success of smart city projects.

2 Literature Review

Two pillars were set as the basis of the chapter's literature review. More specifically, on the one hand, the identification of those characteristics that a city must develop to be characterized as a smart one, under the spectrum of citizens' satisfaction, and, on the other hand, how these characteristics can be quantified and assessed.

2.1 Citizen's Satisfaction in Smart Cities

The "intelligence" of a city is very often characterized without clear criteria (Mozūriūnaitė & Sabaitytė, 2021), and this leads to confusion that tends to be a result of the lack of recognition of a standardized structure of smart cities and their functions. A possible answer to this problem seems to be suggested by Liugailaitė-Radzvickienė and Jucevičius (2014). Although they agree that a city can be qualified as intelligent primarily by being digital, virtual, and technologically innovative, they also emphasize the social point of view (Liugailaitė-Radzvickienė & Jucevičius, 2014). Following such a more complex perspective, the creation of a knowledge-intensive urban environment and its application to decisions about the various city's development initiatives also play an important role in the intelligence of a city.

The successful development of a city can be defined, among other things, by the degree of a city's suitability concerning the human quality of life and the level of satisfaction among the city's inhabitants. These parameters can be measured in several ways, such as those proposed by the economist's approach (The Liveability Ranking and Overview, 2021) or other methods (Lowe et al., 2020), (Khorrami et al., 2021). Satisfaction with city life is at the heart of many urban policies and varies considerably within various cities and across Europe (Weziak-Białowolska, 2016).

Quality of life in cities is also shaped by a variety of individual factors, such as place of birth (Zenker & Rütter, 2014), age, and health (WHO, 2012), as well as by the environment of the urban space (Anastasiou & Manika, 2020) and the duration of residence (Florek, 2010), with a tendency to equate the quality of place with the quality of life (Møller, 2001).

The intelligence of a smart city and its performance, in comparison with traditional cities, affect the quality of life of its residents/inhabitants, making them better-educated, better-informed, and more active participants in the city society (Albino et al., 2015). Moreover, several studies show that there is a strong interaction between ICT-based smart city services and the citizens' overall quality of life (Khorrami et al., 2021).

2.2 *Ranking Systems for Smart City Evaluation*

Many systems have been used to access smart city initiatives, and these systems often consider the correlation between smart and sustainable city assessment (Ahvenniemi et al., 2017). We present in Table 1 some indicative ranking/evaluation systems for smart cities, which have been successfully applied to assess the application of initiatives aiming to transform a city into a "smart" one. We have noticed that each ranking system places a different emphasis on the citizen's parameter. The literature review and the critical evaluation of these systems reveal that the technological possibilities and choices afforded by technology-oriented smart cities positively affect citizens' perceived quality of life (Yeh, 2017). Nevertheless, despite the recognized role of quality of life as one of the goals of a smart city (smart people and smart living parameters), the question arises whether and to what extent the parameters of quality of life and citizen satisfaction are taken into account in a plethora of studies which have been conducted aiming to evaluate a city as smart.

3 Methodology

The bibliometric analysis as a method aims to provide in-depth information on literature statistics (Wang et al., 2021) by visually demonstrating a comprehensive overview of the principal literature, and analyzing the tendencies in the scientific literature on a specific research field (Zolkover & Terziev, 2020). The bibliometric analysis method has been applied in a plethora of urban studies with the aim to identify contemporary research trends in the scientific literature, regarding various research topics and issues, such as urban smart mobility (Tomaszewska & Florea, 2018), the evolutionary path of urban expansion (Xie et al., 2020), the exploitation of big data parameters in urban management and other urban application fields (Allam, 2018), as well as the urban sustainability assessment (Sharifi, 2021). Furthermore, other topics of research have been also analyzed through the bibliometric method,

Table 1 Indicative smart cities' ranking systems with an emphasis on the citizen perspective

Smart cities' ranking systems	Indicators	Developed by source	Emphasis on citizen
Assessment metric used to rank 70 European medium-sized cities	Specific indicators for each of the six identified dimensions of a smart city (smart economy, smart people, smart governance, smart mobility, smart environment, smart living)	University of Vienna (Giffinger & Gudrun, 2010)	Quality of life is considered as a synonym of one of the basic pillars of a smart city, i.e., the "smart living" factor
Smart city reference model	The city is perceived on six levels, layer 0, the city; layer 1, the Green City layer; layer 2, the interconnection layer; layer 3, the instrumentation layer; layer 4, the open integration layer; layer 5, the application layer; layer 6, the innovation layer	Zygiaris (2012)	More emphasis is given to citizens as part of layer 0—The city
Smart city indices	Smart city indices are based on fuzzy logic and fuzzy sets	Lazaroiu and Roscia (2012)	More emphasis is given to independent and aware citizens
Triple helix and performance indicators	Analysis of knowledge-based innovation systems and the relations between university, industry, and government	Lombardi et al. (2012)	Citizens' information

such as issues related to the development of urban education, analyzing the research agenda of energy in urban areas (Cicea, 2020), defining roadmaps for global researchers who focus on collaborative decision-making in urban regeneration (Wang et al., 2021), tracing historical and geographic trends of accessibility in cities (Shi et al., 2020), as well as synthesizing the parameters which define a city's sharing economy (Filser et al., 2020).

More targeted, in the field of smart cities, bibliometric analysis has been also used in some research studies which are mainly focusing on the typical characteristics of smart cities (Guo et al., 2019), understanding smart cities (Li, 2019) and typical development/evolution paths of smart cities (Mora et al., 2019), comparing and benchmarking smart cities in different countries (Dias, 2018), and evaluating the importance that the "smart city concept" has gained through the years (Pérez et al., 2020).

In the following section, this chapter will demonstrate the hot trends that are presented in the bibliographic agenda when several articles on the subject of smart cities or projects evaluation or success are collected and utilized. Data related to each bibliographic reference/item type, its content (keywords, authors, authors collaborations, origin of the sources), and characteristics have been analyzed using the R-based Biblioshiny app, which is freely available from https://bibliometrix.org/ by following the practices described in Aria et al. (2020) and in Aria and Cuccurullo (2017). Emphasis was given on describing the trends of the keywords, sub-topics (timeline and distribution), author countries, and journals of the selected articles. Country scientific production map, trend topic diagrams, thematic evolution time slices, and co-occurrence networks were generated.

4 Results

With the application of the bibliometric analysis software and setting as keywords concepts related to the evaluation of the success factors (or challenges) of smart city projects, 934 documents were identified (after duplicates were excluded, only papers in English were considered and a hand search was performed) in literature published between 2010 and 2021 and used in the analysis of our research work. Web of science was the basis of our research. At this point, we need to recognize the important role of keywords selection for bibliometric analysis to take shape, as different keyword combinations can yield different results. In our research, emphasis was placed on evaluation as a key parameter of the search for the success or not of a smart city. The main information about the data used is described in Table 2. The period from January 2010 to June 2021 was chosen as the reference period because the smart city, as a concept, had already been adopted in the literature and several cities had implemented smart city projects and they had been deemed as smart ones.

It is also interesting to look at the sources of these articles (Table 3), as most of them refer to articles related to technological applications, sensors, or computer science, information, and communication technologies, and much less to urban or social issues. Of course, we should not overlook the fact that the majority of the sources were selected from the sustainability journal (MDPI), which includes also articles that have as their main object the social or demographic dimensions of the phenomenon under examination.

From Fig. 1 it is observed that in the countries with the largest scientific production of this subject, the role of China is primary. South Korea follows and then, with similar sums of relevant articles, the USA, Italy, India, and Spain. Next, scientific production is shown by researchers from Australia, the UK, then Brazil, Canada, Japan, Germany, and Finland. Next are Greece, Pakistan, Saudi Arabia, the Netherlands, and Portugal. We, therefore, observe that the research in the relevant field presents an increased spatial dispersion (Fig. 1).

Regarding the dispersion of the research topic, in individual topics and keywords, we observe that the keywords most frequently used during the period under review are terms related to the internet, cities or city, framework, model, management, performance, systems, challenges, impact, and growth (Fig. 2). To identify the

Table 2 Main information about the data used

Description	Results
Main information about data	
Timespan	2010: 2021
Sources (journals, books, etc.)	396
Documents	934
Average years from publication	2.13
Average citations per document	11.25
Average citations per year per doc	3.091
References	1
Document types	
Article	884
Article; book chapter	1
Article; data paper	5
Article; early access	31
Article; proceedings paper	13
Document contents	
Keywords plus (ID)	1458
Author's keywords (DE)	3533
Authors	
Authors	3085
Author appearances	3743
Authors of single-authored documents	52
Authors of multiauthored documents	3033
Author's collaboration	
Single-authored documents	57
Documents per author	0.303
Authors per document	3.3
Co-authors per documents	4.01
Collaboration index	3.46

differentiation of keywords used per year of examination, we proceeded to time slices (Figs. 3, 4, and 5). We used not only quantitative indicators, such as publication count or country count, but also qualitative indicators, such as citation count and weight index (i.e., the inclusion index weighted by word occurrences was used).

As we see in the first time slice, the major themes (the so-called "motor" themes) "Internet" and "innovation" have a higher impact. The innovation parameter can be related either to innovative solutions to citizens' problems or to innovative ways of operating the public mechanism (Manika, 2020). The role of the "framework" is also important, gaining ground from the period 2016 to 2018. In this period (Fig. 4), the roles of "networks" and "participation" seem to be quite important, while the concepts of "cities" and "big data" are gradually gaining importance. From 2019 onward (Fig. 5), it appears that the contribution of the "city" and its "systems" to the models of their evaluations is recognized.

Table 3 Most relevant literature sources (top 20)

Sources	Articles
Sustainability	69
IEEE access	55
Sensors	34
Sustainable cities and society	28
Energies	23
IEEE internet of things journal	22
Future generation computer systems—The international journal of Escience	19
Journal of cleaner production	17
Applied sciences-BASEL	13
Cities	12
Multimedia tools and applications	9
Technological forecasting and social change	8
IEEE communications magazine	7
Journal of network and computer applications	7
Remote sensing	7
Computer communications	6
IEEE transactions on intelligent transportation systems	6
Pervasive and Mobile computing	6
Smart cities	6
Wireless communications and Mobile computing	6

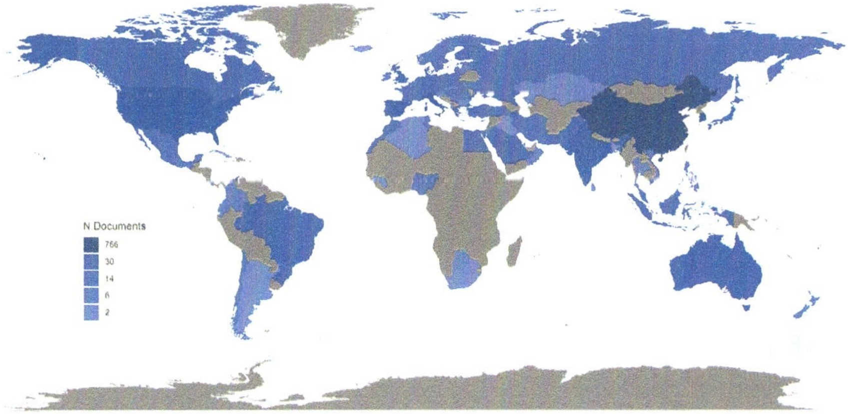

Fig. 1 Country scientific production

Because the network analysis aimed at collecting as much information as possible, keywords plus were used as a field, the minimum number of edges was set at two (2), isolated nodes were removed, and the Louvain clustering algorithm was used (Blondel et al., 2008). In the final co-occurrence network, 49 keywords were visualized. As can be seen in Fig. 6 (keyword co-occurrence network), popular

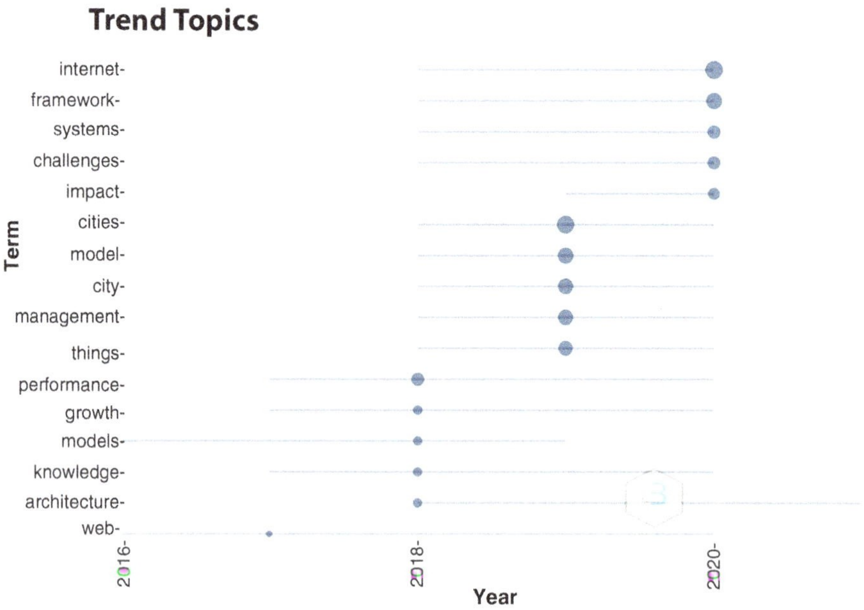

Fig. 2 Trend topics in the literature

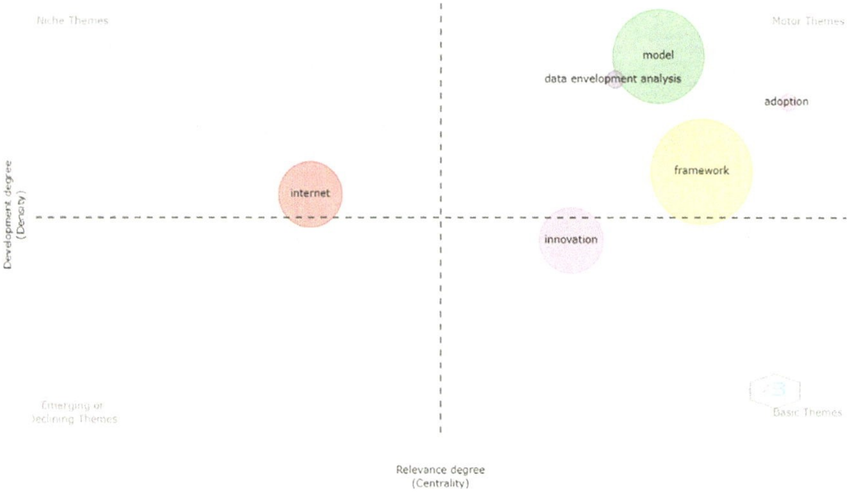

Fig. 3 Thematic evolution, time slice 1 (2010–2015)

research topics are grouped into four (4) dominant clusters. The size of the circles corresponds to the number of occurrences of the represented keywords, while each color corresponds to the respective cluster that is formed.

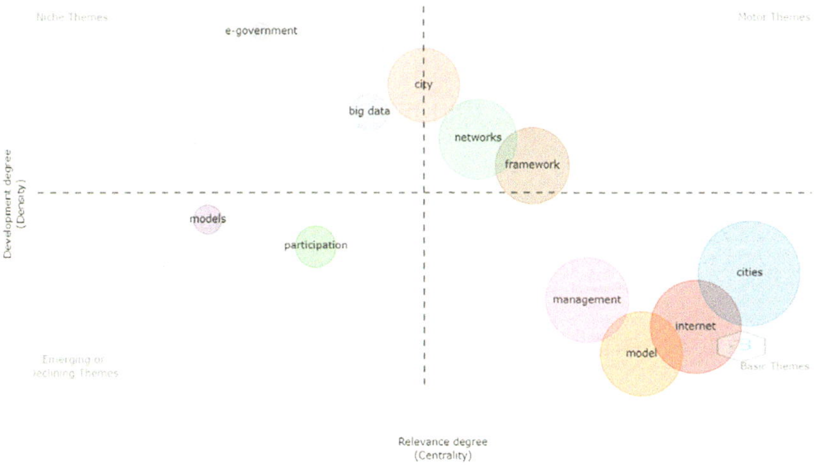

Fig. 4 Thematic evolution, time slice 2 (2016–2018)

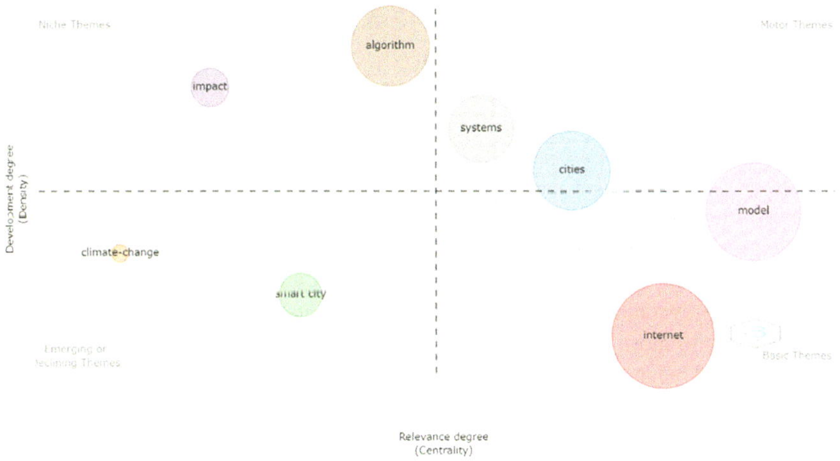

Fig. 5 Thematic evolution, time slice 3 (2019–2021)

The first keyword cluster corresponds to the most technological dimension of the Internet. More specifically, it includes the concepts of "big data" and issues such as "security," "privacy," or "services" that stem from it. The second cluster includes "cities" as "systems" of "innovation," "knowledge," "policies," and "governance." The third cluster focuses on the unity of city management and more specifically on the "framework," "management," and "performance," while in this cluster the concept of "satisfaction" (marked with the black rectangular border) is now distinct.

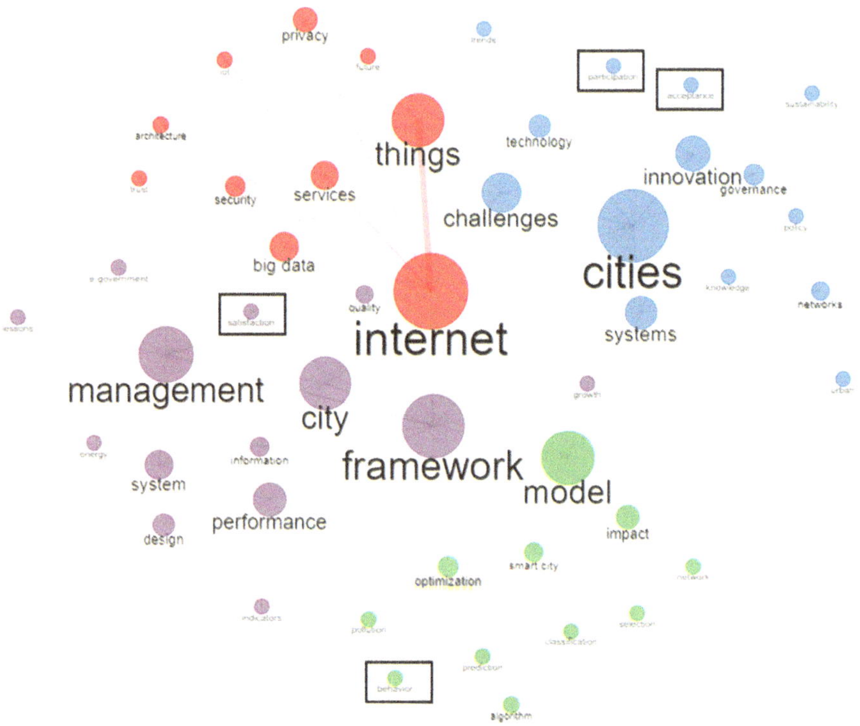

Fig. 6 Co-occurrence network

The fourth cluster includes "model," "smart cities" "impact," "optimization," and "classification." We, therefore, observe that most concepts as they emerge through analysis are related to the perception of evaluation of a more technically structured system and less as a human-centered one. The few concepts that can at first glance be combined with more *anthropocentric* parameters are marked with a black square in Fig. 6 (e.g., "behavior," "participation," and "acceptance"). We notice that not only they are few but also their weight is much lower compared to the other parameters. Nevertheless, we must emphasize that the individual analysis of the concepts presented is necessary to strengthen this argument.

5 Discussion

Evaluating success factors of smart cities through a systematic literature review as has been recently implemented by (Aldegheishem, 2019) has led to a wide range of smart success characteristics including the human-related parameter. As related features of smart people are described, the solution to city problems is based on innovative approaches, public participation, and the provision of access to various education options. However, although this literature review article compares the smart people parameter with the other success parameters of a smart city, it does not proceed to quantitative representations based on their frequency of occurrence.

Smart city evaluation and reporting based on case studies and the city's perspective is also an interesting approach (Caird, 2018). The concept of this case is that the human dimension is highlighted through the programs that promote a strong partnership between the various business or educational structures and the community. At the same time, other research studies reveal evaluation impacts of smart city project programs on people's lives (Buscher & Doody, 2013), and more, specifically, they emphasize the importance of the "citizen value that programs wanted to achieve." There are also research studies that analyzed the various approaches for smart city evaluation by using hybrid research methods, as a combination of literature review and semi-structured interviews to sample smart cities (Shen et al., 2018). The human factor approach, in this case, tends to treat people not only as individuals but also as groups or communities, who often present common needs that can be met by appropriate infrastructure and governance-related factors in promoting among others communication, transparency, participation, and data exchange. To quantify the emergence of smart people, emphasis is placed on proportions, for example, the percentage of the population with higher education or level of access to network facilities by citizens. From another point of view, the principle of humanism is presented in a part of Li et al. (2020) such as evaluation of the friendliness interface, ease of use of the sharable system, and projection of appropriate information.

In general, a more citizen-centered approach can be found in a variety of research papers attempting to evaluate smart city or smart city projects. The current chapter attempted to contribute to this discussion by highlighting the limited role of the human dimension in the prevailing smart city evaluation bibliographic trends.

6 Conclusions

Although many studies have recognized the importance of citizen satisfaction as the goal of smart city's development, this parameter tends to be absent from most measures of city intelligence and the corresponding academic discussions. Based on the results of bibliostatistic analysis, the evaluation of smart cities has become an increasingly prevalent issue in science since 2017 and now seems to be an almost universal concern for the international scientific community. Although the discussion around the evaluation of smart cities over time seems to be related to its more technocratic aspects, such as the Internet and the possibilities that are opened through its management, at the same time, the importance of cities as operating citizen-centric systems is recognized. An important parameter of smart cities, especially in the context of their social and economic development, is the satisfaction of their inhabitants, a rapidly emerging field of study. Of course, as mentioned, the methodology applied in the present research has some limits which are related either to the keywords used or to the time horizon that were set as the basis of the study. Nevertheless, the results are able in the first phase to outline the major bibliographic trends and to highlight the gap of the citizen-based approach in the evaluation of smart cities.

Acknowledgments This work was supported by the Erasmus+ KA2 under the project "DevOps competences for Smart Cities" (Project No.: 601015-EPP-1-2018-1-EL-EPPKA2-SSA). This paper reflects the views only of the authors, and the Commission cannot be held responsible for any use which may be made of the information contained therein.

References

Ahvenniemi, H., Huovila, A., Pinto-Seppä, I., & Airaksinen, M. (2017). What are the differences between sustainable and smart cities? *Cities, 60*, 234–245. https://doi.org/10.1016/j.cities.2016.09.009

Albino, V., Berardi, U., & Dangelico, R. M. (2015). Smart cities: Definitions, dimensions, performance, and initiatives. *Journal of Urban Technology, 22*(1), 3–21. https://doi.org/10.1080/10630732.2014.942092

Aldegheishem, A. (2019). Success factors of smart cities: A systematic review of literature from 2000-2018. *TEMA Journal of Land Use, Mobility and Environment, 12*, 53–64.

Allam, S. (2018). The future of urban models in the big data and AI era: A bibliometric analysis. *International Journal of Creative Research Thoughts, 6*(1), 797–800. https://papers.ssrn.com/sol3/papers.cfm?abstract_id=3821178

Anastasiou, E., & Manika, S. (2020). Perceptions, determinants and residential satisfaction from urban open spaces. *Open Journal of Social Sciences, 8*(6), 1–18. https://doi.org/10.4236/JSS.2020.86001

Aria, M., & Cuccurullo, C. (2017). Bibliometrix: An R-tool for comprehensive science mapping analysis. *Journal of Informetrics, 11*(4), 959–975. https://doi.org/10.1016/j.joi.2017.08.007

Aria, M., Misuraca, M., & Spano, M. (2020). Mapping the evolution of social research and data science on 30 years of social indicators research. *Social Indicators Research, 149*(3), 803–831. https://doi.org/10.1007/s11205-020-02281-3

Blondel, V. D., Guillaume, J. L., Lambiotte, R., & Lefebvre, E. (2008). Fast unfolding of communities in large networks. *Journal of Statistical Mechanics, Theory and Experiment, 2008*(10), P10008. https://iopscience.iop.org/article/10.1088/1742-5468/2008/10/P10008

Bosch, P., Jongeneel, S., Rovers, V., Neumann, H.-M., Airaksinen, M., & Huovila, M. (2017). *CITYkeys indicators for smart city projects and smart cities*. Retrieved from http://nws.eurocities.eu/MediaShell/media/CITYkeysD14Indicators forsmartcityprojectsandsmartcities.pdf.

Buscher, V., & Doody, L. (2013). Global innovators: International case studies on smart cities. Department for Business, Innovation and Skills, *Report of the UK Government, Research Paper 135*. Retrieved from https://www.gov.uk/government/publications/smart-cities-international-case-studies-global-innovators.

Caird, S. (2018). City approaches to smart city evaluation and reporting: Case studies in the United Kingdom. *Urban Research & Practice, 11*(2), 159–179. https://doi.org/10.1080/17535069.2017.1317828

Cicea, C. (2020). Mapping the research agenda of energy in urban areas: A bibliometric analysis. *Theoretical and Empirical Researches in Urban Management, 15*(2), 66–80. https://www.jstor.org/stable/26915291

Dias, G. P. (2018). Smart cities research in Portugal and Spain: An exploratory biliometric analysis. *13th Iberian Conference on Information Systems and Technologies* (CISTI), IEEE, June 2018, 1–6. https://doi.org/10.23919/CISTI.2018.8399295

Filser, M., Tiberius, V., Kraus, S., Spitzer, J., Kailer, N., & Bouncken, R. B. (2020). Sharing economy: A bibliometric analysis of the state of research. *International Journal of Entrepreneurial Venturing, 12*(6), 665–687. https://doi.org/10.1504/IJEV.2020.112234

Florek, M. (2010). No place like home: Perspectives on place attachment and impacts on city management. *Journal of Town & City Management, 1*(4), 346–354. https://eds.b.ebscohost.com/eds/detail/detail?vid=0&sid=a9811d7f-5b76-4607-a6e4-4e9b47a6fcf8%40pdc-v-sessmgr02&bdata=JnNpdGU9ZWRzLWxpdmU%3d#AN=87711049&db=bsx

Giffinger, R., & Gudrun, H. (2010). Smart cities ranking: An effective instrument for the positioning of the cities? *ACE: Architecture, City and Environment, 4*(12), 7–26. https://doi.org/10.5821/ACE.V4I12.2483

Gooch, D., Wolff, A., Kortuem, G., & Brown, R. (2015). Reimagining the role of citizens in smart city projects. In *Proceedings of the 2015 ACM international joint conference on pervasive and ubiquitous computing & proceedings of the 2015 ACM international symposium on wearable computers* (UbiComp & ISWC 2015) (pp. 1587–1594). https://doi.org/10.1145/2800835.2801622

Guo, Y.-M., Huang, Z.-L., Guo, J., Li, H., Guo, X.-R., & Nkeli, M. J. (2019). Bibliometric analysis on smart cities research. *Sustainability, 11*(13), 3606. https://doi.org/10.3390/SU11133606

Ji, T., Chen, J. H., Wei, H. H., & Su, Y. C. (2021). Towards people-centric smart city development: Investigating the citizens' preferences and perceptions about smart-city services in Taiwan. *Sustainable Cities and Society, 67*, 102691. https://doi.org/10.1016/J.SCS.2020.102691

Khorrami, Z., Ye, T., Sadatmoosavi, A., Mirzaee, M., Fadakar Davarani, M. M., & Khanjani, N. (2021). The indicators and methods used for measuring urban liveability: A scoping review. In *Reviews on environmental health*. De Gruyter Open Ltd.. https://doi.org/10.1515/reveh-2020-0097

Lazaroiu, G. C., & Roscia, M. (2012). Definition methodology for the smart cities model. *Energy, 47*(1), 326–332. https://doi.org/10.1016/J.ENERGY.2012.09.028

Li, C., Dai, Z., Liu, X., & Sun, W. (2020). Evaluation system: Evaluation of smart city shareable framework and its applications in China. *Sustainability, 12*(7), 2957. https://doi.org/10.3390/su12072957

Li, M. (2019). Visualizing the studies on smart cities in the past two decades: A two-dimensional perspective. *Scientometrics, 120*(2), 683–705. https://doi.org/10.1007/S11192-019-03134-8

Liugailaitė-Radzvickienė, L., & Jucevičius, R. (2014). Going to be an intelligent city. *Procedia—Social and Behavioral Sciences, 156*, 116–120. https://doi.org/10.1016/J.SBSPRO.2014.11.131

Shen, L., Huang, Z., Wong, S. W., Liao, S., & Lou, Y. (2018). A holistic evaluation of smart city performance in the context of China. *Journal of Cleaner Production, 200*, 667–679., ISSN 0959-6526. https://doi.org/10.1016/j.jclepro.2018.07.28

Lombardi, P., Giordano, S., Farouh, H., & Yousef, W. (2012). Modelling the smart city performance. *Innovation—The European Journal of Social Science, 25*(2), 137–149. https://doi.org/10.1080/13511610.2012.660325

Lowe, M., Arundel, J., Hooper, P., Rozek, J., Higgs, C., Roberts, R., & Giles-Corti, B. (2020). Liveability aspirations and realities: Implementation of urban policies designed to create healthy cities in Australia. *Social Science & Medicine, 245*, 112713. https://doi.org/10.1016/J.SOCSCIMED.2019.112713

Macke, J., Casagrande, R. M., Sarate, J. A. R., & Silva, K. A. (2018). Smart city and quality of life: Citizens' perception in a Brazilian case study. *Journal of Cleaner Production, 182*, 717–726. https://doi.org/10.1016/J.JCLEPRO.2018.02.078

Manika, S. (2020). Mechanisms for innovative-driven solutions in European smart cities. *Smart Cities, 3*(2), 527–540. https://doi.org/10.3390/smartcities3020028

Møller, V. (2001). Monitoring quality of life in cities: The Durban case. *Development Southern Africa, 18*(2), 217–238. https://doi.org/10.1080/037/68350120041910

Monzon, A. (2015). Smart cities concept and challenges: Bases for the assessment of smart city projects. In *2015 International conference on smart cities and green ICT systems (SMARTGREENS)* (pp. 1–11). Retrieved from https://ieeexplore.ieee.org/abstract/document/7297938

Mora, L., Deakin, M., & Reid, A. (2019). Combining co-citation clustering and text-based analysis to reveal the main development paths of smart cities. *Technological Forecasting and Social Change, 142*, 56–69. https://doi.org/10.1016/j.techfore.2018.07.019

Mozūriūnaitė, S., & Sabaitytė, J. (2021). To what extent we do understand smart cities and characteristics influencing city smartness. *Journal of Architecture and Urbanism, 45*(1), 1–8. https://doi.org/10.3846/jau.2021.12392

Perboli, G., De Marco, A., Perfetti, F., & Marone, M. (2014). A new taxonomy of smart city projects. *Transportation Research Procedia, 3*, 470–478. https://doi.org/10.1016/j.trpro.2014.10.028

Pérez, L. M., Oltra-Badenes, R., Gutiérrez, J. V. O., & Gil-Gómez, H. (2020). A bibliometric diagnosis and analysis about smart cities. *Sustainability, 12*(16), 6357. https://doi.org/10.3390/SU12166357

Scuotto, V., Ferraris, A., & Bresciani, S. (2016). Internet of things: Applications and challenges in smart cities: A case study of IBM smart city projects. *Business Process Management Journal, 22*(2), 357–367. https://doi.org/10.1108/BPMJ-05-2015-0074

Sharifi, A. (2021). Urban sustainability assessment: An overview and bibliometric analysis. *Ecological Indicators, 121*, 107102. https://doi.org/10.1016/j.ecolind.2020.107102

Shi, Y., Blainey, S., Sun, C., & Jing, P. (2020). A literature review on accessibility using bibliometric analysis techniques. *Journal of Transport Geography, 87*, 102810. https://doi.org/10.1016/j.jtrangeo.2020.102810

The Liveability Ranking and Overview. (2021). The liveability ranking & overview—the economist intelligence unit. *The Economist*. Retrieved from https://www.eiu.com/public/topical_report.aspx?campaignid=Liveability2011.

Tomaszewska, E. J., & Florea, A. (2018). Urban smart mobility in the scientific literature—Bibliometric analysis. *Engineering Management in Production and Services, 10*(2), 41–56. https://doi.org/10.2478/EMJ-2018-0010

Wang, H., Zhao, Y., Gao, X., & Gao, B. (2021). Collaborative decision-making for urban regeneration: A literature review and bibliometric analysis. *Land Use Policy, 107*, 105479. https://doi.org/10.1016/j.landusepol.2021.105479

Weziak-Białowolska, D. (2016). Quality of life in cities—Empirical evidence in comparative European perspective. *Cities, 58*, 87–96. https://doi.org/10.1016/j.cities.2016.05.016

WHO. (2012). *WHOQOL—Measuring quality of life, the World Health Organization*. Retrieved from https://www.who.int/toolkits/whoqol

Xie, H., Zhang, Y., & Duan, K. (2020). Evolutionary overview of urban expansion based on bibliometric analysis in web of science from 1990 to 2019. *Habitat International, 95*, 102100. https://doi.org/10.1016/j.habitatint.2019.102100

Yeh, H. (2017). The effects of successful ICT-based smart city services: From citizens' perspectives. *Government Information Quarterly, 34*(3), 556–565. https://doi.org/10.1016/j.giq.2017.05.001

Zenker, S., & Rütter, N. (2014). Is satisfaction the key? The role of citizen satisfaction, place attachment and place brand attitude on positive citizenship behavior. *Cities, 38*, 11–17. https://doi.org/10.1016/j.cities.2013.12.009

Zolkover, A., & Terziev, V. (2020). The shadow economy: A bibliometric analysis. *Business Ethics and Leadership, 4*(3), 107–118. https://doi.org/10.21272/bel.4(3).107-118.2020

Zygiaris, S. (2012). Smart City reference model: Assisting planners to conceptualize the building of smart city innovation ecosystems. *Journal of the Knowledge Economy, 4*(2), 217–231. https://doi.org/10.1007/s13132-012-0089-4

Modeling Project Management Complexity in Smart Cities' Projects

Vyron Damasiotis

1 Introduction

Smart city transformation is the new goal for every city in the planet in order to support sustainable growth and prosperity to their citizens. This transformation requires significant changes in how cities are organized, operate, offer services, and interact with their citizens (Thompson, 2016; Giffinger et al., 2007). A smart city transformation is of high complexity due to requirements for long-term planning, interoperability and integration, and operation in a continuous changing environment and due to heterogeneity of people performing or affected by this transformation. Most of the transformation process will be implemented in the form of projects. These projects have an inherited degree of complexity trying to incorporate and implement these changes. If in this complexity it is be added the complexity sourcing from projects itself, such as complexity sourcing from project scheduling, budgeting, and various other project environment uncertainties, then it can easily be end up with projects of high complexity that their success is compromised by this complexity and hence its is undermined the success of smart city transformation. Under these circumstances, an effective project management approach is revealed as an important factor of managing complexity and increasing chances for project success.

This research aims to identify a model that can be used to assess the complexity of smart cities' projects at early stages. Acknowledging the importance of project management in this type of projects (Kaur et al., 2019; Grizhnevich, 2018; Alshahadeh & Marsap, 2018), sets as starting point in this research the identification of a suitable project management complexity typology. Approaching smart cities'

V. Damasiotis (✉)
School of Economics and Business Administration, University of Thessaly, Larissa, Greece
e-mail: damasiotis@uth.gr

P. Fitsilis (ed.), *Building on Smart Cities Skills and Competences*, Internet of Things,
https://doi.org/10.1007/978-3-030-97818-1_10

projects under this prism, a set of complexity factors affecting these projects will be identified, and a relative model will be created. The use of this model will allow the assessment of smart city projects' complexity at their early stages allowing by that way project managers to be prepared and take early the necessary measures to deal or manage the complexity of these projects. By that way, project failures attributed to project complexity are expected to be reduced, resulting in higher rates of successful smart cities' projects. This is going to create a positive attitude about these types of projects both in citizens and cities' authorities and by that way to enhance their eagerness to implement projects that will forward smart city transition.

The proposed model is based on PMBOK management categorization and identifies project complexity factors in each category. Although agile project management approach gains more and more attraction to modern project management and PMBOK (PMI, 2013), it is considered as a traditional project management approach; the Project Management Institute (www.pmi.org) argues that it can be applied to agile projects too and has issued a guide for that (PMI, 2017).

The structure of this paper is as follows. In Sect. 2, the background literature in project complexity and smart cities is presented. In Sect. 3 the methodological steps followed in this research are described, while in Sect. 4, the formation of proposed complexity model is presented. Finally, in Sect. 5 conclusions and implications for future research are discussed.

2 Background Literature

2.1 Project Complexity

Projects, as unique and temporary endeavors, have some characteristics that can challenge their success, either as success considered the delivery of a product or service that fulfills the initial functional, nonfunctional, and quality requirements or the successful execution of project management process (Sudhakar, 2012). Complexity is considered among the main reasons leading to these failures due to the difficulty it imposes to project execution. As such, advance management skills are required for their successful implementation (Nguyen & Mohamed, 2021).

The notion of project success is not clearly defined and has different views for different types of stakeholders, according to the approach they have to project. Initially, success was considered the implementation of a project within its time and cost constraints and quality requirements. However nowadays, there are many more attributes that define project success such as the satisfaction of end users and other project stakeholder, the achievement of the strategic goals of project owner and sponsor, and the increased project performance (Bergmann & Karwowski, 2019).

During the last years, a significant number of studies have been undertaken in order to understand, define, and determine the concept of project complexity, and as a result a number of different definitions exist according to how each researcher approaches complexity, but still a lot of work has to be done (Oehmen et al., 2015). The majority of these studies are empirical, as they are based on the opinions of experts or key point project team members and stakeholders in order to identify factors which affect project complexity. Although there is no consensus on the definition of complexity among the various researchers, there is a general consensus that complexity in projects should contain dynamic and unpredictable elements with uncertainty stemming from ambiguity, lack of clarity, lack of structure, and unpredictable behaviors among project elements to be considered as the most common factor (Ward & Chapman, 2003). Organizational and technological factors are next to uncertainty, the most commonly identified complexity factor among the researchers. The organizational factor is related to project staffing, coordination of stakeholders, contract management project planning and scheduling, organization departments, hierarchy structure, etc. and has received great attention by researchers during the previous years (Nguyen et al., 2015; Lu et al., 2014; Vidal et al., 2011; Bosch-Rekveldt et al., 2011; Xia & Lee, 2005; Baccarini, 1996). Vidal et al. (2011) suggest that organizational complexity is the most significant source of project complexity. The technological factor refers to relationships between technology elements; the variety of technology platforms, technology novelty, newness of project technology, and technology changes has also attracted attention from researchers (Nguyen et al., 2015; Lu et al., 2014; Vidal et al., 2011; Bosch-Rekveldt et al., 2011; Xia & Lee, 2005; Remington et al., 2009; Baccarini, 1996).

The role of project management in handling project complexity is increasingly acknowledged by many researches who studied project management complexity (Damasiotis, 2018; Aydin & Dilan, 2017; Rahman et al., 2016; Stevenson & Starkweather, 2017; Truong & Jitbaipoon, 2016; Tie & Bolluijt, 2014). Project management has a major contribution to project success, and its complexity can significantly affect the project result (Cooke-Davies et al., 2007). Tie and Bolluijt (2014) state that project management and project complexity management are very closely related. Kermanshachi et al. (2016) acknowledging the relationship between project complexity and project management identified a set of complexity indicators and the corresponding management strategies. Damasiotis (2018) proposed an integrated framework for assessing software project complexity based on project management aspects and technical aspects of software development process.

Another division of project complexity is based on its origin. As such, it is identified either as descriptive complexity or as perceived complexity (Floricel et al., 2018; Schlindwein & Ison, 2004). Descriptive complexity describes complexity as a property of a system, while perceived complexity describes complexity as the subjective complexity that someone experiences through the interaction with the system. Based on the second approach, researchers argue that the perception of complexity is dependent on the cognitive level (knowledge, experience, background, personality) of the people involved (Remington et al., 2009; Fioretti & Visser, 2004) and that the subjectivity in evaluation of factors affecting project complexity is an inherent characteristic of this process (Montequín et al., 2018). This means that the perception of complexity can vary by time and according to the cognitive level of the observer, and can change as someone gains experience and knowledge by performing repeatedly similar type of projects or by managing more ambitious projects (Chapman, 2016).

This approach is significant as it entails that project complexity can be handled more efficiently as experience of project manager and project stakeholders is increasing and more effective and efficient management methods such as agile methods are applied. As enhancement to this approach, Sohi et al. (2016) identified a number of researches stating that complexity affects projects in such a way that makes changes in various project aspects inevitable, and as a result project management should be evolved to a direction of flexibility by providing structured management methods and be aware and prepared for changes and alignments.

2.2 Smart Cities

The world population residing in cities are continuously increasing during the last decades, and this trend is going to continue for the next decades. Cities trying to deal with this situation are changing and transformed to a more flexible and sustainable form called "smart city." Although the concept of smart cities exists for some years now, there is still no common definition about its definition (Chourabi et al., 2012). Initially the concept of smart city was very close to the use and adoption of information systems and communication technology. However, nowadays the concept of a smart city has been enriched with the concepts of sustainable growth, efficiency, and people aspects such as quality of life, improved services, etc. (Kaur et al., 2019; Daneva & Lazarov, 2018). However, according to Tsoutsa et al. (2021), information and communication technologies (ICT) still remain the foundation for all key themes related to a smart city.

Giffinger et al. (2007) proposed six "smart" dimensions of smart city that can be measured and can be used to assess the smartness of it. These are environment, governance, economy, people, living, and mobility. This approach indicates the multidimensional notion of a smart city and the multidimensional approach that someone should have when trying to implement a smart city and achieve smart city goals. On the other hand, Ekman (2018) state that the emergence of smart cities in network societies reflects the challenges of urban planning that call for complexification due to a large number of interactions and dynamical systems of the urban and environment.

Based on the above, this study defines smart city as the concept of a continuously growing and dynamically changing organization of high complexity relations that tries through the adoption of ICT technologies to improve the well-being of its citizens in domains of natural environment, sustainable growth, quality of services, accessibility, governance, entrepreneurship, mobility, and healthcare.

3 Research Methodology

This research is aiming to approach complexity of smart cities' projects, as defined in the previous section, from the project management perspective. To implement the objectives of this research, initially a literature review will be performed in order to identify the challenges of smart cities' projects that are sources of complexity. The appropriate literature was determined through e-resources such as e-databases and web search engines such as ScienceDirect, IEEE Xplore, ACM Digital Library, Google Scholar, Google Search, etc. The relevant papers were identified using criteria such as "smart cities' projects," "smart cities' project challenges," "smart cities' project complexity," etc. Through literature a list of smart cities' project challenges that are sources of complexity were identified.

This research adopts the complexity model proposed by Damasiotis (2018) that studied project management complexity of software projects and set it as the base model for this research. Building on the existing model, this research adds to it another dimension of complexity, namely, "specific features of smart city projects", resulting in a 12-dimensional model and matches the challenges identified in the complexity dimensions of the model as described in the next section. The specific framework was selected because it is the only framework that approaches project management complexity through a structured way, as it is built on PMBOK's (PMI, 2013) management knowledge areas, namely, scope, integration, risk, time, cost, quality, communication, human resources, procurement, and stakeholder. Furthermore, it emphasizes the complexity of software projects and identifies a set of complexity factors related to technical aspects of software projects that makes it suitable for smart cities' project as they incorporate a significant degree of ICT components. In addition, its structure is modular allowing the addition or removal of dimensions and the alignment of their weighting according to project needs.

4 Smart Cities' Project Complexity Framework

4.1 Smart Cities' Project Challenges

The implementation of a smart city is based on a set of projects aiming to fulfill the priorities and objectives set and require a huge amount of project management (Kaur et al., 2019). Smart cities' projects are usually not simple due to their nature as they have to deal with environmental challenges, political changes, social issues, ethic problems, innovative approaches and solutions, and technological challenges that inherent complexity to them and make their management a challenging task. According to Alshahadeh and Marsap (2018), the complexity of smart cities' projects is twofold as beyond the classic projects' challenges, there are challenges arising from the nature and complexity structure of smart cities. Another characteristic of smart cities' projects is that they are usually long term, but are implemented in incremental stages in the form of smaller projects, as the cost for one stage implementation is too high and/or other conditions (e.g., political, economic, social, technological) do not always allow it. This may lead to differences in perception of their obligation and benefits about project between project stakeholders which can affect their attitude toward the project. Furthermore, a smart city project requires the cooperation and involvement of stakeholders with high diversity in their interests and in many cases in their vision for project success and ROI.

In literature, a set of challenges that a project manager has to face during implementation of smart cities' projects are identified. Alshahadeh and Marsap (2018) identified technological challenges, financial constraints, stakeholder collaboration issues, governmental restraints, social challenges, and various managerial and organizational challenges as the main challenges of a smart city project. Grizhnevich (2018) identified as the most challenging aspects for a smart city the requirements for long-term planning, huge investments in field components, smart city platform flexibility, suitable connectivity, citizen engagement, and integration to existing infrastructure. Chourabi et al. (2012) identified eight categories of smart city challenges, namely, managerial and organizational, technological, government, policy context, people and communities, economy, built infrastructure, and natural environment. Pierce and Andersson (2017) identified challenges as technical and nontechnical and proposed six categories of them, namely, collaboration, financial, governance, awareness, interoperability, and privacy. Kakarontzas et al. (2014) in their study for a conceptual enterprise architecture framework for smart cities identified critical issues in smart city management, organization structure, and funding.

This study based on the above researches identified seven main categories of challenges that are sources of complexity, namely, technological, financial, stakeholders, administration, managerial and organizational, social, and ethic and provided a taxonomy of these challenges as presented in Table 1.

4.2 Project Management Complexity Model

According to Kaur et al. (2019), "*The establishment of a smart city requires a huge amount of project management.*" On the other hand, managing a smart city project is a challenging task as it has to deal with tasks of high complexity (Grizhnevich, 2018; Alshahadeh & Marsap, 2018). As that, the handling of complexity of project management of smart cities' project is a crucial factor for the successful implementation of them. As already mentioned, Damasiotis (2018) proposed a framework for assessing project management complexity of software projects and proposed a model of 11 dimensions and 35 complexity factors. The complexity model proposed by Damasiotis can be seen in Table 2.

In the next section, the formation of complexity model is presented.

Table 1 Smart cities' project challenges

Categories	Challenges identified	Identified in
Technological	Lack of appropriate knowledge of ICT systems Compatibility issues Security and privacy problems Operating and maintaining issues Flexibility and scalability of selected architecture Data collection and storage Unclear vision of IT management	da Silva et al. (2013), Grizhnevich (2018), Kakarontzas et al. (2014), Chourabi et al. (2012), Pierce and Andersson (2017)
Financial	Cost of installing ICT systems Cost of hiring and/or training IT specialists New system adaptation to existing infrastructure Who is the financier How to avoid huge long-term investments and replace them with short term and smaller Slow investment payoff Lack of business model	da Silva et al. (2013), Pierce and Andersson (2017)
Stakeholders	Stakeholder collaboration Communication Coordination Stakeholder diversity Conflicting stakeholders Clear aims and objectives—Vision	da Silva et al. (2013), Grizhnevich (2018), Chourabi et al. (2012)
Administration	Support from government and city administration Leadership Accountability Transparency Outdated rules and regulations	da Silva et al. (2013), Chourabi et al. (2012), Pierce and Andersson (2017)
Managerial and organizational	Project size and scope Manager skills Existence of qualified teams Project marketing issues Long-term planning through iterative short-term implementation	da Silva et al. (2013), Kakarontzas et al. (2014), Chourabi et al. (2012)
Social	Citizen engagement Common vision and understanding Motivate and involve citizens positively toward project Align citizens' behavior and attitude with smart city approach Quality of life Accessibility Political agenda	da Silva et al. (2013), Grizhnevich (2018), Chourabi et al. (2012)
Ethic	Data sharing policies Data ownership Data access level by various stakeholders Data storage Data privacy	da Silva et al. (2013), Kakarontzas et al. (2014), Pierce and Andersson (2017)

Table 2 Project management complexity factors (Damasiotis, 2018)

Complexity areas/dimensions	Complexity factor code	Complexity factor name
Time	TM1	The density of project activities
	TM2	Project activities' resource constraints
	TM3	The density of project schedule
	TM4	Protracted project/activities' duration
	TM5	Organization time management immaturity
Cost	CM1	Organization cost management immaturity
	CM2	Complicated financial structure and processes
	CM3	Long project duration
Quality	QM1	Inadequacies in quality management design
	QM2	Organization quality management immaturity
	QM3	Rigorous quality control procedures
Communication	COM1	Organization communication management immaturity
	COM2	Communication constraints due to project structure and staffing
	COM3	The density of project communication
Human resources	HRM1	Project team cohesion
	HRM2	Organization HR management immaturity
	HRM3	HR management constraints due to team structure
	HRM4	Project team size and skill diversity
Procurement	PM1	The density of procurement process
	PM2	Organization procurement management immaturity
	PM3	External barriers in project procurement process
Risk	RM1	Organization risk management immaturity
	RM2	Project risk density
Scope	SM1	The density of project requirements
	SM2	Quality of requirements
	SM3	Organization scope management immaturity
Integration	IM1	Integration constraints due to project characteristics
	IM2	Organization integration management immaturity
	IM3	The density of deliverables
Stakeholders	STM1	The density of stakeholders' management
	STM2	Organization stakeholders' management immaturity
Software development (technical) factors	SD1	Organization technological immaturity
	SD2	Product development constraints
	SD3	Product quality requirements
	SD4	Software size

4.3 Modeling Smart City Project Complexity

The next step of this research is the alignment of identified project challenges which are sources of complexity with the components of the complexity model. As smart cities' projects face the same challenges with other projects (Alshahadeh & Marsap, 2018), it was decided to keep the proposed model as it is and enriched it with special characteristics of smart cities' projects.

Concerning the identified challenges, most of the technological challenges are already included in technical dimension of complexity model, as "organizational technological immaturity" factor incorporates components such as appropriate knowledge and expertise in the domain. The factor "product development constraints" incorporates components related to compatibility, scalability, and flexibility of selected architecture and other product development requirements and constraints such as storage requirements, while issues related to security and privacy are included in "product quality requirements." What is missing from this dimension is "the existence of integrated and clear IT management support." This factor is important as the development of a smart city will lead to the development of a high number of different ICT systems with high heterogeneity between them that can easily cause operating and maintaining issues. Also, a common management approach of ICT systems can affect systems' interoperability.

Considering the financial challenges of a smart city project, there are a lot of different characteristics identified in relation to traditional cost management. Usually smart cities' projects are long term and require substantial investments in field components which in turn raises questions about investment payoff period and how cost reduction can be achieved without reduction in functional performance (Alshahadeh & Marsap, 2018; Grizhnevich, 2018). Furthermore, smart cities' projects are not always part of a clear business model that makes clear the marketing strategy, the payoff of investment, and the beneficiaries of the projects that are usually the financiers. This may lead to disagreements between stakeholders and decrease their willingness to patriciate or actively engage to project. In addition, a smart city project may create immediate costs due to needs for hiring and/or training staff while the repayment period is unclear. As that, the inclusion of a project in the context of a broader smart city business model can facilitate investors finding process and motivate management and staff.

Stakeholder challenges such as stakeholder's diversity, conflicts, and collaboration and communication issues are satisfactorily covered by the existing model's stakeholders and integration complexity dimensions. Emphasis should be placed in engagement of a specific type of stakeholders the citizens, as active engagement of them allow governmental authorities to tackle new concepts (Kogan & Lee, 2014) and forward the concept of smart city. On the other hand, the nonengagement of some group of citizens in project planning process will create problems in project execution, as they will not be included in requirement elicitation process and probably will not be part of the project beneficiaries. This will result in the creation of a gap between those groups of citizens and other citizens that can endanger not only current but also future projects due to different perception they will receive about project benefits and gains. Stakeholder engagement is also identified as a social challenge. The creation of common vision and understanding of project aims among citizens is also a social issue as it is the mean that will motivate citizens positively toward project and facilitate the change of citizens' thinking toward a "smart direction" (Woods et al., 2016; Kogan & Lee, 2014). As that, the existence of a strategy aiming to citizens "smart thinking" that can support and be supported by the project can increase citizens' engagement and hence reduce the complexity sourcing from citizens' opposition or indifference toward project.

Managerial and organizational challenges are also covered satisfactorily by model's scope, integration, and human resource management complexity dimensions. However, due to the diversity of project stakeholders and their different project views, it may be difficult to be agreed the main roles in project, e.g., who is the leader, the vendor, the beneficiary, etc. As such, the identification and agreement about the project role should have each one of the main project stakeholders is significant for smooth project execution (Alshahadeh & Marsap, 2018).

Significant challenges of smart cities' projects arise from citizens and public sector involvement and from the fact that their administration is strongly exposed to political and governmental changes. For example, changes in project administration can occur during its execution due to political changes. Even if new administration is in favor of the project, it may ask for different requirements and priorities that can endanger or entirely stop the project. The situation is even worse if new administration is opposed to the project. As such, the existence of consensus of all or major political parties about the project is significant. Another source of complexity is outdated rules and regulations. Major project stakeholders are local government, various central government authorities, and other public organizations that are usually slow in making decisions and changing outdated rules, laws, and procedures and hence are the cause for critical delays that can hinder project execution (Kogan & Lee, 2014). Another complexity factor is the requirements for accountability and transparency that in a smart city project is greater than in a typical project, not only because it may involve public funding but mainly because it affects the living of city's citizens and they require to be informed for any step.

Complexity sourcing from ethic challenges of smart cities is a dimension that usually is not vital to other projects generally. However, in smart cities projects are critical factor and can impose significant complexity as usually they involve the collection and/or processing of personal and business data, which set matters of privacy, confidentiality, ownership, access, and sharing. Complexity can occur not only from the need to satisfy these constraints due complex requirements settings but also from the need to define the necessary policies and reach to an agreement about these policies. Specifically, citizens need to be aware about which parts of their lives are being monitored, recorded, and exposed to authorities or other third parties and how they benefit from this situation. To achieve a consensus on these policies citizens, authorities, lawyers, and various organizations need to participate and reach to a consensus which in many times can be a complex issue.

Table 3 Complexity factors related to smart city special characteristics

Complexity factor code	Complexity factor name
SC1	Existence of integrated and clear IT management support
SC2	Existence of a broader business model that includes project
SC3	Unclear and protracted project payoff investment period
SC4	Existence of special strategy aiming at citizens' informing and engagement
SC5	Main project roles are clear and unquestionably acceptable by all major project stakeholders
SC6	Existence of outdated rules and laws affecting project
SC7	Density of requirements for accountability and transparency
SC8	Existence of consensus of all or major political parties about project's aim and objectives
SC9	Density of ethic constraints related to project
SC10	Existence of clear definition of ethic policies

Fig. 1 Proposed complexity typology

In Table 3 the final set of complexity factors sourcing from smart city characteristics is presented.

4.4 *Proposed Complexity Assessment Framework*

The final form of proposed complexity framework for smart cities can be seen in Fig. 1. It is composed of twelve complexity dimensions/categories, from which eleven are derived from the work proposed by Damasiotis (2018) and one dimension that was added in the context of this research.

The complexity model will be in the form of questionnaire, and each factor can be assessed separately based on a Likert scale question as can be seen in Fig. 2.

For calculating complexity, the formula proposed in the base complexity framework will be used in this case too. However, the full implementation of the model requires weights to have been assigned to complexity factors. To implement that, a

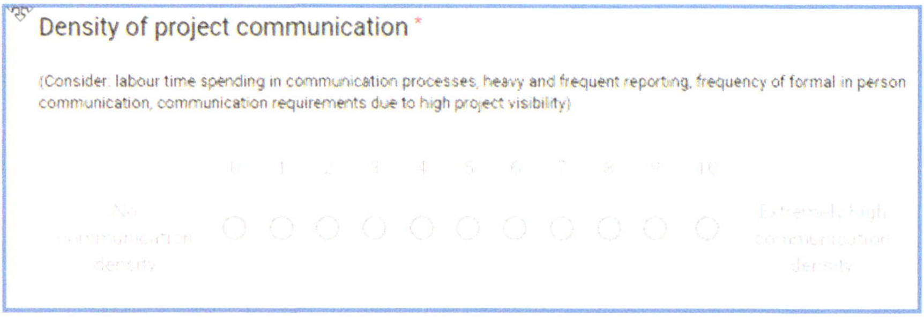

Fig. 2 Example of complexity factor assessment question

multicriteria decision method such as AHP can be used, but as this is beyond of the scope of this research, it will be done in a future research.

5 Conclusions and Implications for Future Research

Smart cities are the vision for cities of the future; however, the "smart" transformation of a city is a challenging and multidimensional process of high complexity. Transformation is usually made in the form of iterative projects as onetime projects require a huge amount of investments that are impossible to be found. This research acknowledges the importance of project management in smart cities' project success and focusing on project management aspects identifies complexity factors in these projects that affect their management. It provides project managers with a model that can help them to identify the sources of complexity in a smart city project and to assess the level of the complexity of the project. This research is based on an existing project complexity model that focuses on management complexity of software development projects and enriches it with a set of complexity factors sourcing from the special characteristics of smart cities' projects, resulting in a new model with 12 complexity dimensions and 45 complexity factors, 10 of which were identified in this research and concern the special characteristics of smart cities' projects.

It is profound that not all complexity factors neither dimensions are of equal importance. As that, assignment of weights to 12 complexity dimensions and to 45 complexity factors is a significant implication for future work and a requirement for full model implementation. Furthermore, model validation is another implication for future work. Validation process can reveal weaknesses and be a guide for model improvements. However, the proposed model consists of a solid base for future enhancements and research.

Acknowledgments This work was supported by the Erasmus+ KA2 under the project "DevOps competences for Smart Cities" (Project No.: 601015-EPP-1-2018-1-EL-EPPKA2-SSA).[1] This paper reflects the views only of the authors, and the Commission cannot be held responsible for any use which may be made of the information contained therein.

References

Alshahadeh, T., & Marsap, A. (2018). Smart cities, smarter management: Developing a smart framework for smart city projects management in Europe. *Ge-International Journal of Management Research, 6*(9), 41–73.

Aydin, M. N., & Dilan, E. (2017). Project management method adoption: A service industry case study. *International Journal of Information Technology Project Management, 8*(2), 17–33.

Baccarini, D. (1996). The concept of project complexity—A review. *International Journal of Project Management, 14*(4), 201–204.

Bergmann, T., & Karwowski, W. (2019). Agile project management and project success: A literature review. In J. I. Kantola, S. Nazir, & T. Barath (Eds.), *Advances in Human Factors, Business Management and Society. AHFE 2018* (pp. 405–414). Springer International Publishing.

Bosch-Rekveldt, M., Jongkind, Y., Mooi, H., Bakker, H., & Verbraeck, A. (2011). Grasping project complexity in large engineering projects: The TOE (technical, organizational and environmental) framework. *International Journal of Project Management, 29*(6), 728–739.

Chapman, R. (2016). A framework for examining the dimensions and characteristics of complexity inherent within rail megaprojects. *International Journal of Project Management, 34*(6), 937–956.

Chourabi, H., Nam, T., Walker, S., Gil-Garcia, J. R., Mellouli, S., Nahon, K., Pardo, T. A., & Scholl, H. J. (2012). Understanding smart cities: An integrative framework. In *Proceedings of the 45th Hawaii international conference on system science (HICSS), Maui, HI, 4–7 January 2012* (pp. 2289–2297). IEEE.

Cooke-Davies, T., Cicmil, S., Crawford, L., & Richardson, K. (2007). We're not in Kansas anymore, Toto: Mapping the strange landscape of complexity theory, and its relationship to project management. *Project Management Journal, 38*(2), 50–61.

da Silva, W. M., Alvaro, A., Gustavo, H. R. P. T., Afonso, R. A., Dias, K. L., & Garcia, V. C. (2013). Smart cities software architectures: A survey. In *Proceedings of the 28th annual ACM symposium on applied computing (SAC '13)* (pp. 1722–1727). Association for Computing Machinery. https://doi.org/10.1145/2480362.2480688

Damasiotis, V. (2018). *Modelling software Project Management complexity-an assessment model* (Stoke on Trend). Staffordshire University.

Daneva, M., & Lazarov, B. (2018). Requirements for smart cities: Results from a systematic review of literature, In *12th International conference on research challenges in information science (RCIS),* 2018 (pp. 1–6), doi: https://doi.org/10.1109/RCIS.2018.8406655.

Ekman, U. (2018). Smart city planning: Complexity. *International Journal of E-Planning Research, 7*(3), 1–21. https://doi.org/10.4018/IJEPR.2018070101

Fioretti, G., & Visser, B. (2004). A cognitive interpretation of organizational. *E:CO Special Double Issue, 6*(1–2), 11–23.

Floricel, S., Piperca, S., & Tee, R. (2018). Strategies for managing the structural and dynamic consequences of project complexity. *Complexity, 2018*, 3190251, p. 17.

[1] SmartDevOps project w1ebsite is https://smartdevops.eu

Giffinger, R. et al. (2007). *Smart cities—Ranking of European medium-sized cities*. [Online] Retrieved July 12, 2021, from http://www.smart-cities.eu/download/smart_cities_final_report.pdf

Grizhnevich, A. (2018). *Smart city project management: Avoiding common pitfalls*. [Online] Retrieved July 16, 2021, from https://www.smartcitiesworld.net/opinions/opinions/smart-city-project-management-avoiding-common-pitfalls

Kakarontzas, G., Anthopoulos, L., Chatzakou, D., & Vakali, A. (2014). A conceptual enterprise architecture framework for smart cities. In *Proceedings of the 11th International Joint Conference on e-Business and Telecommunications—Volume 2 (ICETE 2014)* (pp. 47–54). SCITEPRESS—Science and Technology Publications, Lda. https://doi.org/10.5220/0005021400470054

Kaur, S., Akre, V., & Arif, M. (2019). SMART project management for SMART cities: Analyzing critical factors affecting trust among virtual project teams. *Sixth HCT Information Technology Trends (ITT), 2019*, 65–72. https://doi.org/10.1109/ITT48889.2019.9075131

Kermanshachi, S., Dao, B., Shane, J., & Anderson, S. (2016). An empirical study into identifying project complexity. *Procedia Engineering, 145*, 603–610.

Kogan, N., & Lee, K. J. (2014). Exploratory research on the success factors and challenges of smart city projects. *Asia Pacific Journal of Information Systems, 24*(2), 141–189. https://doi.org/10.14329/apjis.2014.24.2.141

Lu, Y., et al. (2014). Measurement model of project complexity for large-scale projects from task and organisation perspective. *International Journal of Project Management, 33*(3), 610–622.

Montequín, V. R., Balsera, J. V., Fernández, S. M. C., & Fernández, F. O. (2018). Exploring project complexity through project failure factors: Analysis of cluster patterns using self-organizing maps. *Complexity, 2018*, 9496731, p. 17.

Nguyen, A. T., Nguyen, L. D., Long, L., & Dang, C. N. (2015). Quantifying the complexity of transportation projects using the fuzzy analytic hierarchy process. *International Journal of Project Management, 33*(6), 1364–1376.

Nguyen, T., & Mohamed, S. (2021). Mediation effect of stakeholder management between stakeholder characteristics and project performance. *Journal of Engineering, Project, and Production Management, 11*(2), 102–117. https://doi.org/10.2478/jeppm-2021-0011

Oehmen, J., Thuesen, C., Ruiz, P., & Geraldi, J. (2015). Complexity management for projects, programmes, and portfolios. An engineering systems perspective. In *PMI global congress 2015—EMEA*. Newtown Square, Project Management Institute.

Pierce, P., Andersson, B. (2017). Challenges with smart cities initiatives—a municipal decision makers' perspective. *HICSS*.

PMI. (2013). *A guide to the project management body of knowledge (PMBOK® Guide)*—Fifth Edition". sl.: Project Management Institute, Inc.

PMI. (2017). *Agile Practice Guide*. sl.: s.n.

Rahman, N., Wittman, A., & Thabet, S. (2016). Managing an engineering project. *International Journal of Information Technology Project Management, 7*(1), 1–17.

Remington, K., Zollin, R., & Turner, R. (2009). A model of project complexity: Distinguishing dimensions of complexity from severity. In *The 9th international research network of project management conference (IRNOP-IX)*. IRNOP.

Schlindwein, S. L., & Ison, R. (2004). Human knowing and perceived complexity: Implications for systems practice. *Emergence: Complexity & Organization, 6*(3), 27–32.

Sohi, A. J., Hertogh, M., Bosch-Rekveldt, M., & Blom, R. (2016). Does lean & agile project management help coping with project complexity? *Procedia—Social and Behavioral Sciences, 226*, 252–259.

Stevenson, D., & Starkweather, A. (2017). IT project success: The evaluation of 142 success factors by IT PM professionals. *International Journal of Information Technology Project Management, 8*(3), 1–21.

Sudhakar, G. P. (2012). A model of critical success factors for software projects. *Journal of Enterprise Information Management, 25*(6), 537–558.

Thompson, E. M. (2016). Smart city: Adding to the complexity of cities. In A. Herneoja & T. O. A. P. Markkanen (Eds.), *Complexity & simplicity—Proceedings of the 34th eCAADe conference—Volume 1* (pp. 651–660). University of Oulu. sl.: University of Oulu.

Tie, B. N., & Bolluijt, J. (2014). Measuring project complexity. In *9th International conference on system of systems engineering (SOSE)* (pp. 248–253). IEEE.

Truong, D., & Jitbaipoon, T. (2016). How can agile methodologies be used to enhance the success of information technology projects? *International Journal of Information Technology Project Management, 7*(2), 1–16.

Tsoutsa, P., Fitsilis, P., Anthopoulos, L., et al. (2021). Nexus services in smart city ecosystems. *Journal of the Knowledge Economy, 12*, 431–451. https://doi.org/10.1007/s13132-020-00635-3

Vidal, L. A., Marle, F., & Bocquet, J. (2011). Using a Delphi process and the analytic hierarchy process (AHP) to evaluate the complexity of projects. *Expert Systems with Applications, 38*(5), 5388–5405.

Ward, S., & Chapman, C. (2003). Transforming project risk management into project uncertainty management. *International Journal of Project Management, 21*(2), 97–105.

Woods, E., Alexander, D., Rodriguez Labastida, R., & Watson, R. (2016). *UK smart cities index. UK: Commissioned by Huawei From Navigant Consulting, Inc.* [online:] Retrieved from July 16, 2021, from https://www-file.huawei.com/-/media/CORPORATE/PDF/News/Huawei_Smart_Cities_Report_FINAL.pdf?la=en

Xia, W., & Lee, G. (2005). Complexity of information systems development projects: Conceptualization and measurement development. *Journal of Management Information Systems, 22*(1), 45–83.

Part III
Citizen Engagement

Learning to Engage Citizens to Make Smarter Cities

Kleanthis Sirakoulis and Stella Manika

1 Introduction

The priority for the development of smart cities has been included in the "Digital Agenda 2020" of the European Union, which has created the European Innovation Partnership on Smart Cities and Communities (EIP) (European Commission, 2021) intended to disseminate knowledge and innovative solutions in European cities. Among the issues that dominated the Digital Agenda 2020, the highlight was the role of environmental issues, such as energy-saving or mitigation of traffic problems. EIP working papers highlight the need for common infrastructures, with an emphasis on information and communications technology (ICT) tools and methods that collect, circulate, store, and process data for measurable and smart city management (e.g., big data analytics tools, information search systems, and applications using location information and smart subscriptions). Among these tools, we can include tools that aim to make available to citizens information produced through participatory applications that utilize different sources of information and improve the response of citizens and the quality of life in the city. Via this way, citizens, as end users, involve actively in urban planning. Recognizing that governing is better through the evidence-informed decision- and policymaking, this chapter seeks to approach civic engagement in the context of achieving the intelligence of a city, using, on the one hand, the bibliometric analysis and on the other hand the pilot study of a specific representative spatial unit. It, therefore,

K. Sirakoulis (✉)
Department of Business Administration, University of Thessaly, Thessaly, Greece
e-mail: sirakoul@uth.gr

S. Manika
Department of Planning and Regional Development, School of Engineering, University of Thessaly, Thessaly, Greece
e-mail: smanika@uth.gr

P. Fitsilis (ed.), *Building on Smart Cities Skills and Competences*, Internet of Things,
https://doi.org/10.1007/978-3-030-97818-1_11

proceeds to "map" the civic engagement in order for this mapping to be the starting point of a municipality or a community that wishes to increase the levels of civic engagement.

2 Literature Review

Although in the academic community the smart city is often criticized as overly technocratic (Rong, 2019), nowadays, most urban policies in smart cities tend to adopt more citizen-centric practices and an experimental city approach. At the same time, discussions are constantly coming to the fore which highlight the issue of public sector modernization and the need to move toward a more customer-focused approach. In order for this to happen, new public services and new performance or evaluation metrics of current services must be developed.

According to (Giffinger et al., 2007), smart city concept has been developed based on the aspects of the economy: smart economy, transport systems; smart mobility, natural resources; smart environment, social and human capital; smart people, quality of life; smart living; and administration and participation (Smart Governance) (Orlowski & Romanowska, 2019). Smart governance is an important smart city pillar since smart cities are complex sociotechnical systems (Razaghi & Finger, 2018) and require many interactions between governments, citizens, and other stakeholders (Pereira et al., 2018).

In order to strengthen the smart governance aspect, it is advisable to give innovative access to government for citizens by cocreating the smart city projects, so that citizens can participate in decision-making processes/cycles. Smart cities' potential can support community development purposes (Stratigea, 2012). Of course, in any case, the way the residents participate in the civic engagement processes is related to their special characteristics, such as their education (Evans et al., 2018) or their knowledge (Stratigea, 2015), p. 50).

At the same time, even in the case of actions promoting public participation, they are treated with suspicion and lack of trust on the part of citizens and many discrepancies between technocracy and democracy (Hartley, 2021). Regarding the relationship between smart city and civic engagement, a strong correlation seems to prevail through innovative systems (Manika, 2020b) and digital participatory platforms (De Filippi et al., 2019), civic crowdfunding choices (Carè et al., 2018), codesign and cocreation activities (Santos et al., 2018) (Manika, 2020a), or developing active self-decisive citizenry (Ojo et al., 2016). Activation policies and governance of activation correspond to a new service provision model (van Berkel & Borghi, 2008), promote governance based on the conception of responsibility (Bonvin, 2008), and produce self-governing and responsible subjectivities (Haikkola, 2018). The intensity of the relationship between smart city and civic engagement is confirmed by the large number of articles that formed the basis of bibliometric analysis. At the same time, various evaluation frameworks of citizen participation in smart cities have been developed, as shown in Table 1.

Table 1 Evaluation framework of citizen participation in smart cities

Three different elements form the basis of citizen participation: (1) citizens as democratic participants, (2) citizens as cocreators, and (3) citizens as ICT users	Simonofski et al. (2017)
Four assessment perspectives are described as follows: Perspective 1—public value generation ↔ community building and management ↔ vision and strategy formulation Perspective 2—vision and strategy formulation ↔ community building and management ↔ assets management Perspective 3—assets management ↔ community building and management ↔ financial and economic sustainability Perspective 4—public value generation ↔ community building and management ↔ financial and economic sustainability	Castelnovo et al. (2015)
Emphasis on the mobile app's impact on civic participation (fulfillment of mobile app usability)	Hartmann (2019)
Emphasis on the relation between the type of governance and the effects of citizen participation	Gaventa and Barrett (2010)
Comparative measures used as comparison base, among Talk London. My Ideal City Bogota, MyGov.in.Goal and objectives Ownership and operation level of governance Intended users Outcomes	Praharaj et al. (2017)

While, to determine the degree to which a city can increase levels of citizen agency and public participation can be utilized various scaffolds such as of Sherry Arnstein's Ladder of Citizen Participation (Arnstein, 1969), as one of the most influential models in this field. This model applied in the case of Dublin (Cardullo & Kitchin, 2018) showed that most "citizen-centric smart city initiatives come from *stewardship, civic paternalism, and a neoliberal conception of citizenship*." At the same time, studies based on the same model and having as a key study Poznan, Poland and Portland, Oregon, USA highlighted the importance of extending this model to new categories such as *«disorder, awakening, radicalisation, civil disobedience and rebel action»* (Kotus & Sowada, 2017).

3 Methodology

Civic engagement in smart cities, as a shift from a top-down, largely hierarchical planning (Stratigea et al., 2015), is a cutting-edge field of study, and this article explores the dominant bibliographic and research approaches to this topic via bibliometric analysis. A variety of surveys have used bibliometric analysis to capture the smart city concept (Janik et al., 2020) (Winkowska et al., 2019) or certain aspects of it, like smart cities' infrastructure (Kasznar et al., 2021). The bibliometric analysis used in this paper aims to identify the dominant trends of civic engagement, as they are formed and evolve by year, by geographical unit, and by their specific theme, either through solely theoretical bibliographic references or through applied examples. This methodological framework provides a holistic approach to the issue of civic engagement as well as evolving trends in the field.

The analysis carried out was based on the Web of Science database and as the basis of the search was placed on the one hand keywords related to "smart city" or "intelligent city" or "digital city" concept and on the other hand keywords related to "citizen" or "governance" or "engage citizens" or "participation" or "public participation" or "participatory design" or "civic engagement". Of course, it should be noted that this method is very sensitive in terms of the keywords chosen and the results they lead to. For this reason, several tests have been performed to include the largest and most representative range of relevant articles available.

The result of the first search was 1,395 documents. In the next step, the time range of study was set, the period from 2010 to 2020 (the last full year), and English was chosen as the language of publication. The final set of documents included (after duplicates were excluded and a hand search was performed) 817 related articles published during that period in scientific journals. Data on document type, document content (keywords, authors, authors' collaborations), and their characteristics were analyzed using the R-based Biblioshiny app, which is freely available from https://bibliometrix.org/ and as described in (Aria et al., 2020) and (Aria & Cuccurullo, 2017). Table 2 shows in detail the main information about the data used.

4 Results

Authors created the following graphs and tables by compiling the selected articles. Graph 1 demonstrates the number of articles about smart cities published each year. One can observe a positive trend in the number of published articles related to civic engagement in smart cities that peaks in 2020. The annual growth rate is equal to 62.57%. This supports the article's initial argument that civic engagement in smart cities is an emerging topic that is gradually gaining the interest of researchers.

More specifically, Graph 2 demonstrates that keywords such as "e-government," "participation," "public participation," and "citizen participation" appear with increasing frequency in these articles. The time course of the emergence of the concepts is increasing, with a particular upward trend from 2018 onward, demonstrating once again the topicality of these issues.

Based on Graph 3, we can group the countries of origin of the main volume of authors into three categories. The first category includes the countries of origin of most articles, more specifically, the United Kingston, Italy, China, the United States, Spain, the Netherlands, Australia, and Canada. The participation of articles from Germany, Brazil, India, Korea, Greece, and Sweden is smaller. Meanwhile, in the last category, to which corresponds the smallest contribution to the objects under study, are Belgium, Finland, Ireland, Austria, Poland, and Portugal. It is also worth mentioning the differentiation that occurs when articles are divided into

Table 2 Main information about data used

Description	Results
Main information about data	
Timespan	2010:2020
Sources (journals, books, etc.)	358
Documents	817
Average years from publication	2.8
Average citations per document	18.34
Average citations per year per doc	3.889
Document types	
Article	817
Document contents	
Keywords plus (ID)	1175
Author's keywords (DE)	2616
Authors	
Authors	2235
Author appearances	2550
Authors of single-authored documents	162
Authors of multiauthored documents	2073
Authors' collaboration	
Single-authored documents	184
Documents per author	0.366
Authors per document	2.74
Coauthors per documents	3.12
Collaboration Index	3.27

multiple-country publications (MCP) and single-country publications (SCP) with the landscape changing significantly in almost all categories.

Combining the countries of origin of the authors, the keywords and the sources results in Graph 3. Graph 4 shows the three-field plot (Sankey diagram) as analyzed in Riehmann et al. (2005) and in Fatehi et al. (2020). In the left-field (entitled "DE"), we see the keywords (number of items: 10); in the middle field, entitled "AU_CO" (number of items: 10), we see the countries of origin of the source, while in the right field, "entitled SO," the sources are observed (number of items: 10). This plot was created to depict the proportion of research topics for each country and the recency of the papers that they cited.

We observe that in all the countries of the first category, i.e., countries of origin of most articles, in addition to the concepts of smart cities, the keywords that appear most often are "smart governance" and "e-government," the "Internet of Things," the "citizen participation," the "governance," "big data," and "sustainability." In countries of smaller origin of articles (third category), the appearance of the predominant keywords focuses beyond the "smart city" and "smart cities," mainly in "smart governance" and "e-government." The authors considered it appropriate to keep both phrases, either "smart city" or "smart cities," rather than to combine

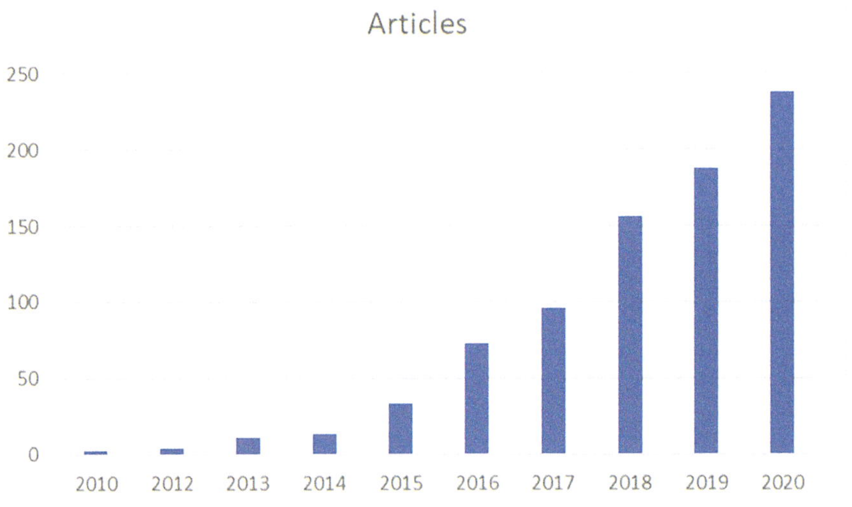

Graph 1 Articles published per year

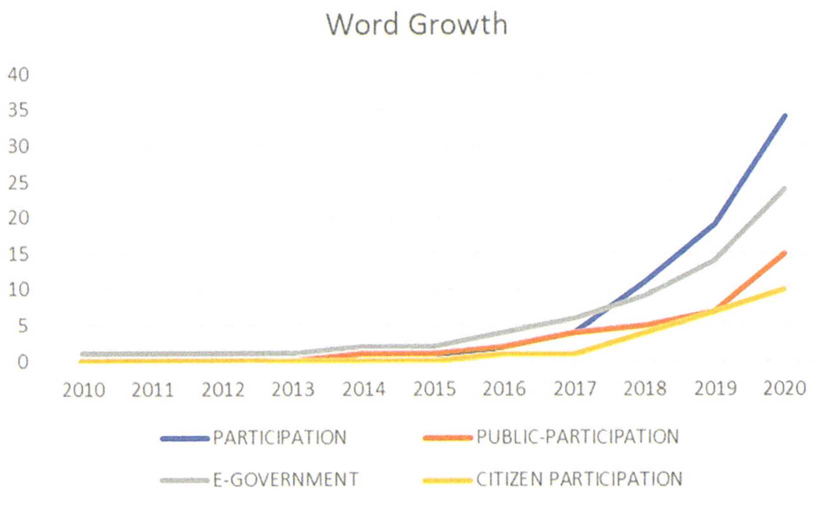

Graph 2 World growth per year

them, as the latter phrase often corresponds to city networks or groups of cities of the common country of origin.

To give the correlations and interfaces of the main concepts and their dominant tendencies, the network map (Graph 5), connecting the main concepts of the chapter, was created. Authors can distinguish three clusters. In the larger cluster, which

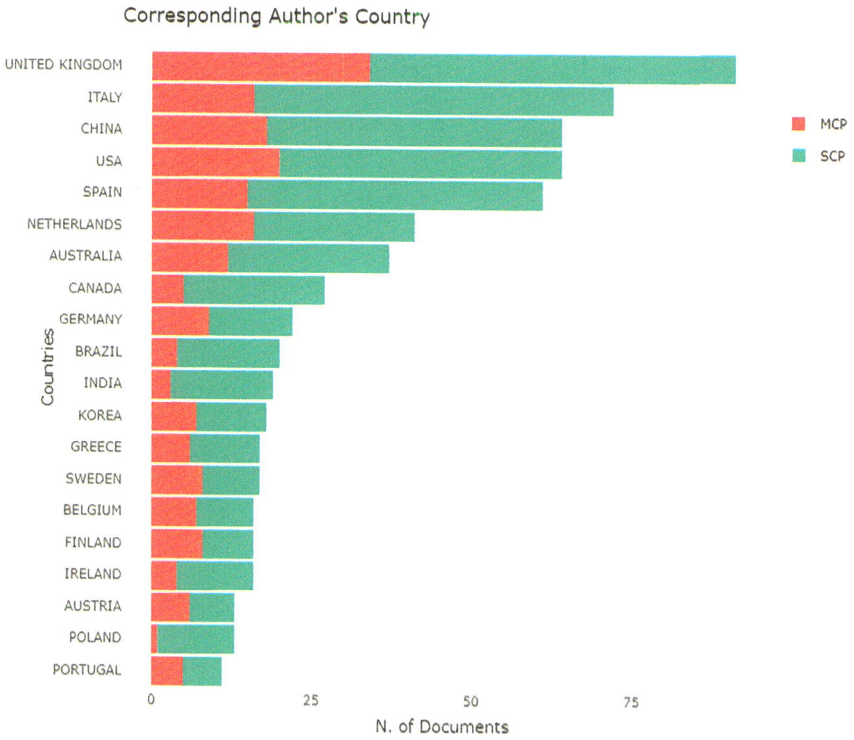

Graph 3 Corresponding author's country

is shown in red, we see the correlation of "smart cities" and "governance" with "innovation," "knowledge," "networking," "urban sustainability," and their achievement "policies." The second cluster recognizes the correlation between "smart city," "management," and "information," while the decisive link between them is the contribution of "participation" (citizen participation, public participation) and "e-government", their perspective and design.

5 Civic Engagement in Greek Smart Cities: The Case of the Thessaloniki Metropolitan Area, Greece

Given that Graph 3 demonstrates that Greece is the country of origin for a significant number of writers yet does not appear in the three-field plot, authors found it appropriate to identify the applications of civic engagement as they are adopted in recent years by illustrative examples of Greek cities.

The basis for choosing these examples was set to move in the same direction as the bibliographic searches. This section aims to provide a brief outline of the Greek

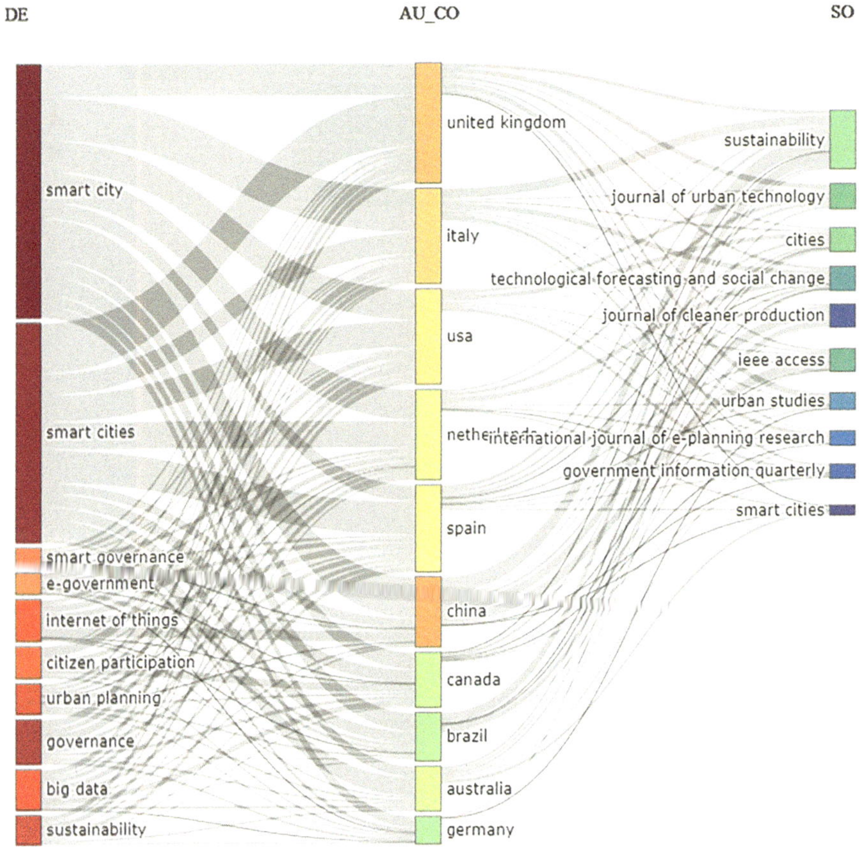

Graph 4 Three -field plot

reality and then to use the features as they are found in the bibliography as a road map to achieving even greater civic engagement.

The case of the Thessaloniki metropolitan area, one of the regional units of Greece, presents a key study. Thessaloniki is the capital of the Region of Central Macedonia, and the Thessaloniki regional unit is divided into 14 municipalities. It was selected as the municipalities present an increased diversity in terms of their particular geographical and morphological characteristics and on the other hand their levels of intelligence in the part of the civic engagement under consideration.

The Municipality of Thessaloniki is the second largest municipality in Greece by population (325,182 inhabitants, 2011 census). As part of the municipal government's efforts to increase the percentage of civic engagement, it has developed an e-services program that allows all citizens to access the policy decisions and vote counts of the municipal councils online.

Regarding the development of intelligence in the Municipality of Kalamaria, one of the largest municipalities in the Thessaloniki regional unit, it has been

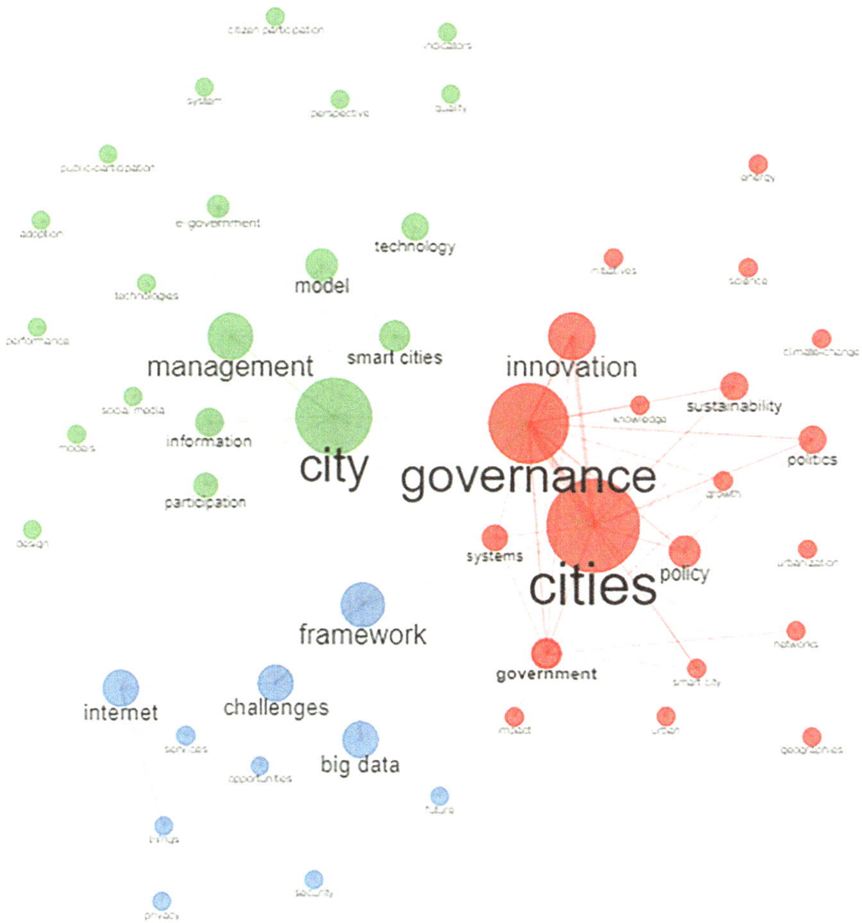

Graph 5 Network map

focused on transparency for budget data and interaction between citizens and municipal authorities. In this context, it is worth mentioning that the Municipality of Oreokastro is exploring the possibility of broadcasting municipal council meetings in real time. Detailed aggregated information about the other municipalities is provided in Table 3. Table 3 presents on the one hand the municipalities and on the other hand the participatory potentials through the official municipal website (data source). At this point, it should be emphasized that the authors attempted to identify the features that are part of the participatory planning and smart governance pillar and not in all the smart city actions.

Based on the information from Table 3, one can see that, in general, the Peripheral Unit of Thessaloniki has already implemented a fairly satisfactory set of policies meant to achieve smart governance and the promotion of civic engagement,

Table 3 Municipalities and participatory potentials

Municipal	e-services (issuance of certificates), e-poleodomia	Form for performance or improvement comments	GIS	Job announcements, actions for the unemployed	Online municipal meeting, promoting deliberation actions	Online municipal services website (data source)
Ampelokipi-Menemeni	x		x		x	https://www.ampelokipi-menemeni.gr/
Chalkidonos	x	x	x	x		https://dimoschalkidonos.gr/
Delta	x	x	x	x		https://www.dimosdelta.gr/
Kalamaria	x	x	x	x	x	https://kalamaria.gr/
Kordelio-Evosmos	x	x	x	x	x	http://www.kordelio-evosmos.gr/
Lagada	x				x	https://www.lagadas.gr/
Neapolis-Sikeon	x	x			x	http://www.dimosneapolis-sykeon.gr/
Oraiokastro	x					https://oraiokastro.gr/
Pavlou Mela	x	x	x	x	x	https://pavlosmelas.gr/
Pilea-Hortiatis	x	x	x	x	x	https://www.pilea-hortiatis.gr/web/guest/home
Thermaikos	x	x		x	x	https://www.thermaikos.gr/
Thermi	x	x				https://www.thermi.gov.gr/
Thessaloniki	x	x	x	x	x	https://thessaloniki.gr/?lang=en
Volvi	x					https://www.dimosvolvis.gr/

through concepts and tools, as shown in Graph 5, that include, but are not limited to, the promotion of innovation, information, citizen participation, and big data management. However, there is no common line from the municipalities in terms of all the possibilities they provide to the citizens. The municipalities' sizes by population (or even other parameters related to the individual characteristics of each municipality, such as location or economic conditions) do not determine the choices of each local government to achieve higher or lower percentages of smart governance. For example, small municipalities, such as the Municipality of Pylaia Chortiatis, have created more innovative actions compared to other larger ones, such as the Municipality of Kalamaria.

At this point, it is worth mentioning that it would be even more interesting to see whether the opportunities to access information provided by the municipalities are used in practice if their environments encourage the use and if they are properly supported by the municipalities, i.e., what information is collected by the citizens or if they receive answers to their questions immediately

6 Conclusions

The bibliometric analysis demonstrates that civic engagement in smart cities is a rapidly emerging issue, especially in terms of its integration and promotion in the context of the intelligence of a city. From the previous analysis, it seems that there is a direct correlation between the manifestations of participation (citizen participation or public participation) with innovation, knowledge, and networking.

From the analysis of the basic aspects of smart governance, as they were used in the case of Thessaloniki, it seems that the vision of each municipality is unrelated to its size in physical area or population, as many smaller municipalities are more encouraging of citizen participation than larger ones.

Utilizing the findings of the bibliometric analysis and the examination of a Greek metropolitan area, authors conclude that in the case of Greece, municipal governments have taken important steps to promote civic engagement. However, it is important to place, according to bibliometric analysis, even greater emphasis on the following:

1. Increasing the amount of information provided by municipal government websites and dissemination of this information to all citizens (e.g., through social media)
2. Promoting innovation and networking between municipalities and citizens, efficient management and immediate update of information, and real-time communication
3. Adoption of civic engagement approach regardless of the size of the municipality, and on a common national basis, as the available technology tools can now significantly strengthen in this direction

These findings illuminate a theoretical and practical puzzle and advance a dialog regarding the next steps for municipalities to take in the effort to achieve a better smart governance quota for their citizens.

References

Aria, M., & Cuccurullo, C. (2017). bibliometrix: An R-tool for comprehensive science mapping analysis. *Journal of Informetrics, 11*(4), 959–975. https://doi.org/10.1016/j.joi.2017.08.007

Aria, M., Misuraca, M., & Spano, M. (2020). Mapping the evolution of social research and data science on 30 years of social indicators research. *Social Indicators Research, 149*(3), 803–831. https://doi.org/10.1007/s11205-020-02281-3

Arnstein, S. R. (1969). A ladder of citizen participation. *Journal of the American Planning Association, 35*(4), 216–224. https://doi.org/10.1080/01944366908977225

Bonvin, J.-M. (2008). Activation policies, new modes of governance and the issue of responsibility. *Social Policy and Society, 7*(3), 367–377. https://doi.org/10.1017/S1474746408004338

Cardullo, P., & Kitchin, R. (2018). Being a 'citizen' in the smart city: up and down the scaffold of smart citizen participation in Dublin, Ireland. *GeoJournal, 84*(1), 1–13. https://doi.org/10.1007/S10708-018-9845-8

Carè, S., Trotta, A., Carè, R., & Rizzello, A. (2018). Crowdfunding for the development of smart cities. *Business Horizons, 61*(4), 501–509. https://doi.org/10.1016/J.BUSHOR.2017.12.001

Castelnovo, W., Misuraca, G., & Savoldelli, A. (2015). Citizen's engagement and value co-production in smart and sustainable cities. *International Conference on Public Policy*, 1–16.

De Filippi, F., Coscia, C., Cocina, G. G., Lazzari, G., & Manzo, S. (2019). Digital participatory platforms for civic engagement: A new way of participating in society? *International Journal of Urban Planning and Smart Cities, 1*(1), 1–21. https://doi.org/10.4018/ijupsc.2020010101

European Commission. (2021). *Smart cities*. Retrieved from https://ec.europa.eu/info/eu-regional-and-urban-development/topics/cities-and-urban-development/city-initiatives/smart-cities_en

Evans, B. J., Marsicano, C. R., & Lennartz, C. J. (2018). Cracks in the bedrock of American democracy: differences in civic engagement across institutions of higher education. *Differences in Civic Engagement, 48*(1), 31–44. https://doi.org/10.3102/0013189X18809053

F. Fatehi, F. Hassandoust, R.K.L. Ko, S. Akhlaghpour, General data protection regulation (GDPR) in healthcare: hot topics and research fronts. Stud. Health. Technol. Inform. **270**, 1118–1122 (2020). https://doi.org/10.3233/SHTI200336

Gaventa, J., & Barrett, G. (2010). So what difference does it make? Mapping the outcomes of citizen engagement. *IDS Working Papers, 2010*(347), 01–72. https://doi.org/10.1111/J.2040-0209.2010.00347_2.X

Giffinger, R., Kramar, H., Haindlmaier, G., & Strohmayer, F. (2007). *European smart cities*. Retrieved from http://www.smart-cities.eu/?cid=2&ver=3

Haikkola, L. (2018). Shaping activation policy at the street level: Governing inactivity in youth employment services. *Regional Science and Urban Economics, 62*(3), 334–348. https://doi.org/10.1177/0001699318784341

Hartley, K. (2021). Public trust and political legitimacy in the smart city: A reckoning for technocracy: doi: https://doi.org/10.1177/0162243921992864

Hartmann, S. (2019). Citizen relationship management for civic participation: How smart cities use 311 to involve citizens. *Public Administration and Information Technology, 34*, 59–77. https://doi.org/10.1007/978-3-319-89474-4_4

Janik, A., Ryszko, A., & Szafraniec, M. (2020). Scientific landscape of smart and sustainable cities literature: A bibliometric analysis. *Sustainability, 12*(3), 779. https://doi.org/10.3390/SU12030779

Kasznar, A. P. P., Hammad, A. W. A., Najjar, M., Qualharini, E. L., Figueiredo, K., Soares, C. A. P., & Haddad, A. N. (2021). Multiple dimensions of smart cities' infrastructure: A review. *Buildings, 11*(2), 73. https://doi.org/10.3390/BUILDINGS11020073

Kotus, J., & Sowada, T. (2017). Behavioural model of collaborative urban management: Extending the concept of Arnstein's ladder. *Cities, 65*, 78–86. https://doi.org/10.1016/J.CITIES.2017.02.009

Manika, S. (2020a). Transforming vacant commercial spaces: From localized hotspots of urban shrinkage to 'smart' co-working places. *Open Journal of Social Sciences, 8*, 86–97. https://doi.org/10.4236/jss.2020.86009

Manika, S. (2020b). Mechanisms for innovative-driven solutions in European smart cities. *Smart Cities, 3*(2), 527–540. https://doi.org/10.3390/smartcities3020028

Ojo, A., Dzhusupova, Z., & Curry, E. (2016). Exploring the nature of the smart cities research landscape. *Public Administration and Information Technology, 11*, 23–47. https://doi.org/10.1007/978-3-319-17620-8_2

Orlowski, A., & Romanowska, P. (2019). Smart cities concept: Smart mobility indicator. *Smart Cities, 50*(2), 118–131. https://doi.org/10.1080/01969722.2019.1565120

Pereira, G. V., Parycek, P., Falco, E., & Kleinhans, R. (2018). Smart governance in the context of smart cities: A literature review. *Information Polity, 23*(2), 143–162. https://doi.org/10.3233/IP-170067

Praharaj, S., Han, J. H., & Hawken, S. (2017). Innovative civic engagement and digital urban infrastructure: Lessons from 100 smart cities mission in India. *Procedia Engineering, 180*, 1423–1432. https://doi.org/10.1016/J.PROENG.2017.04.305

Razaghi, M., & Finger, M. (2018). Smart governance for smart cities. *Proceedings of the IEEE, 106*(4), 680–689. https://doi.org/10.1109/JPROC.2018.2807784

Riehmann, P., Hanfler, M., & Froehlich, B. (2005). *Interactive Sankey diagrams* (pp. 233–240). IEEE Symposium on Information Visualization, 2005. INFOVIS 2005. https://doi.org/10.1109/INFVIS.2005.1532152

Rong, H. H. (2019). *Designing with data: collective intelligence in urban design*. Retrieved from https://dspace.mit.edu/handle/1721.1/123601

Santos, I., Nobre, A. C. B., Ibiapina, J. C., Oliveira, P. R. M., Carvalho, Z. V. N. De, & Oliveira, A. D. De. (2018). Strategies and methodologies for civic engagement and social empowerment. In *2017 IEEE 1st summer school on smart cities, S3C 2017—Proceedings* (pp. 157–160). doi: https://doi.org/10.1109/S3C.2017.8501396

Simonofski, A., Asensio, E. S., De Smedt, J., & Snoeck, M. (2017). Citizen participation in smart cities: Evaluation framework proposal. In *Proceedings—2017 IEEE 19th conference on business informatics, CBI 2017* (Vol. 1, pp. 227–236). doi: https://doi.org/10.1109/CBI.2017.21

Stratigea, A. (2012). The concept of 'smart cities'. Towards community development? *Open Edition Journals, Networks and Communication Studies, 26-3*(4), 375–388. https://doi.org/10.4000/NETCOM.1105

Stratigea, A. (2015). *In Greek: Θεωρία και Μέθοδοι Συμμετοχικού Σχεδιασμού* (HEAL-Link). www.kallipos.gr

Stratigea, A., Papadopoulou, C.-A., & Panagiotopoulou, M. (2015). Tools and technologies for planning the development of smart cities. *Journal of Urban Technology, 22*(2), 43–62. https://doi.org/10.1080/10630732.2015.1018725

van Berkel, R., & Borghi, V. (2008). Introduction: The governance of activation. *Social Policy and Society, 7*(3), 331–340. https://doi.org/10.1017/S1474746408004302

Winkowska, J., Szpilko, D., & Pejić, S. (2019). Smart city concept in the light of the literature review. *Engineering Management in Production and Services, 11*(2), 70–86. https://doi.org/10.2478/EMJ-2019-0012

Developing Smart City Ambassadors in Oman

Judy Backhouse and Laila al Hadhrami

1 Introduction

This chapter examines the case of the Smart City Ambassadors program in Oman (2019–2021) as a model for developing skills as well as raising awareness of smart city possibilities.

When the Sultanate of Oman embarked on a strategy for smart cities, one of the challenges was related to capacity, both in the specific technology skills and in the general levels of knowledge about smart cities that enables people to envisage smart solutions. In particular, they faced the challenge of training staff in a large number of government entities, including in different locations. At that point, the country drew on a small pool of experts for training and projects that was inadequate to meet their ambitious plans.

The Smart Cities Ambassadors program was developed in response to these challenges. It sought to expand the pool of experts and draw a wider range of participants into the process while, at the same time, spreading knowledge about smart technologies more broadly throughout the country. The program uses skilled volunteers to conduct lectures and workshops on smart city technologies and to champion smart city initiatives. As the program has evolved, it has built a smart community of practice that can support innovation by having smart city experts available across the country.

J. Backhouse (✉)
United Nations University Operating Unit on Policy-Driven Electronic Governance, Guimarães, Portugal
e-mail: backhouse@unu.edu

L. al Hadhrami
National Energy Center, Knowledge Oasis Muscat, Muscat, Sultanate of Oman
e-mail: laila.alhadhrami@nec-oman.com

P. Fitsilis (ed.), *Building on Smart Cities Skills and Competences*, Internet of Things,
https://doi.org/10.1007/978-3-030-97818-1_12

This chapter takes the form of a case study, describing the program, how it is structured, and how it operates. It discusses some of the results of the program as well as some of the challenges and reflects on the potential for replicating and scaling the program.

2 Smart City Skills and Learning

One of the main challenges in progress toward smart cities in the next decade is a lack of skills (UNCTAD, 2016), and the rapidly changing nature of the skills needed makes it difficult to plan their development. Smart cities need highly educated, ICT-literate individuals to implement technology solutions, but they also depend on less ICT-literate individuals to perform support functions. Smart cities also need to involve citizens with varying levels of ICT skills, in planning and contributing to city development. At the same time, digital skills are needed by researchers, innovators, entrepreneurs, employees, and leaders in the public and private sectors. This section examines the literature to understand what skills are needed and how they can be developed.

2.1 The Skills Needed

Identifying the skills needed for smart cities to thrive is complicated by the great variety of smart city domains and specializations, as well as the range of stakeholders.

There have been attempts to catalog the competencies needed for smart cities and digital transformation (Bakhshi et al., 2017; Michelucci et al., 2016; European Commission, 2020) and to define appropriate learning pathways (Iatrellis et al., 2021). These attempts are useful, but they focus on what is needed for existing and anticipated occupations, while smart city stakeholders include residents and organizations not directly employed in smart city roles, who also need skills to live in smart cities and contribute to their development.

Researchers have categorized the skills needed for smart cities into technology skills, data skills, and community skills (David & McNutt, 2019), and this serves as a useful way to structure our discussion.

2.1.1 Technology Skills

Smart cities need ICT workers with technical skills to innovate and originate new products and services (David & McNutt, 2019). Researchers have identified priority technical skills such as developing and assembling electronic systems, developing automated learning algorithms, programming embedded systems, and managing big data (Madureira et al., 2020).

Businesses (including entrepreneurs) and public administrators need to know what technologies are available, to be able to conceptualize how to use them in enterprises or cities, as well as the limitations of technology. They also need to understand the cycles of technology adoption. Government employees need to be able to anticipate the technology impacts, including negative impacts on issues such as equality and sustainability (David & McNutt, 2019; Mukhametov, 2019). Individuals need sufficient skill to be able to use digital services and sources of information. They also need to understand the risks technology presents and how to mitigate them (Kritzinger, 2017).

2.1.2 Data Skills

Data skills include how to collect, clean, curate, store, protect, and archive data as well as the analysis of data using modeling to extract meaning. Data skills also include the ethics and legal considerations of data collection, storage, and use. Data literacy also includes being able to interpret data to understand problems and to create stories about issues and as evidence in decision-making.

Business owners need to be aware of the power of data in developing new products and markets, as well as the limitations (Al Harthy et al., 2019). Similarly, public administrators need to understand how they can use data, particularly the predictive capabilities of data, effectively and justly (David & McNutt, 2019). Both businesses and public administrators need to be aware of legal and ethical issues surrounding data collection, storage, and use. Public administrators in particular have a responsibility to act in the best interests of those they govern. Citizens need to be data literate in order to shape and communicate their expectations of cities (Wolff et al., 2019) and make use of the available open data sets (David & McNutt, 2019).

2.1.3 Community Skills

The complexity of smart solutions requires multiple inputs, and the range of city stakeholders means that understanding and accommodating their needs is important in developing good smart solutions. As a result, community skills are a key part of smart cities. Community skills go beyond the skills of networking and collaboration that are common in business and government to include cultural intelligence (Faraji et al., 2021) as well as how to engage with communities, understand their needs, communicate effectively, and work together to cocreate solutions that enhance the city (David & McNutt, 2019).

In a study of hackathons, researchers found that participants particularly valued "learning about other participants' views, improving their ability to work in a diverse team, and gaining new perspectives on the city's problems" (Jaskiewicz et al.,

2019:23). Engaging multiple stakeholders in planning and development of smart cities ensures that efforts are directed to needs. Citizens benefit through increased self-esteem, skills development, and community connections (Wolff et al., 2019). Some authors argue that the skills to critique and challenge smart city developments are also important (Perng & Maalsen, 2020).

2.2 How to Develop Skills

Skills development needs to be targeted to the many different constituencies involved in smart cities. Approaches identified include making changes to existing formal education and training, adding digital elements to continuing professional development, the use of various forms of just-in-time learning (including how such learning is discovered), and informal ways of learning (David & McNutt, 2019). Researchers have called for "the centrality and continuous development of ICT literacy, knowledge, talents and skills" (Curseu et al., 2021:668).

Some have examined how existing formal learning programs in schools, colleges, and universities can be altered to incorporate digital skills (David & McNutt, 2019; Manjrekat & Deshmukh, 2020). Other programs supplement formal learning, such as the Urban Data School in Milton-Keynes which teaches skills in "asking and answering questions from data," to critique data, use data as evidence to tell stories, and to recognize "opportunities for using data" (Wolff et al., 2019: 161).

High-level technical skills for knowledge workers are addressed through education and through professional programs with training often provided by the large technology companies (Cisco, Microsoft, Siemens, SAP, Oracle, etc.). Technical skills can also be developed through volunteers in civic tech movements (David & McNutt, 2019). Tech entrepreneurs learn through formal education and through incubator programs and hackathons. In China, creative entrepreneurs have been supported to improve economic as well as social and cultural outcomes (Jiang et al., 2019).

Cities have made use of hackathons to initiate and develop smart projects. Participants learn both technical skills and new perspectives on city problems from each other (Jaskiewicz et al., 2019). Hackathons can provide the learning found in a community of practice (Wenger, 1998), but they are short-lived events and there needs to be ongoing support to sustain learning beyond the events (Jaskiewicz et al., 2019).

For employees and leaders in public administration, digital skills learning can be more effective when solving "living problems" (Kranjac et al., 2021). Formal programs for public administrators are unlikely to keep pace with changes in technology, and academic staff may lack the skills to teach digital competencies. One possibility is to develop partnerships with technology companies to augment professional learning (David & McNutt, 2019).

Developing digital skills among citizens more broadly is challenging. Authors suggest involving people in public space and play (Manjrekat & Deshmukh, 2020), citizen participation in smart city governance (Curseu et al., 2021), using open data to engage people in city plans, and making resources available for citizen-initiated smart city projects (Wolff et al., 2019). People can also participate in community innovation platforms or living labs to find solutions to city problems (Anttiroiko, 2016) although significant challenges have been noted in running such programs.

Smart cities depend on innovation ecosystems to take advantage of emerging technologies, and they themselves function as ecosystems (Fernandez-Anez et al., 2018). Such ecosystems include diverse actors and goals and are notoriously difficult to manage effectively (Ooms et al., 2020), although research has identified conditions for effective ecosystems. For smart cities, this includes having strong business and research sectors, engaged citizens, effective institutions, opportunities for these actors to meet and work together, infrastructure for experimenting, and access to funding mechanisms to develop projects (Fernandez-Anez et al., 2018). Strong ecosystems support the development of digital skills because there is greater awareness of the need for skills, demand for those skills, and more opportunities for people to learn them.

3 Background and Research Questions

Oman Smart City Platform was launched in March 2017 as a collaboration between the Research Council, the Supreme Council for Planning, Muscat Municipality, and the Information Technology Authority. In addition to the four government entities, three private companies (Omantel, Omran, and Nama Holdings) are partners and sponsors. The program arises out of the Digital Oman Strategy of 2003 to transform Oman into a knowledge-based society by developing ICT skills; the electronic provision of government services; developing the local ICT sector; putting in place governance, standards, and regulations for a knowledge society; and developing the national ICT infrastructure. Progress toward these goals has already been observed, with the 2020 e-government survey by the United Nations showing that Oman was, for the first time, ranked as "very high" in the E-Government Development Index (UNDESA, 2020).

The Research Council describes the Oman Smart City Platform as "a knowledge-sharing consortium to drive smart city initiatives in Oman"[1]. The objectives of the program are (1) to provide a knowledge-sharing, collaborative, and networking environment between smart city stakeholders, (2) to facilitate and enable smart city innovations through funding research and carrying out innovation competitions, (3) to promote and test smart solutions to city challenges, (4) to create awareness of the importance of smart city solutions and good practices, and (5) to enable a path toward a unified national smart city vision and strategy.[2]

[1] https://www.trc.gov.om/trcweb/topics/research/programs/6651

[2] http://en.smartoman.om/AboutSmartCityPlatform.aspx

In 2018 the Smart City Platform issued a call for project and research proposals related to smart cities in Oman. In that call, they explain that "we know too little about existing smart solutions in Oman, smart city ecosystems, the challenges and opportunities towards building sustainable smart city, as well as potential solutions" (SCP, 2018:4). In addition to providing funding, Smart City Platform runs hackathons in different cities in Oman[3] and has held a competition for software developers (ONA, 2020). They also arrange events and stakeholder engagements to raise awareness of smart cities.

Oversight of the Smart City Platform is carried out by a steering committee consisting of nominated members from each partner, with the head of the committee on the level of Minister undersecretary. Day-to-day management of the program is done by an executive team who oversees all the activities, including the Smart City Ambassador program. The executive team contracts a small Omani company to manage marketing and social media.

The Oman Smart Ambassadors program, which this chapter examines, arose out of two challenges that the Smart City Platform faced.

Firstly, in order to grow awareness of and skills in smart city technologies, they needed to offer training to government employees across the country as well as to expand awareness among businesses and citizens of the potential of smart technologies. However, the same small pool of experts was being repeatedly drawn on for capacity building around smart cities. This meant it was sometimes difficult and costly to secure them and raised concerns that the knowledge shared was not covering all aspects of the smart city and not up to date. It was also difficult to offer training in different cities.

Secondly, on a more practical level, the members of the Smart City Platform executive team undertake these responsibilities alongside full-time appointments in their organizations. As a result, they needed a team of motivated individuals to drive activities.

In response to this situation, Smart City Platform, in collaboration with Government Innovation (a project of the Information Technology Authority which leads the national project of digital transformation), launched the Smart City Ambassadors program to build a network of volunteer smart city ambassadors.

The program has been supported by a research initiative at the United Nations University Operating Unit on Policy-Driven Electronic Governance (UNU-EGOV) to examine the program's results and relevance as a potential good practice for other countries and contexts. Consequently, this chapter addresses the following two research questions:

1. How did the Smart City Ambassadors program operate and what were the key success factors?
2. How has the program contributed to skills development in Oman?

[3] http://en.smartoman.om

4 Theoretical Framing and Methods

We take a constructivist view of learning, meaning that individuals construct their own knowledge based on what they already know and their social interactions with the external environment (Fosnot, 1996). Constructivism stresses the learning process as individual and social, the role of personal experience, and the self-directed nature of learning (Doolittle & Hicks, 2003). Through this lens, personal reports of learning are a reliable source of information about that learning.

A case study was considered appropriate as we were examining a social process as it unfolded, in a context where we were not able to control events (Yin, 1994). To describe the case, we used desk research that collected available materials about the program. We consulted online records including the main program website and social media records,[4] as well as a survey of the ambassadors. We also made use of news reports and reports by other government agencies about the program activities.[5]

The survey of the Smart City Ambassadors professional community was used to understand the experience of participating in the program. It was conducted online, in May 2020. All ambassadors were invited to participate, and 17 responded which is considered adequate for small-scale case study research (Boddy, 2016). Questions were presented in both English and Arabic, and responses could be in the language of their choice. Most of the responses were in Arabic and were translated to English using Google Translate. The translations were checked for accuracy and interpretation by the second author who is fluent in Arabic.

The respondents included 5 women and 12 men. Two were business owners, 14 were employed, and 1 was a job seeker. Three respondents had each conducted three workshops, five had conducted two workshops, and the remaining nine respondents had conducted one workshop each. Respondents were asked their reasons for getting involved in the program and about the impacts. These responses are analyzed below, referenced by respondent numbers (R1 to R17).

The documents and survey responses were analyzed using qualitative content analysis to identify themes. This was carried out by the first author and the meanings checked with the second author, who was closely involved in the program and in touch with the ambassadors.

[4] Instagram (www.instagram.com/smartcity_oman) and Twitter (https://twitter.com/SmartCity_Oman)

[5] For example, the Oman Research Council (https://www.trc.gov.om/trcweb/topics/research/programs/6651)

5 Results: Program Operation

5.1 *Structure and Activities*

The Smart Cities Ambassadors program was inspired by two of the Sustainable Development Goals, namely, Goal 4: Ensure inclusive and equitable quality education and promote lifelong learning opportunities for all and Goal 11: Make cities and human settlements inclusive, safe, resilient, and sustainable. It was designed to enable the Smart City Platform to deliver on its mandate with limited resources. Although discussed as early as 2017, the program officially launched in November 2019.

The program brings together volunteer ambassadors with expertise in different aspects of smart cities, selected for their profile and proven experience. The ambassadors are recruited from different cities in Oman so as to ensure that sills are widely available.

The program is a collaboration between government agencies and private companies, and the financial model combines private sector sponsorship with voluntary work by the ambassadors. The companies contribute to the funding of program activities, salaries of the executive team, and marketing costs. In return, they receive exposure for their smart solutions and services. The arrangement is mutually beneficial and makes the program sustainable.

The activities of the program include sharing knowledge, sharing smart solutions, exchanging experience, leading innovative projects, and supporting Smart City Platform initiatives. The program started because of the need to train government employees but expanded to include other city stakeholders and communities. Initially, in 2019 and early 2020, the activities focused on knowledge sharing, particularly around technologies. Over time, more projects were initiated, and the focus shifted to sharing experience gained. This evolution was influenced by the COVID-19 pandemic but also resulted from the growing maturity of the program. Projects became important examples and sources of knowledge, and as ambassadors improved their professional standing, their own stories became important examples and inspiration for other ICT professionals.

5.2 *The Role of the Ambassadors*

Smart City Ambassadors is a bottom-up model driven by the volunteer ambassadors who run knowledge-sharing sessions with support from the executive team. Ambassadors propose lectures and workshops based on their expertise and identified demand and prepare appropriate content. Topics were informed by planned activities in government agencies and community requests. Initial workshops focused on smart technologies with topics such as blockchain, big data, Internet of Things, autonomous vehicles, drones, artificial intelligence, machine learning, and 3D printing.

Ambassadors also arrange venues for their sessions. To keep costs down, they approach institutions and organizations to provide venues free of charge. Sessions have been held at universities, colleges, schools, technology centers, training institutions, government facilities, and business premises. Organizations that provide or sponsor a venue are acknowledged in social media and sent a certificate of appreciation. Proposed sessions and arrangements are reviewed and approved by the executive team who initiates marketing and manages registrations.

The program aimed to balance quality content with inclusiveness, and so recruitment was broad. Academic experts, researchers, industry experts, consultants, business owners, and students were invited to participate, particularly those with appropriate qualifications and experience. The broad approach to participation paid off. For example, a student who graduated in 2019 joined with the necessary formal training in her field, but no experience of presenting and sharing knowledge. Although she struggled at first, she grew in confidence and went on to become a well-known ambassador, contributing her expertise (in blockchain) to the program and to projects.

The ambassadors are volunteers who receive no payment. Since the program depends on volunteers, it is important to understand why they participate as this would be key to replicating the program. For most, the motive for participating was a sense of community responsibility (11 respondents), followed by the desire to exchange knowledge and experiences (7 respondents). Other motivations mentioned were career development and skills development and to spread awareness and culture regarding technology (one respondent each).[6] When the program started, some academics questioned the lack of payment. However, once the program was operational and they saw that the ambassadors got exposure on television, radio, and social media, they were willing to volunteer.

5.3 Developments since COVID-19

As the COVID-19 pandemic unfolded in early 2020, face-to-face events became impossible and the initial Smart City Ambassadors program threatened to stall. The team responded in two ways. First, they organized online events and switched the knowledge-sharing sessions to online. This maintained the momentum and kept ambassadors engaged and committed to the project. One of the formats adopted was virtual "knowledge cafes" where people could share experiences of smart projects.

Second, they launched the "Smart City Ambassadors Professional Community." This initiative provided the ambassadors with a hub through which to exchange knowledge and experiences. Making use of WhatsApp as the primary communi-

[6] Some respondents cited more than one motivation.

cation tool, the ambassadors use the professional community to share news and information about smart technologies, projects, and opportunities.

Ambassadors are recruited from cities around Oman in order to address the need to develop expertise across the country. In the first month of operations, activities were held in three cities, but the pool of ambassadors quickly expanded. There are currently more than 300 ambassadors registered from across the country. With the advent of the professional community, people from neighboring countries, including Qatar, Saudi Arabia, and Egypt, have joined.

5.4 Key Success Factors

The Smart City Ambassadors program appears easy to replicate, particularly given the minimal infrastructure and resourcing needed. The elements that were found to have been important for the success of the program include:

1. High-level support
2. A clearly defined executive structure
3. A clear set of goals with measurable targets
4. Careful targeting of topics to local needs
5. A small and effective communications team
6. A pool of expertise in universities, research institutes, and businesses to draw on
7. A culture of community service and goodwill among the experts

While point 6, the existence of experts, may appear to be an obstacle (as indeed it did at the beginning of this program), the program was able to find these experts by using an inclusive approach to recruitment to develop the skills of willing volunteers. The lesson learned is that much expertise is hidden or nascent and could be made available. The program identified and surfaced that expertise.

6 Result: Skills Development

6.1 The Ambassador Database

The program now has a database of over 300 smart city experts in Oman from different sectors with expertise in a wide range of technologies and application domains including government, private sector, small-medium enterprises, and agriculture. This national resource is widely used and is the most significant outcome of the program. Organizations regularly request the contact details of ambassadors with specific qualifications or with expertise in particular smart city technologies or domains.

The ambassadors are enthusiastic, identifying program improvements such as clearly identifying the level of workshops (R5); unifying the program effectively with common virtual platforms, slide templates, and themes (R12); and expanding the topics covered and archiving the courses delivered (R13). Some ambassadors asked that the Smart City Platform team communicate with their employers to negotiate more time to devote to the program (R6, R9).

6.2 Knowledge Sharing

When the Smart City Ambassadors program was initiated, the goals were to recruit ambassadors, run events in all central states of Oman, and increase public awareness of and engagement with the Smart City Platform. These goals were supported by measurable targets. Between the launch in November 2019 and the end of April 2020, 44 lectures and workshops were conducted with a total of 3052 participants from 11 cities attending. The targets that had been set for the first year were exceeded after only 4 months of operation.

The program has resulted in greater knowledge and awareness of smart technologies and their potential and pitfalls. It has also developed specific technical skills. For example, two technical workshops were conducted on the Internet of Things. Ambassadors received positive feedback on the sessions (R14, R16, R17) and reported great enthusiasm among the attendees with many returning for subsequent sessions. While those that attended have not been surveyed, the popularity of the program suggests that attendees found the sessions worthwhile.

Many of these knowledge-sharing events focused on developing awareness and skills related to technologies and data, both at a technical level and in terms of their applications in government and business.

6.3 Empowering the Ambassadors

Ambassadors report personal benefits from the program. They have developed their own skills as they prepared content and engaged with and learned from workshop participants (R1, R5, R10, R11, R17). They also developed skills in presentation and knowledge sharing (R12) and in photography and video editing (R11) and experienced unspecified self-development (R1, R17) and the satisfaction of serving the community (R12, R13).

One ambassador explained that the program provided a platform to showcase their knowledge (R6), and as a result, they have become known in their communities and have been given opportunities to consult on projects (R3, R11, R17). They have also been given opportunities to collaborate in projects including implementing blockchain and drones in business, as well as projects in municipalities and agriculture.

The ambassadors say that they learned from each other (R1, R5, R10, R11, R13) and from workshop participants (5, R10, R13) about aspects of smart cities and were able to identify and submit project proposals (R11). One respondent explained that the experience provided valuable insights into the levels of understanding and acceptance of smart technologies within the community (R11) and hence what solutions might be accepted and adopted.

The program has thus developed many of the community skills described above, including understanding problems, communication, and developing solutions. Over time, the program has created a community of experts that exchanges innovative ideas and how to implement them.

6.4 Developing an Ecosystem

The program has developed a number of aspects of the smart city ecosystem in Oman.

Firstly, the database of ambassadors and their areas of expertise has become a useful resource. Ambassadors are in demand to run tailored training sessions for organizations and have been invited to participate in high-level meetings with regional and international organizations. The program has increased the visibility of research and development activities through the database which includes researchers working in smart education, health, and government.

Ecosystems can be influenced by judicious governance, and government can serve to bring the actors in an ecosystem together (Ooms et al., 2020). This program connects people across sectors, bringing together smart city experts, government employees, the organizations that host events, researchers, the private sector, and individuals. Knowledge cafes, competitions, and hackathons also bring different stakeholders together, confirming the view of David and McNutt (2019) that skills can be developed through volunteers.

The program created a conversation around smart cities, raising awareness and enriching the ecosystem (Fernandez-Anez et al., 2018) for smart city development in Oman. In addition, the Smart City Platform team gained experience and expertise in the process of managing the Smart City Ambassadors project. While the team members had previous experience, they also learned from high-level experts participating in the program who helped to establish a systematic method for running activities.

6.5 Impact on Digital Transformation

The ambassadors impact innovation in government through specific interventions for government entities as well as the attendance of government employees at other events. Several municipal projects have been initiated that draw on the ambassador's expertise as well as local businesses. The ambassadors have become change champions supporting government innovation in cities around Oman.

The influence of the ambassadors extends beyond smart cities to general ICT expertise as well. More recently, the program is being integrated with the Oman IT Society, as the value of the intervention becomes apparent beyond the initial smart city focus.

The impact on participants has also extended to the school level. One woman reported that "two students from the 11th grade" attended a session she ran and that "the knowledge of the Internet of Things opened wide horizons for them and helped them to form an idea about their future specializations" in higher education (R5). At a College of Technology, the academic head of technology knowledge was inspired by the program to create a similar initiative among students.

7 Replication and Scalability

This approach to developing skills appears relatively easy to replicate and is suited to any organization or city that has limited resources to develop skills for digital transformation. In particular, it provides an innovative way to develop capacity among government employees. The Smart City Ambassadors added value to the Smart City Platform team, enabling it to scale activities and increase its geographic reach. It also enabled the program to expand into new activities and widen its objectives. As such, this kind of program could complement other smart city initiatives by adding a capacity development dimension.

Funding is needed to cover events, staff costs, and marketing. In this case, the program was funded by private sector partners, but it could be funded out of a training or development budget if that exists. Similarly, marketing might be handled by an in-house marketing department. Replicability would depend on the context and would in particular need a source of community-minded volunteers available to act as ambassadors.

The program has proven relatively easy to scale. First, when events were face-to-face, the program exceeded expectations as a result of the enthusiasm of ambassadors. Second, the shift to virtual events enabled greater participation and reach. Finally, the launch of the Smart City Ambassadors Professional Community has increased the number of ambassadors involved as well as the range of activities they engage in.

The experience of the Oman Smart City Ambassadors program has been shared as good practice in building capacity regionally and internationally at the Arab Regional Innovation Forum and the International Telecommunications Union (ITU) Global Innovation Forum. It was one of the winners in the ITU 2020 Innovation Challenge, in the category Ecosystem Best Practice.

8 Limitations and Future Research

This chapter describes the case of a program implemented in Oman to address the specific needs of the Smart City Platform for a wider pool of trainers and support for the work of the executive team. As such, the results are specific to the context and cannot be assumed to apply in other contexts. While we have identified key factors of the success of this program, it would be interesting to compare these with similar programs or programs with similar objectives.

9 Conclusion and Recommendations

The Oman Smart City Ambassadors was launched in response to the challenge of developing smart city skills among government employees, as well as raising awareness of smart city technologies among businesses and citizens. It harnessed existing pockets of skill within the country and drew on the goodwill and community-mindedness of individuals with smart competencies who were willing to freely share their expertise. The project has resulted in a database of more than 300 ambassadors.

With private sector sponsorship covering limited costs, the program was able to deliver events, lectures, and workshops reaching more than 3000 participants before the COVID-19 pandemic halted face-to-face engagements early in 2020. At that point, the program was forced to change to online events. The focus also shifted to creating a virtual professional community, providing more technical training and positioning the ambassadors as expert advisors for smart city initiatives.

The program created a pool of expertise, benefitting the volunteer ambassadors by developing their own skills, professional standing, and contacts. As a result, the Oman Smart City Platform, despite limited resources, had a greater impact than would otherwise have been possible. The ideas and technologies of smart cities have been widely publicized, and the ambassadors are now a national resource, drawn on by business and government to support smart city initiatives. The program is consolidating this pool of expertise as a professional community of practice which will be a cornerstone in the ecosystem needed to support a smart future for Oman.

Given the low resource base needed to launch and sustain the Smart City Ambassadors, it is a model that could be replicated in other settings. Seven key elements were identified that contribute to the success of the program. Most importantly, the program relies on having a base of smart city expertise in universities, research institutes, and businesses, and these people are willing and able to volunteer their time and energy. In the case of Oman, a strong culture of community service and goodwill made the program possible.

This model does not replace traditional forms of training, particularly high-level technical training, although it can supplement it. Rather, it raises awareness of the power of smart technologies. By making young people aware of these technologies, it increases their interest in pursuing technical studies. Raising awareness within government entities and businesses of the potential of smart technologies raises the demand for smart solutions, creating jobs and supporting the development of a thriving ICT sector. Consequently, such a program can add value to the ecosystem needed to support digital transformation and sustainable smart cities.

The real innovation in the program is the coordinating role played by the Smart City Platform, bringing together stakeholders across the country to share knowledge and build skills. The program illustrates very well the powerful role that government and private sector partners, working together, can play in developing a comprehensive smart ecosystem. That the collaboration owes its success more to human energy and goodwill, and less to physical and financial resources, makes it worth further exploration.

References

Al Harthy, K., Puteh, H. F., & Al Harthy, K. (2019). Investigates review of leadership challenges in a smart environment. In *Proceedings of the 4th MEC international conference on big data and smart city (ICBDSC), Muscat, Oman*, Jan 15–16 (pp. 135–138). Retrieved from https://ieeexplore.ieee.org/abstract/document/8645566

Anttiroiko, A. V. (2016). City-as-a-platform: The rise of participatory innovation platforms in Finnish cities. *Sustainability, 8*(9), 922. https://doi.org/10.3390/su8090922

Bakhshi, H., Downing, J. M., Osborne, M. A., & Schneider, P. (2017). *The future of skills, employment in 2030*. Pearson and Nesta. https://futureskills.pearson.com/research/assets/pdfs/technical-report.pdf

Boddy, C. R. (2016). Sample size for qualitative research. *Qualitative Market Research, 19*(4), 426–432. https://doi.org/10.1108/QMR-06-2016-0053

Curseu, P. L., Semeijn, J. H., & Nikolova, I. (2021). Career challenges in smart cities: A sociotechnical systems view on sustainable careers. *Human Relations, 74*(5), 656–677. Special issue, Article: 0018726720949925.

David, N., & McNutt, J. (2019). Building a workforce for smart city governance: Challenges and opportunities for the planning and administrative professions. *Informatics-Basel, 6*(4), 47.

Doolittle, P. E., & Hicks, D. (2003). Constructivism as a theoretical foundation for the use of technology in social studies. *Theory and Research in Social Education, 31*(1), 72–104. https://doi.org/10.1080/00933104.2003.10473216

European Commission. (2020). *European skills/competences, qualifications and occupations*. Retrieved from https://ec.europa.eu/esco/portal/home?resetLanguage=true&newLanguage=en

Faraji, S. J., Nozar, M. J., & Arash, M. (2021). The analysis of smart governance scenarios of the urban culture in multicultural cities based on two concepts of "cultural intelligence" and "smart governance". *GeoJournal, 86*(1), 357–377. https://doi.org/10.1007/s10708-019-10074-6

Fernandez-Anez, V., Fernandez-Guell, J. M., & Giffinger, R. (2018). Smart City implementation and discourses: An integrated conceptual model. The case of Vienna. *Cities, 78*, 4–16.

Fosnot, C. T. (1996). *Constructivism: Theory, perspectives, and practice*. Teachers College Press.

Iatrellis, O., Panagiotakopoulos, T., Gerogiannis, V. C., Fitsilis, P., & Kameas, A. (2021). Cloud computing and semantic web technologies for ubiquitous management of smart cities-related competences. *Education and Information Technologies, 26*(2), 2143–2164. https://doi.org/10.1007/s10639-020-10351-9

Jaskiewicz, T., Mulder, I., Morelli, N., & Pedersen, J. S. (2019). Hacking the hackathon format to empower citizens in outsmarting "smart" cities. *Interaction Design and Architectures, 43*, 8–29, Special Issue SI.

Jiang, Y., Qian, Q., Zhang, X., & Chen, Y. (2019). The formation of government-oriented creative community and its driving mechanisms: A case study of the 39° space art creative community in Foshan, China. *Sustainability, 11*(3), 1–20. https://doi.org/10.3390/su11030625

Kranjac, M., Tomic, S., Salom, J., Ilic-Kosanovic, T., Ilic, D., & Bulajic, S. (2021). New methodology of teaching a smart city concept: A case study. *International Journal of Engineering Education, 37*(1), 204–214.

Kritzinger, E. (2017). Cultivating a cyber-safety culture among school learners in South Africa. *Africa Education Review, 14*(1), 22–41. https://doi.org/10.1080/18146627.2016.1224561

Madureira, C. R., Amorim, M., Rodrigues, M., Ferreira-Dias, M., Souza, A., & Lucas, M. (2020). The digital transformation in Aveiro region (Portugal): Formalize full skillset development through specialized training. In *Proceedings of the 14th international technology, education and development conference (INTED)*, Valencia, Spain, March 2–4 (pp. 8698–8703).

Manjrekat, S. M., & Deshmukh, M. (2020). Exploring societal participation in making cities smart. *E3S Web of Conferences, 170*, 06011. https://doi.org/10.1051/e3sconf/202017006011

Michelucci, F. V., De Marco, A., & Tanda, A. (2016). Defining the role of the Smart City manager: An analysis of responsibilities and skills. *Journal of Urban Technology, 23*(3), 23–42. https://doi.org/10.1080/10630732.2016.1164439

Mukhametov, D. (2019). Smart city: From the metaphor of urban development to innovative city management. *TEM Journal, 8*(4), 1247–1251. https://doi.org/10.18421/TEM84-20

ONA (2020). *Smart city platform launches competition to create smart application to monitor people infected with COVID-19*. Oman News Agency 31-03-2020 03:13 PM. Retrieved July 17, 2021, from https://omannews.gov.om/NewsDescription/ArtMID/392/ArticleID/10959/Smart-City-Platform-Launches-Competition-to-Create-Smart-Application-to-Monitor-People-Infected-with-COVID-19

Ooms, W., Caniels, M. C., Roijakkers, N., & Cobben, D. (2020). Ecosystems for smart cities: Tracing the evolution of governance structure in a dutch smart city initiative. *International Entrepreneurship and Management Journal, 2020*(16), 1225–1258. https://doi.org/10.1007/s11365-020-00640-7

Perng, S.-Y., & Maalsen, S. (2020). Civic infrastructure and the appropriation of the corporate smart city. *Annals of the American Association of Geographers, 110*(Special Issue 2), 507–515.

SCP. (2018). *Surveying and developing Smart city solutions in Oman*. Smart city platform call for proposals. Oman Smart City Platform. Downloaded July 2021 from https://www.trc.gov.om/trcweb/topics/research/programs/6651

UNCTAD. (2016). Smart cities and infrastructure report of the secretary-general economic and social council. *United Nations Conference on Trade and Development, 03074*, 18. https://doi.org/10.1017/S0020818300006640

UNDESA. (2020). *E-government survey 2020: Digital government in the decade of action for sustainable development*. United Nations Department of Economic and Social Affairs. Retrieved from https://publicadministration.un.org/egovvkb/es-us/Reports/UN-E-Government-Survey-2020

Wenger, E. (1998). *Communities of practice: Learning, meaning, and identity*. Cambridge University Press.

Wolff, A., Gooch, D., Cavero, J., Rashid, U., & Kortuem, G. (2019). Removing barriers for citizen participation to urban innovation. In M. de Lange & M. de Waal (Eds.), *The hackable city*. Springer. https://doi.org/10.1007/978-981-13-2694-3_8

Yin, R. K. (1994). *Case study research: Design and methods*. Sage.

At the Root of the Smart Cities: Smart Learning Ecosystems to Train Smart Citizens

Carlo Giovannella

1 Which Smart City?

The use of the adjective *smart* referred to an ecosystem, and in particular to a *city*, is predominantly associated with a more or less intelligent infrastructural backbone whose purpose is the fluidification of both material and immaterial flows (goods, people, data, etc.) in order to optimize processes and the use of resources, time included (n.b. sometimes also intellectual and social capitals are included as part of the city resources) (Lee et al., 2008; Deakin & Al Waer, 2011; Giffinger & Gudrun, 2010; Anonymous, n.d.-a). This vision of a smart city, usually, is also associated to models of territorial development like the one based on six soft factors called also "six pillars" (smart economy, smart people, smart government, smart mobility, smart living, and smart environment) (Hollands, 2008) or like the so-called triple helix model that considers university, industry, and government as engines of the territorial innovation and growth (Ezkowitz, 2008; Leydensdorff & Deakin, 2011; Kourtit et al., 2013). Both models are deemed to provide methods to determine the smartness of a city. Indeed, most of the attempts to bench-mark smart cities and produce smart city rankings, by means of a top-down approach, have been based on the six pillar model (Giffinger et al., 2007). Such rankings, however, should not be taken very seriously because, as we have shown in the past (Giovannella et al., 2014; Giovannella, 2013), they are affected by a significant number of problems: (a) the choice of factors that are not always very meaningful; (b) strong correlations among factors (usually not investigated); and (c) the scarce consideration for the individuals—their opinions are never collected. In alternative we can consider a

C. Giovannella (✉)
Association for Smart Learning Ecosystems and Regional Development—ASLERD, Rome, Italy

University of Rome Tor Vergata—Dip. SPFS, Rome, Italy
e-mail: gvncrl00@uniroma2.it

P. Fitsilis (ed.), *Building on Smart Cities Skills and Competences*, Internet of Things,
https://doi.org/10.1007/978-3-030-97818-1_13

different perspective—that we have defined as *people centered* (Giovannella et al., 2014, 2013; Giovannella, 2013; The origin of the "Person in Place Centered Design" vision is documented in Giovannella C, 2008) and others *human centered* (The Human Smart Cities Cookbook, 2014; Concilio & Rizzo, 2016)—where the citizen is at the center of the city and, at large, of the territory of reference. Accordingly to this alternative perspective, an ecosystem, like a city, is smart if "the individuals living and operating in the ecosystem own not only competences and a skill level adequate to carry on the activities in which they are involved, but are also strongly motivated by continuous and adequate challenges, while their primary needs are reasonably satisfied" (Giovannella et al., 2014). Such alternative vision of smart cities, although not the dominant one, continues to represent, still nowadays, the conceptual basis for numerous initiatives (i.e., Anonymous, n.d.-b, n.d.-c, n.d.-d, n.d.-e).

The smartness of the ecosystems, thus, can be represented by a multidimensional construct (Giovannella, 2014, 2015) (see Fig. 1) that can be obtained by integrating the Maslow's pyramid of needs (Maslow, 1943) with the concept of *flow* (Czisikszentmihalyi, 1990) applied to an ecosystem. It represents the measure of the well-being reached by the various categories of actors involved in the processes implemented within a given territory of reference and should be co-evaluated by all actors, through a bottom-up participatory process. Such participatory evaluation, when deemed relevant, can be integrated with a more top-down traditional one, based on predefined quantitative parameters, provided that the problems discussed in (Giovannella et al., 2014) are taken carefully in consideration. It is interesting to note that in this alternative vision the word *resilience* should not be understood, as done by the majority, as a synonym of resistance or the ability of an ecosystem to return to an initial state (possibly of apparent equilibrium) but, rather, as a dynamic quality associated to the ability of an ecosystem to maintain its trajectory toward the development of a higher degree of smartness.

In a scenario in which the smart city is *people centered*, all citizens must consider themselves as an active agent capable to contribute to the process that leads the ecosystem toward a progressive increase of its smartness, an increase that would not be possible without *smart citizens*. In such *people-centered* scenario, being smart citizen does not simply mean being able to get used to and use at best the technological backbone of the city but to participate consciously in the development of the city smartness and coevolve with the ecosystem. To be a smart citizen means also to be a competent citizen, i.e., not only *to know* and not only *to do* but, rather, *to be*. In other components, *knowledge* and *skills* should serve as basis for the development of *competences*. The identification of the competences needed to participate in the evolutionary process of the cities requires, necessarily, a critical analysis of the existing frameworks of competence and the identification of an adequate *space of competences* that can be taken as a reference.

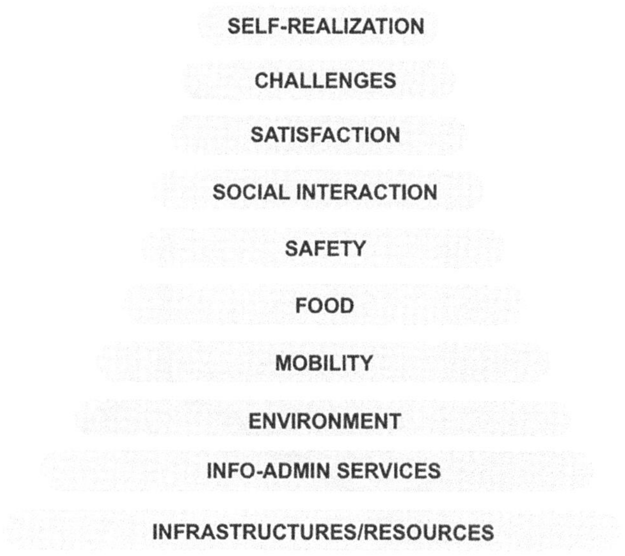

Fig. 1 The dimensions that contribute to define the *smartness* of an ecosystem

2 The Smart Citizens' Space of Competencies

Looking for the most suitable space of competences for a smart citizen, we cannot restrict ourselves exclusively to a given set of digital skills, but we should enlarge our perspective and consider a wider space such that of Fig. 2, represented by three meta axes, each corresponding to a subspace of competences that can interact with the other two.

The axis of the hard competences represents the subspace of the specialized competences that everyone must develop to fulfill a professional/specialized role and therefore, since it has to be contextualized case by case, will not be covered in this article. A second axis is that of the transversal or soft individual competences that can be described, as we will see in the following, by re-elaborating the *LIFE skills* framework (Anonymous, n.d.-f, n.d.-g, n.d.-h). Finally, the third axis is that of the soft digital competences that could be relevant for each individual/citizen. The specialized ICT competences are not included in this subspace because they fall within the subspace of the hard skills and are relevant only for professional roles. Soft digital competences cannot be considered as a separated and independent subspace because, actually, as we will understand better later, their main function is to strengthen the *LIFE skills* within a digitally enhanced context.

In Fig. 3, we show a matrix that organizes the "LIFE skills" (Giovannella, 2016) in three main macro-areas and in a certain number of subareas. The three main areas refer to (a) the competences that can be mobilized at individual level, some of which have been aggregated around particularly relevant competences—such as problem

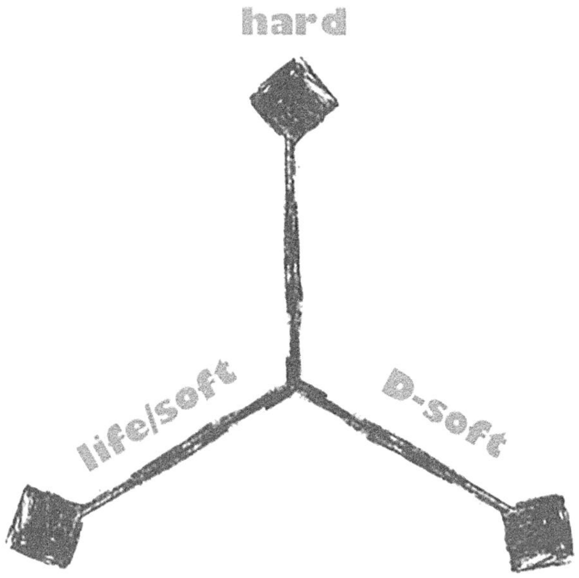

Fig. 2 Meta representation of the complete space of competences of reference

setting and problem-solving—because these latter could represent also phases of a process; (b) the socio-relational competences that are mobilized in the relationship with other individuals and/or with contexts; and (c) the competences that emerge during the development of processes, i.e., the activities that are carried on by the individuals and communities within a given ecosystem.

At this point it is also worthwhile to underline how a competence differs from a skill: the latter concerns the ability *to do* and reproduce procedures, while in the first case a full autonomy is required together with the ability to evaluate the border conditions and mobilize the competence even in unknown situations. It is only the competence that allows individuals to produce new *objects of knowledge* that have the possibility to become part of our *cultural DNA*.

Most of the competences listed in the matrix of Fig. 3 contribute to what we can call *design literacy*, and this latter is fundamental in all participatory processes and for the development of a proactive approach both as an individual and as a citizen.

Once that the competences of Fig. 3 are fully developed, we expect that the *design literacy* could undergo a transformation to become *meta-design literacy*, i.e., enriching itself with the ability to modify and adapt on-fly the processes, taking into account possible evolution of the border conditions and of the needs/expectations with the aim to optimize the outcomes of the processes.

On a robust framework like the one of Fig. 3—composed by individual, socio-relational, and procedural-managerial competences—we can graft what we have defined *soft-digital competencies* with the aim to support "the confident, critical, and creative use of ICT to achieve goals related to work, learning, leisure, inclusion, and/or participation in society" by individuals/citizens. The most popular framework of reference for the soft digital competencies is the DigiComp 2.1 (Carretero Gomez

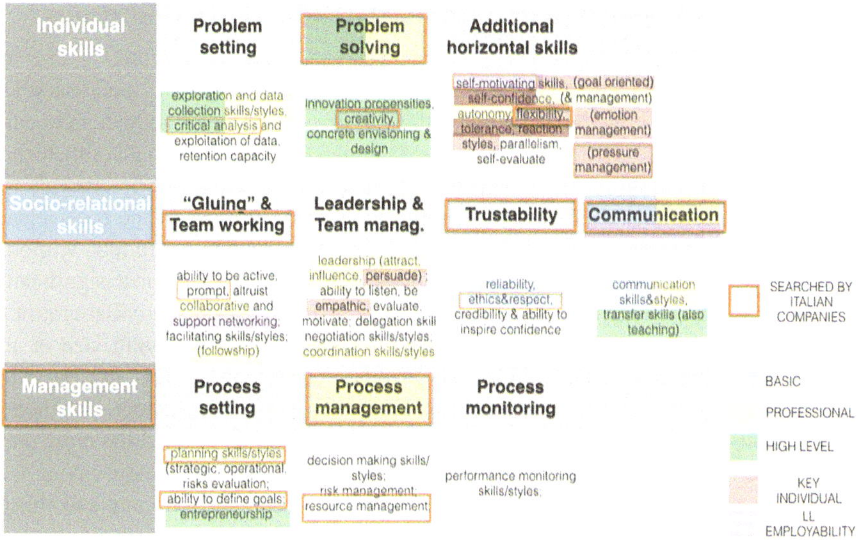

Fig. 3 Taxonomic table of essential LIFE skills organized by macro-areas—individual, socio-relational, and management skills—and micro areas (e.g., problem setting, leadership and team management, process monitoring, etc.). The colored backgrounds highlight those LIFE skills—basic, professional, high-level, key individual, and LL employability—considered relevant for employability by ref. (Chatterton & Rebbeck, 2015). The orange frames, on the other hand, identify the LIFE skills required at most by Italian companies (Anonymous, 2019)

et al., 2017). Basically, it is a framework that describes the competences needed to (a) take advantage of the potentialities of the web considered as an information and communication amplifier; (b) expand the individual propensity for collaborative activities and digital citizenship, by respecting the others; (c) strengthen the ability in the production of digital content (from consumer to prosumer); (d) increase the awareness about safety issues; and, finally, (e) develop further the problem-solving ability taking advantage of digital resources and tools. As a warning, we would like to note that some aspects of the DigiComp 2.1 tend to cross the border with the subspace of the hard digital skills—as in the case of programming and coding skills—but, as stated above, they are not considered here because they are not relevant for the aim of this paper.

The key point to take away is that the soft digital competences are not competences that one can develop on their own from scratch but they need to grow on a fertile ground, for example, that of the previously described "LIFE skills" that, thus, become digitally amplified.

3 Interdependencies and Technology-Induced Modifications

To refine the picture emerged from the previous paragraph, it is necessary to elaborate further on how the elements of the various subspaces of competences can interact and influence each other. Take, for example, the case of digital competence in finding information on the web. It is very clear that it would not be possible to take advantage of this competence without the mastery of most of the transversal competences belonging to the problem setting subarea (see Fig. 3). In fact, without an adequate critical ability, one would not be able to select the most significant information, and it would be impossible to exploit it at best during collaborations or communication exchanges, because one would not be able to synthesize in an original way the information collected. In other words, one would not be able to provide a personalized synthesis capable to emerge with respect to others' elaborations produced by the representatives of what can be defined the *copy and paste generation*. Elaborations that unluckily tend to increase the background noise and make more difficult to distinguish between competent individuals and those who own simply basic knowledge and skills.

The transversal competences become even more relevant in highly technological contexts, like the smart ecosystems, in which the production of data is continuously growing at an impressive rate. The competence in knowing how to collect, analyze, and read data becomes fundamental in order not to delegate to others the reading and interpretation of the reality and, at the same time, to contribute as an active actor to the development of one's own context of reference. With the development of the big data sector, in fact, it becomes essential to develop, both as a citizen and individual, a minimum level of *data literacy*, although specialized data analysis has to be delegated, anyway, to professionals owning adequate hard competences.

Alongside interdependencies among competences, another phenomenon that must be taken into due account is the transformation of competences and skills that may be induced by the advent of new technologies and in particular by AI. To fully understand the relevance of this issue, let's take two situations that might have been experienced by most of the readers.

The first one concerns the ability to write: with the advent of electronic writing and the use of keyboards, many of us tend to lose naturalness in writing with the pen, to the point that sometimes writing with accurate handwriting may require a greater effort of concentration than in the past.

The second one concerns the ability to orient oneself, for example, within a city. Until the advent of navigators, the usual habit was the progressive development and memorization of a map of the territory in our mind. Today the easiness with which one is guided by the navigator slows down and reduce the extent of the process of memorization, to the point that many individuals tend to feel lost without a navigator and loose familiarity with the use of paper maps.

Most likely are the premillennial generations, rather than by the millennials, that are more aware of such ongoing transformations.

During the evolution of our civilization, we have witnessed continuous evolutions of human skills—e.g., most of us no longer know how to ride a horse, but we know how to drive a car—but the second example highlights the delicate boundary and balance existing between the transformation of skills and competences (especially hard ones) and the delegation of one own skills to external artificial intelligences. For sure, it is not be possible for individuals to analyze the infinity of resources available on the web—this is why search engines have been designed—but at the same time no individual can give up to critically analyze the list of results ranked by a search engine or by an aggregator and, as well, to try to identify the factors that have been used to draw up any ranking (just think to the more trivial example of the smart cities ranking we have referred to in the first section).

A corollary to the above considerations is the concerns about the aging of the individuals. Each person represents on the one hand a historical memory while on the other hand, like any machine, becomes obsolete with respect to contexts populated by artificial superstructures and technologically augmented processes that continuously evolve. In very broad terms, this implies the need for a continuous retraining of the individuals (also in terms of competences) and for a continuous transformation of their role with the age, especially if we consider the consistent lengthening of the average human life and the need to support as much as possible active aging.

4 At the Origin of the Smart Citizens: Smart Learning Ecosystems

The next step is to provide an answer to the following question: from where the smartness of the citizens should origin?

The answer to this question is, all in all, simple (although the corresponding actions are not so easy to implement): from schools, considered as engines of *smart learning ecosystems* centered on students, whose growth should be supported in a participatory manner by all territorial stakeholders; the learning ecosystems, moreover, are expected to foster territorial development and social innovation.

The concept of a *learning ecosystem* from its definition (Anonymous, n.d.-i) is gaining ground and momentum (Anonymous, n.d.-j) as well as that of a territorial community. It is no coincidence that during the pandemic the Italian government gave impetus to a new tool that can be seen as the basis for the establishment and development of learning ecosystems: the so-called territorial pact (Anonymous, n.d.-k). The territorial pacts—which during the pandemic were sustained by the search for spaces and resources useful to counteract the confinement and the social distancing imposed by the lockdown—have actually their main goal in the activation of local communities to find shared solutions to serious problems that affect, for example, the most disadvantaged areas of a given territory: such as the dropout at school, the high delinquency rate, etc.

However, looking forward, beyond the first intentions, the pacts can also be seen as a tool to foster the achievement of the excellence, to spread the culture of active and digital citizenship, and to support SDGs (Anonymous, n.d.-l), social innovation and territorial development, and, therefore, the increase of smartness of territories and cities. They are the right tool to stimulate the sensitivity of communities and involve them in the processes of co-evaluation, codesign, and co-responsibility, with the aim to support the growth of students' competences and, as well, those of other citizens belonging to such communities, in a coevolutionary perspective.

The development of the competences and the skills needed to become smart citizen is fully aligned with the objectives of the territorial pacts although it requires, obviously, the design and the delivery of specific training activities that ideally should be integrated with those foreseen by the standard curricula dedicated to the development of the basic skills. Among the possible actions, already included in the regulatory frameworks of many European countries (Anonymous, n.d.-m, n.d.-n; Chatzichristou et al., 2014; Scuola, 2014), there are the *dual education* and the so-called alternation schemes aimed at allowing young people to get more familiar with job environments and working processes and, as well, at making them reflect on the challenges posed by either their home territory or the sustainable development goals (SDGs) at global level.

Taking advantage of the opportunity offered by the alternation schemes, in the recent past, we have designed training activities based on the *simulation of innovation processes* for the students of the last 3 years of few vocational and high schools (Giovannella, 2017, 2021a, 2021b; Giovannella et al., 2018). During such alternation schemes, the students got in contact with all phases of a design process (problem setting, problem-solving, communication of innovation, and, in the case of the last year students, prototyping). Due to the specific nature of the innovation processes, the students were able to develop an adequate set of skills (which in some cases have turned into truly competences) and to experiment themselves in solving social problems and, sometime, in enhancing the level of smartness of their own territory and city.

In Fig. 4 are shown the logos of some applications for which the students have designed a preliminary project and, sometime, low profile prototypes (see Fig. 5). In many cases, the projects were based on the outcomes of processes of participatory evaluation, as in the case of projects aimed at improving the processes put in place by the learning ecosystem (the school itself). In all the other cases, the projects were inspired by an accurate analysis of the territorial ecosystem and from the identification of problems of interest to the whole community.

In the last 2 years, then, due to the pandemic, the processes have been carried out in *smart working* and, thus, from the working point of view, have been enriched by an additional experiential dimension. As one may expect, it allowed also to foster a more robust development of soft and hard digital skills in the students—i.e., the future citizens. Details on the design and analyses of the alternation schemes are available in Ref. (Giovannella, 2021a, 2021b).

Finally, during this last year, to provide more relevance to the training process experienced by the students and to enrich it with an element of greater concreteness

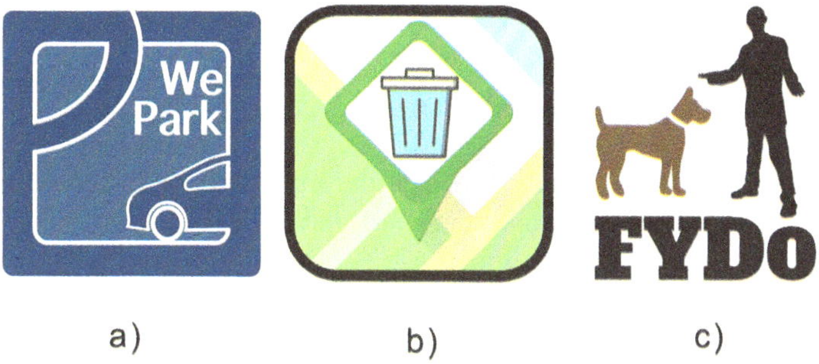

Fig. 4 Examples of logos: (**a**) WePark, an app intended to support the social parking; (**b**) Cestinami, an app intended to support the separate waste collection; (**c**) FYDo (Find Your Dog) an app intended to support the identification of the most suitable faithful friend among those abandoned in a kennel

Fig. 5 Example of navigation tree for the app Cestinami (low profile prototype)

for the future of the students, also as digital citizens, we have introduced a voluntary process of certification of the *role* and of the *competences* developed during the alternation scheme. It led to the release of a blockchain anchored e-certificate, i.e., what we may define an *e-currency* that the students can spend in following up of their studies and/or during a job-placement phase.

5 Final Thoughts and "Healthy" Pessimism

It is worth noting that this kind of training activities and processes requires continuous adjustments which, year after year, lead to a better identification and to the strengthening of the areas in which the smart citizens of tomorrow can be trained. This continuous improvement must necessarily go hand in hand with a parallel cultural change that is impossible to achieve in a few months. It is also very important to stress that the "best practices" should be disseminated to favor their spread out within the territory of reference, and to other territories, and their assimilation as part of the cultural baggage of the learning ecosystems of a city or of wider territories. All this would imply a commitment also at political level and a related investment of resources, which is not easy to obtain. For this reason we wish to stress here, again, that it is not difficult to find an answer to the question posed at the beginning of the previous section, but it is very complicated to start and keep growing actions that can lead to a truly cultural change of a community.

Because of this, as a corollary to above reflections dedicated to the training of future smart citizens, we want to spend few words on the need to train also the political and managerial classes, i.e., those categories that can provide guidelines and resources to drive and sustain a cultural change.

In the past we have also designed and delivered a university master's degree in *Design of People-Centered Smart Cities* (Anonymous, n.d.-o). What have been the results of this experience? Double-face ones. Extremely positive as regards the few executives who motu proprio have taken steps to ensure that their administrations could support them in this process of re-skilling. It is totally insignificant if we consider the large audience that could have been attracted by the initiative. We think that this is a result of two facts: (a) the dominant vision of smart cities that is not a people-centered one and (b) the interest of politicians in achieving results that can be spent on a short term rather than investing in initiatives whose outcomes would be visible on a medium or long term and that, thus, will be probably "monetized" by others who will come after them.

To conclude, the main message is that the theoretical grounds, conceptual tools, and even some best practices—which are by large more numerous with respect to the few examples presented in this article—are available, but the biggest obstacles for citizens to become smart and, thus, for cities to start a resilient path toward a progressive increase of their smartness are overall cultural and economic. Indeed, one would need the future vision of veritable statesmen joined to a high level of corporate social responsibility by veritable entrepreneurs, especially those involved in the development of "smart cities."

The alternative would be the affirmation of "smart cities" where the lever of economic interests will support the optimization of flows and resource consumption and where the citizen will grow digitally only as function of the satisfaction of their real or induced needs, as it happened in the last decade with the use of the "social networks" that are now absorbing a large part of the time and potentialities of the youngest: a step forward Matrix (Anonymous, n.d.-p).

References

Anonymous. (2019). Unindustria, Lazio, Italy. Private communication.

Anonymous (n.d.-a). Retrieved August 2021, from https://ec.europa.eu/info/eu-regional-and-urban-development/topics/cities-and-urban-development/city-initiatives/smart-cities_en

Anonymous (n.d.-b). Retrieved August 2021, from https://www.nesta.org.uk/feature/10-people-centred-smart-city-initiatives/

Anonymous (n.d.-c). Retrieved August 2021, from https://unhabitat.org/programme/people-centered-smart-cities

Anonymous (n.d.-d). Retrieved August 2021, from https://www.resite.org/stories/design-and-the-city-human-centered-smart-cities-with-marianthi-tatari

Anonymous (n.d.-e). Retrieved August 2021, from https://www.euronews.com/next/2021/01/18/japan-s-human-centred-smart-cities-enhancing-well-being

Anonymous (n.d.-f). Retrieved August 2021, from https://apps.who.int/iris/handle/10665/63552

Anonymous (n.d.-g). Retrieved August 2021, from http://disco-tools.eu/disco2_portal/terms.php

Anonymous (n.d.-h). Retrieved August 2021, from http://en.wikipedia.org/wiki/Life_skills

Anonymous (n.d.-i). Retrieved August 2021, from https://en.wikipedia.org/wiki/ASLERD

Anonymous (n.d.-j). Retrieved August 2021, from https://www.wise-qatar.org/topic/?topic=learning-ecosystems-leadership

Anonymous (n.d.-k). Retrieved August 2021, from https://www.miur.gov.it/documents/20182/2467413/Le+linee+guida.pdf/4e4bb411-1f90-9502-f01e-d8841a949429

Anonymous (n.d.-l). Retrieved August 2021, from https://en.unesco.org/education2030-sdg4/targets

Anonymous (n.d.-m). *Council recommendation of 22 April 2013 on establishing a youth guarantee 2013/C 120/01*. Retrieved August 2021, from http://eur-lex.europa.eu/legal-content/EN/ALL/?uri=CELEX:32013H0426%2801%29

Anonymous (n.d.-n). *Europe 2020 a strategy for smart sustainable and inclusive growth*. Retrieved August 2021, from http://eur-lex.europa.eu/legal-content/EN/TXT/PDF/?uri=CELEX:52010DC2020&from=EN

Anonymous (n.d.-o). Retrieved August 2021, from http://www.mifav.uniroma2.it/inevent/events/pcst_master/index.php?s=185

Anonymous (n.d.-p). Retrieved from https://en.wikipedia.org/wiki/The_Matrix

Carretero Gomez, S., Vuorikari, R., & Punie, Y. (2017). *DigComp 2.1: The digital competence framework for citizens with eight proficiency levels and examples of use. EUR 28558 EN*. Publications Office of the European Union. Retrieved August 2021, from https://publications.jrc.ec.europa.eu/repository/handle/JRC106281

Chatterton P. and Rebbeck G. (2015). Technology for Employability,. JISC.

Chatzichristou, S., Ulicna, D., Murphy, I., Curth, A. (2014). *Dual education: A bridge over troubled water EU DG for internal polices—Culture and Education*. Retrieved August 2021, from http://www.europarl.europa.eu/RegData/etudes/STUD/2014/529072/IPOL_STU(2014)529072_EN.pdf

Concilio, G., & Rizzo, F. (Eds.). (2016). *Human smart cities. Rethinking the interplay between design and planning*. Springer.

Czisikszentmihalyi, M. (1990). *Flow - the psychology of optimal experience*. Harper & Row.

Deakin, M., & Al Waer, H. (2011). From intelligent to smart cities. *Journal of Intelligent Buildings International: From Intelligent Cities to Smart Cities, 3*(3), 140–152.

Ezkowitz, H. (2008). *The triple helix: Universtiy, industry and government*. Routledge.

Giffinger, R., Fertner, C., Kramar, H., Pichler-Milanovic, N., & Meijers, E. (2007). *Smart cities. Ranking of European medium-sized cities*. Retrieved August 2021, from http://www.smart-cities.eu/press-ressources.html

Giffinger, R., & Gudrun, H. (2010). Smart cities ranking: An effective instrument for the positioning of cities? *ACE: Architecture City & Environment, 4*(12), 7–25.

Giovannella, C. (2013). *Territorial smartness and emergent behaviors. ICSCS 2013* (pp. 170–176). IEEE Publisher.

Giovannella, C. (2014). *'Smartness' as complex emergent property of a process. The case of learning eco-systems. ICWOAL* (pp. 1–5). IEEE Publisher, ISBN: 978-1-4799-5739-2.

Giovannella, C. (2015). *Territorial smartness and the relevance of the learning ecosystems ICS2 2015* (pp. 1–5). IEEE Publisher. https://doi.org/10.1109/ISC2.2015.7366220

Giovannella, C. (2016). Schools as driver of social innovation and territorial development: A systemic and design based approach. *IJDLDC, 6*(4), 64–74.

Giovannella, C. (2017). Incubator of projectuality: An innovation work-based approach to mitigate criticalities of the Italian massive alternance scheme for the school-based educational system. *IJDLDC, 8*(3), 55–66.

Giovannella, C. (2021a). An analysis of alternation schemes to increase student employability and the smartness of secondary schools. In *Ludic, co-design and tools supporting smart learning ecosystems and smart education* (pp. 39–51). Springer Publisher.

Giovannella, C. (2021b). Smart alternation schemes and design practices during pandemics. In *Smart learning ecosystems and the regional development after the pandemic*. Springer publisher, in press.

Giovannella, C., Crea, I., Brandinelli, G., Ielpo, B., & Solenghi, C. (2018). Improving massive Alternance scheme: The paradigmatic case history of the incubator of projectuality at the Ferrari school of Rome. In *The interplay of data technology, place and people for smart learning* (pp. 3–14). Springer Publisher.

Giovannella, C., Dascalu, M., & Scaccia, F. (2014). Smart city analytics: State of the art and future perspectives. *Interaction Design and architecture(s) (IxD&A) Journal, 20*, 72–87.

Giovannella, C., Gobbi, A., Zhang, B., Elsner, J., Del Fatto, V., Pérez-Sanagustín, M., Avouris, N., & Zualkernan, I. (2013). *Villard-de-Lans: A case study for collaborative people-centered Smart City learning design. ICALT 2013* (pp. 459–460). IEEE Publisher.

Hollands, R. (2008). Will the real smart city please stand up? *City, 12*(3), 303–320.

Kourtit, K., Deakin, M., Caragliu, A., Del Bo, C., Nijkamp, P., Giordano, S., & Lombardi, P. (2013). An advanced triple -helix network framework for smart cities performance. In M. Deakin (Ed.), *Smart cities:Governing, modelling and analysing the transition*. Routledge.

Lee, S. Ho, Han, J. Hoon, Leem, Y. Taik, & Yigitcanlar T. (2008). Towards ubiquitous city: Concept, planning, and experiences in the Republic of Korea. In *Knowledge-based urban development: Planning and applications in the information era* (pp. 148–169).

Leydensdorff, L., & Deakin, M. (2011). The triple helix of smart cities: A neoevolutionist perspective. *Journal of Urban Technology, 18*(2), 53–63.

Maslow, A. H. (1943). A theory of human motivation. *Psychological Review, 50*(4), 370–396.

Scuola, L. B. (2014). Retrieved August 2021, from (in Italian). https://labuonascuola.gov.it/documenti/La%20Buona%20Scuola.pdf.

The Human Smart Cities Cookbook. (2014). *Planum n. 28*. Retrieved August 2021, from https://issuu.com/planumnet/docs/cookbook_planum

The origin of the "Person in Place Centered Design" vision is documented in Giovannella C. (2008). Person-in-Place Centered design: educare 'istructional designer' e operatori dei futuri "learning places". In Didamatica 2008, Andronico, A., Roselli, T. & Lamborghini, B. (eds.) Bari, 2008 (p. 973) and later in Giovannella, C. & Graf. S. (2010). Challenging technologies, rethinking pedagogy, being design-inspired: The grand challenge of this century. eLearn Magazine, ACM ed., 25 Feb. 2010, http://elearnmag.acm.org/archive.cfm?aid=1734044

Toward E-Deliberation 2.0

Vassilis Tsakanikas, Georgia Rokkou, and Vassilis Triantafyllou

1 Introduction

Democracy is a system of government where citizens exercise power through a voting process. In direct democracy, the citizens form a governing body and vote on every issue. In representative democracy, citizens elect representatives from among the electorate (those who are eligible to vote). Representatives form a government structure, such as parliament, which is responsible for passing laws that apply in a country, while laws are proposed by the government. In constitutional democracy, the dynamics of the majority restrict the majority and protect the minority through the exercise of specific rights, e.g., freedom of speech, freedom of assembly. In recent years, however, trust in politicians and political organizations has been put to test. Citizens perceive politics as an arena where conflicting forces are involved in the pursuit of their own interests that do not reflect the needs of the majority. This has given rise to political cynicism, i.e., the feeling of lack of public confidence in politicians and the political system. Citizens perceive political cynicism as distrust of the sincerity, integrity, and intentions of political authorities, political personnel, and political institution, which do not live up to the expectations of the electorate, leading to a lack of trust.

Some political theorists (Habermas, 1996, 1987) argue that a more deliberative democracy would be an antidote to the problems of a conventional electoral democracy. Citizens who are invited to participate in a deliberation to formulate public policies, under an environment that ensures equal participation, mutual understanding, and at the same time the presentation of substantiated views, are very likely to contribute positively to bridging different views. In addition, in this

V. Tsakanikas (✉) · G. Rokkou · V. Triantafyllou
NetDit Lab, Electrical and Computer Engineering Department, University of Peloponnese, Tripoli, Greece
e-mail: vtsakanikas@uop.gr

P. Fitsilis (ed.), *Building on Smart Cities Skills and Competences*, Internet of Things,
https://doi.org/10.1007/978-3-030-97818-1_14

way, it is very likely to produce political decisions that are more consensual, logical, and fair.

One definition of public deliberation could be:

> *. . . is the dynamic process of dialogue between individuals or groups, based upon a genuine exchange of views, with the objective of influencing decisions, policies, or programmes of action.* (Institute, 2004)
>
> *. . . a local attempt to seek the views of a broad constituency of persons. User involvement is a local attempt to include organized groups of service users in the planning, and occasionally the management, of such services.* (Harrison & Mort, 2003).

1.1 What Is and What Is Not E-Deliberation

E-deliberation is an online deliberation process that uses Internet to sense public opinion on one or more specific issues, to enable and enhance discussion among citizens, and to shape consent among citizens. E-deliberation introduces the use of computers and communication technologies in meeting procedures and complements existing practices. E-deliberation can be an effective tool for encouraging participation and gathering answers to consultation papers and social policy issues as part of a wider range of methodologies (Triantafyllou, et al., 2019).

E-participation is an umbrella term for a set of e-democracy actions, like e-petitions and e-voting. Yet, it is important to clear out that e-deliberation refers to the process of formulating a common statement, rather than selecting predefined solutions.

2 E-Deliberation Systems

2.1 System Categories

Depending on the subject of the deliberation and the objectives, the profile, and the experience of the participants and coordinators in relevant processes, but also the general characteristics of the environment in which all this takes place, different categories of e-deliberation systems can be used.

2.1.1 Online Forums

It is used for large-scale public discussions and consultations, especially when it is desirable to involve many people in a discussion at the same time. They can be used for information and for discussion—debate on an issue as well as for deciding by voting. These methods and tools apply to a relatively large number of people, and the technology they use includes e-voting, text messaging, online polls, etc. Thus, it is possible to engage many people in the same place and at the same time.

2.1.2 Deliberative Poll

The selection of samples is random but representative of the citizens. There is a discussion, during which all opinions are presented. The positive aspect is that there is a thorough presentation of views and that the community is also involved through the transmission of media. The disadvantage, however, is that it does not create the feeling of wide participation but also requires a lot of time from those involved. In the poll process, there is no need for special anonymity as it can also be done in real time and constantly monitor the results. The deliberative poll is used specially to form an opinion.

2.1.3 Votes

E-deliberations voting is usually part of deliberations where, after information is given and since they can highlight the most interesting and relevant views on the issues, it is achieved to combine voting and deliberation. This has the effect of creating an audience with a high level of interest and better information about the subject under discussion than the actual citizens (Fishkin & Luskin, 2004). In most deliberations votes, the impact on the course of public policy is indirect and difficult to exploit. The hope here is to influence public opinion, to have an impact on policymaking beyond those who show interest in such issues and processes (Luskin, et al., 1999).

2.1.4 Discussion

The deliberative discussion method was developed with the aim of creating means to attract people and communities for dialog with each other. In essence, a deliberative discussion asks participants to discuss and weigh the costs and implications of a variety of solution options to a public problem (Goodin & Stein, 2008).

2.1.5 Questionnaires (E-Surveys)

By creating structured questionnaires by experts, depending on the topic, it is possible for everyone to participate, anonymously or by name. Bulk participation is a positive of e-surveys as structured discussions are achieved in large groups but also easy to implement in a multilingual environment. The analysis of the results is easy and immediate but can also give long-term results. Great attention should be paid to the questions, but the answers should be relatively simple; it is also possible that not all opinions are necessarily heard.

2.1.6 E-Petitions

A public e-petition refers to a petition published on a public network. The actual petition text can also be amended with additional background information concerning the petition issue and/or the different procedural steps related to the submitted petition. In addition, the final decision, important in terms of transparency, may be published.

2.2 Main E-Deliberation Systems

The most widely used e-deliberation systems deployed worldwide can be summarized in the following section.

2.2.1 E-Dialogos

The project (e-dialogos, 2012) concerns the provision of an innovative and fully developed methodology and e-democracy platform for the citizens of the Municipality of Trikala who will participate in the decision-making processes of the city, combining online discussions and voting procedures. The innovation of the project lies in its holistic and integrated approach to e-democracy and e-participation.

2.2.2 CrowdLaw

With proper planning, participation could help improve both the legitimacy and effectiveness of the legislative process at every stage, by introducing more data and ensuring that legislation is better informed (Noveck, 2018). The potential benefits at each stage are as follows:

- Setting the agenda: when parliaments decide what issues to take on and legislate
- Proposal: when legislative and regulatory bodies come to the substance
- Drafting: when legislators announce solutions through legislation, regulation, or drafting of constitutions
- Implementation: when the legislature instructs administrative bodies or staff to translate the law into practice
- Evaluation: when the public can assist in overseeing and monitoring the results of legislation

2.2.3 Cornell E-Rulemaking Initiative

Groups such as the Cornell E-Rulemaking Initiative (CeRI) have focused on research into how technologies such as Web 2.0 can help enhance public participation in the political process, specifically in Federal Service rules. In 2009, CeRI launched the RegulationRoom.org website. The site, hosted by the Legal Information Institute, is an independent nongovernmental online community that allows users to read, comment, and discuss proposed regulations from federal agencies (CornelleRulemakingInitiative, 2017).

2.2.4 MiSenado

It is a mobile phone app, available on iOS and Android devices. Via the app, users have access to attendance and voting records for all their elected representatives. Push notifications alert users to when live plenary sessions are scheduled to occur so that they can provide their feedback and vote on bills in real time (Senado, 2018).

2.2.5 ParlementetCitoyens

It is a platform that enables the French public to provide data on legislation through a multistage, online consultation process. From the platform, delegates can host a consultation consisting of three to five different participation opportunities. The consultation processes, while funded by the representatives present throughout the process, are run by volunteers (Rozenberg & Viktorovitch, 2014).

2.2.6 E-Democracia

It is a participation platform managed by HackerLab that provides the Brazilian public with three participation opportunities:

- Collaborative drafting of legislative texts through WikiLegis
- Involvement with members in a discussion board (Expressao)
- Open to the public through online conferences with representatives

Through its tools, E-Democracia allows citizens to propose and draft legislation with multiple opportunities for government. The final reports ensure that citizens understand how and when their platform activity updated the bill (E-Democracia, 2016).

2.2.7 LiquidFeedback

An open-source software that supports Internet platforms, LiquidFeedback is a unique democracy software used by municipalities, political parties, associations, social movements, private organizations, and companies. It facilitates a comprehensive consultation process to empower citizens, members, and employees to participate in democratic decisions important to their organizations. LiquidFeedback promotes democratic participation and self-organization to redefine the future of society. LiquidFeedback offers a completely transparent decision-making process. Predefined rules and times ensure that all steps in the process are made public in real time. Decisions are made by registered ballot only, and all relevant data is available to all participants in human and computer readable formats (Behrens, et al., 2014).

2.2.8 DebateHub

DebateHub is an open, online, collaborative tool to support deliberation and democratic decision-making. DebateHub has a very simple user interface, which may look like a common web forum, but is enhanced by a semantic data model. This allows for better up-to-date brainstorming support, as well as the development of advanced analytics for a team, discussion, and challenge data, which is delivered with a visualization control panel. This system also has moderator functions to reduce the repetition of ideas in an online chat, which is one of the main weaknesses of existing online platforms for ideas and discussions (Quinto, et al., 2021).

2.2.9 OnlineConsultationPlatform (GR0059)

The online consultation platform will serve as a key reference point for the participation of citizens in the decision-making processes of the Municipality of Thessaloniki and will support the implementation of the organization's strategy for the transformation of Thessaloniki into an integrated, participatory platform. It will launch a new framework for discussion with citizens, empowering and empowering them throughout the project, and will be a key element of the broader initiative on disseminating the benefits of e-democracy (Anon., 2018).

2.2.10 Decidim

Decidim (decidim, 2019) is a digital infrastructure for participatory democracy, built entirely on free software. More specifically, Decidim is a web environment produced in Ruby on Rails, which allows users to create and configure a website platform or portal, which will be used in the form of a social network for democratic participation. The portal allows each organization (local council, association,

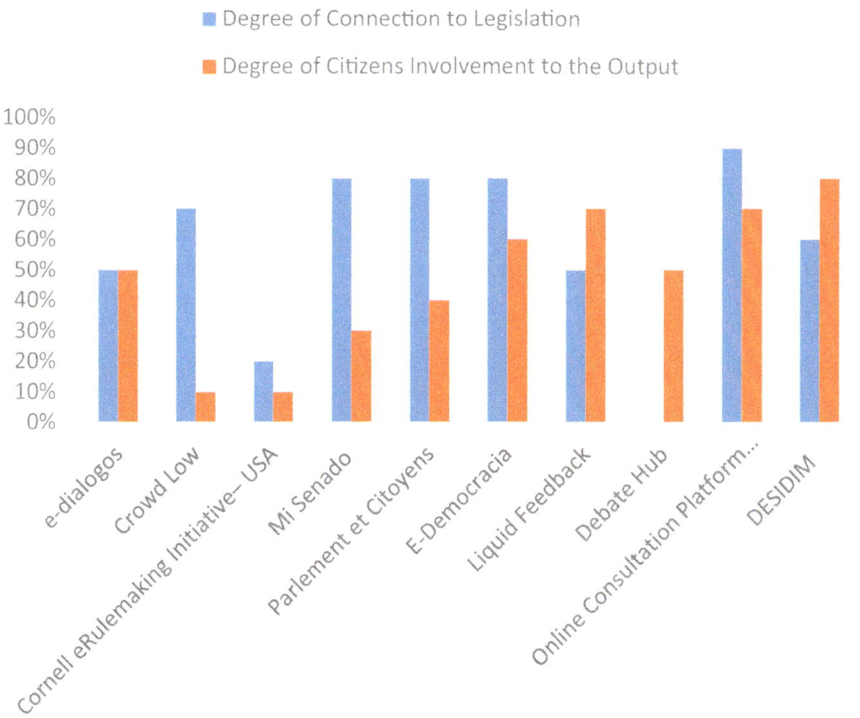

Fig. 1 Comparison among common e-deliberation systems

university, NGO, trade union, neighborhood, or cooperative) to create democratic processes for strategic planning, participatory budgeting, collaborative planning, urban planning, and elections. It also allows personal meetings to be organized and signed, minutes to be posted, agenda items to be proposed, and results notifications to be received.

There are three levels of authentication that indicate how anonymous users are or are not. The first concerns systems that are completely open to authentication and can be accessed as many times as users want. The second concerns the systems with which they somehow identify the users, but without much information, but to have access only once in this way (e.g., by phone number). And the third concerns the systems with which they identify users, with a lot of important information. Degree of connection with the legislation has to do with how much each system is related to the legislation. The rating is from 0 to 100.

The comparison of the aforementioned systems, in terms of citizen identification and connection to legislation, is presented in Fig. 1. About the degree of connection to legislation, we have taken under consideration two aspects: (i) the capacity of the system to produce a final e-deliberation report which can act as an enabler to a legislation process and (ii) the number of e-deliberation projects that reached, at some point, to a legislation form. For example, e-dialogos is equipped with a final report functionality, but no e-deliberation projects (to the best of our knowledge) have reached to a legal form. On the other hand, MiSenado's final reports have affected many documented legislation processes.

As far as the degree of the citizens to the final output metric, it refers to both tools (e.g., discussion forums) and e-deliberation models which can facilitate the evolvement of a proposal. More specifically, a system which just presents proposals and citizens can only vote their preferable proposal will receive a score 0 at this specific metric, while a system which promotes comments discussion and integration, proposals' merging, and other collaboration actions will receive a score 100. Therefore, Decidim has received a high score in this metric, while Crowd-low a low one.

3 "Smart Cities" Need E-Deliberation and E-Deliberation Needs "Smart Citizens"

3.1 *"Smart Citizens"*

In the past few years, the term "smart city has taken over research, academia, policy, and industry by building and deploying digital technologies, networks, and urban governance (Marvin, et al., 2015). While it has been initially proposed as an agenda for designing, developing, and deploying ICT products, "smart city" embraced more broad concepts for digitally enabled urbanism, equipped for nowadays city challenges like mobility, sustainability, and green policies. However, materializing and deploying concepts for smart cities, public authorities and other relative stakeholders need to come to a "moral" agreement with the citizens: in what extent citizens are familiar with the "smart city" technologies, their rights, and their obligations in this digitalized world (Kitchin, et al., 2019). Subsequently, a smart city requires a "smart citizen": a willing subject in digitally shaped urban governance, infrastructure, and services (Joss, et al., 2017). A smart citizen is usually registered to a set of networks and services while utilizing hardware devices (e.g., sensors, webcams, etc.) (Cardullo & Kitchin, 2018). Yet, the participation of smart citizens to e-participation platforms (such as e-deliberation, e-voting, e-comment, etc.) has the potential for the citizens not only to consume digital goods but also produce utilization, in terms of engagement with the public administration. A growing market in e-participation platforms, initiated by public authorities, multinationals and academia enable citizens to debate, propose, comment, vote, and contribute to urban strategy and local plans. Yet, criticism has emerged that such initiatives implemented within an incurably neoliberal smart citizenship are mostly tokenistic (Kitchin, et al., 2019).

Smart citizenship or the smart city bordering of the involved technology has been reported widely (Rabari & Storper, 2015). For example, usage of sensing devices across city infrastructures and integration of platforms that connect and utilize these data sources (also produced by citizens via their cell phones) by smart city managers can produce useful analytics, like visualization of urban phenomena in real time. Governance transforms mainly to a managerial process, i.e., digital tools and services provide neutral means for meeting seemingly universal and legitimate actions of efficiency for sustainable and modest cities. While the smart city challenge is mostly technical, transforming complex urban processes to platform functionalities (Marvin & Luque-Ayala, 2017), the smart citizen challenge is more complex. Citizens not only have to acquire the necessary digital literacy but also to possess the knowledge of the power the e-participation platforms have.

Recently, various authors have argued that one of the criticisms of smart city discourse concerns how e-governance is effectively ceded to public-private corporations conquered by the trade technology interests who install, own, and deploy e-participation platforms and whose authoritative presence imposes a particular computational logic upon the city (Vanolo, 2016). Criticism of platform technocracy explores the assumptions stored into digital platforms, and it challenges the values privileged in so-called "technical" decisions that obfuscate the real politics of those decisions (Gillespie, 2010). The smart city is projected as the latest brand for neoliberal urban political economy, deploying digital technologies to materialize competitiveness, inward investment, economic productivity, and efficiency (March & Ribera-Fumaz, 2014). Smart services are criticized for enhancing corporate technology priorities, urban entrepreneurship, and imperatives in capital accumulation, at the expense of citizenship rights and democracy to the smart city (Kitchin, et al., 2019).

Smart citizenship promotion can be realized, even partly, as legitimacy-seeking responses to smart city criticism. Citizen e-participation platforms become an appealing road for smart city managers seeking to cultivate active citizens. A business market has been set up, developing online citizen participation services. These industry providers expose participation as a circumscribed service: running citizen e-deliberation and voting services for clients, who receive data analysis and reports (Graeff, 2018). Consistent with the smart city, vendors of these services are contracted to provide a uniform technological template for citizen participation. There have been some research works though, which indicates that the democratic capacity of these platforms needs scrutiny in terms of their ability to actively challenge power (Cardullo & Kitchin, 2019). Studies of smart citizen policies and projects find citizen inclusion to be a shallow invitation (Vanolo, 2016). Initiatives envisage citizens as either passive, compliant participants in each process or, at best, entrepreneurial contributors to smart services (Cardullo & Kitchin, 2018). Active, autonomous citizenship is largely absent. Unfortunately, most of the e-participation platforms treat citizens passively, by developing functionalities which steer end users to specific pathways. If there is civic engagement, it is in the form of a participant who provides feedback or suggestions, rather than being a proposer, cocreator, decision-maker, or leader. Motivation is not considered as citizenship, let alone democracy, but rather self-interested acknowledgement from developers of the benefits of user-centered design in the successful implementation of digital technology projects. Questions about control, representation, participation, and democracy remain unaddressed (de Hoop, et al., 2019).

3.2 *Models of Participatory Democracy*

Governments and politicians recognize the worth of e-participation (Bryson, et al., 2013), but usually their narrative usually fails to define what the outcome of participation should be. Several models and frameworks have been proposed to frame the output of e-participation processes and more specifically e-deliberation processes. (Held, 2006) identifies almost ten different models of democracy. While these models differ from each other, they all converge on a set of democratic principles: free elections, freedom of speech, inclusive citizenship, freedom to form and become members of organizations, and the rule of law. (Päivärinta & Sæbø, 2006) presented four models of democracy, categorizing e-participation into two pillars: agenda setting (citizens or public authorities set the agenda) and decision-making (citizens or public authorities have the final decision). Authors stress the importance of clearing out the model of democracy being followed at the beginning of a citizen e-deliberation (or e-participation in general) project and that this should be communicated to participants to avoid confusion about the outcome of the discussion.

Attempting to become "smart cities," public authorities are embracing various technologies that promise opportunities for increasing participation by expanding access to public comment and deliberation. Yet, stakeholders have encountered the problem of defining participation, determining who is able to participate through technology-enhanced public engagement.

3.3 *Benefits of E-Participation in Smart Cities*

The increasing penetration of ICT technologies has enabled e-participation platforms to be integrated in several smart cities and for several purposes. As part of e-democracy, e-participation can offer the following benefits:

- Focus on citizen needs: e-participation, mostly through e-deliberation platforms, provides to citizens the opportunity to express their opinions and formulate their actual needs.
- Government transparency: through open government initiatives, public authorities offer citizens access to government information.
- Citizen involvement: digital democracy platforms provide a tangible mean for citizens to actively involve to public issues and to public authorities to provide systematic information updates and official meetings and/or ask for citizens' involvement in activities regarding the local government.

For e-deliberation (and e-participation platforms in general) to act in the heart of a smart city and constitute the base of every governing activity, the remarks should be enhanced. Thus, both government policies and the legislative decision-making process should be transformed properly, aiming to increase the online participation and the relative data and information being shared with a significant reduced cost and in an easier manner (Matei & Savulescu, 2014). Along with this process, the role that the citizens have regarding public services' delivery can be changed from one of a passive service beneficiary to one of an active informed partner (Sherriff, 2015).

4 PODS: A Generic Approach for E-Deliberation in Smart Cities

The formulation of an integrated approach to online deliberation led to the design and implementation of the PODS (Public On-line Deliberation System). As described in (Triantafyllou, et al., 2019), the system comprises four functional modules, which materialize the components of the different PODS models.

- **Information base**. The information base is the primary functional module where all data are stored and retrieved from, through the deliberation process. These data include personal information of the participants (proposers and citizens), such as name, address, gender, age, identity, expertise, social profile, and other relevant information which are used to infer (during the deliberation process) the most prevalent proposals/suggestions/comments.
- **Discussion forum**. The deliberation utilizes a type of *electronic forum* through which participants can express their opinions and engage in the dialog and exchange of opinions between them. Certain rules of civilized dialog are applied to ensure equal opportunities of participation, encourage involvement of citizens who are reluctant to express themselves, and limit those who tend to monopolize the debate. These rules are known ex ante, and compliance to these rules is the responsibility of an independent adjudicator, the debate moderator.
- **Evaluation process**. Evaluation of the proposals is based on a compound procedure that highlights *winning proposals* such as positive, negative, neutral votes, and social factors (citizens' digital reputation, participation frequency and innovation, geolocation, etc.). After a number of consequent rounds, where proposals with low acceptance are discarded or proposals with similar context can be merged, a set of "winning proposals" are surfaced.
- **Deliberation authority**. A governing body monitors the deliberation process, and it is responsible for setting up the criteria and coefficients to be used by the evaluation procedure to facilitate the screening process toward awarding the "winning proposals." The criteria and coefficients introduce the "political view" of local authorities.

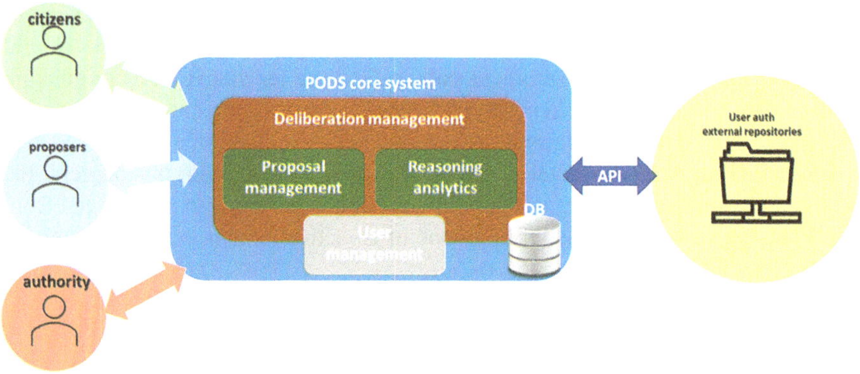

Fig. 2 PODS conceptual architecture

As depicted in Fig. 2, the PODS involves three types of users:

- *Deliberation authorities*. These include all authorities that introduce a deliberation. The PODS is mainly addressed to authorities (local council, public bodies or groups, management executives, etc.) that need to deliberate on an issue allowing the participants/citizens to express their opinions and evaluate possible proposals.
- *Proposers*. Any citizen can present a proposal about the issue introduced for deliberation. The proposers present their proposal including any documentation needed to support their thesis and are responsible to work with other proposers to merge their proposition because of the evaluation procedure.
- *Citizens*. The citizens formulate a closed set of individuals that participate in the deliberation process based on their identity (citizens, interested people, members of a specific group, etc.). Each participant logs in using credentials, and the user inserts all data at relevant fields that correspond to relevant criteria and factors the authority body believes should be a part of the evaluation procedure.

Compared to the work published in (Triantafyllou, et al., 2019), in the past 2 years, several developments have been introduced to the PODS system, aiming to improve the deliberation process. These developments mainly include analytics for encouraging proposers to take under consideration the citizens' comments, thus emphasizing on building proposals with higher levels of consensus. More specifically, two analytics are automatically calculated for each proposal, facilitating the deliberation process:

- *Proposal plasticity index*. This analytic refers to the degree a proposal has changed compared to the initial submission of the proposal. A document comparison algorithm (hybrid implementation of the Jaccard score (Temma, et al., 2019) and the BERT model (Nogueira, et al., 2019)) is applied to the working version of the proposal, compared to the initial proposal, and the result is presented to the citizens. Following this rationale, citizens can "judge" a proposal for its evolvement through the deliberation process. Thus, PODS encourages the proposers to make changes, alterations, and improvements to their proposal, facilitating the integration of opinions and aspects not included at the initial proposal.

- *Comments integration index*. The comments integration index is a quantification of how many comments have been taken under consideration by the proposer. The index is affected by two actions. The first action is the reply of the proposer to a comment, until the communication chain (replies on comments by both the proposer and a citizen) is characterized as resolved by the citizen. The second action is the incorporation of a comment into a proposal, after it is verified by the citizen who posed the comment. It is obvious that the more comments are either addressed or incorporated in the proposal, the higher this analytic will be.

These analytics are visible for all citizens who participate in the deliberation process, providing quantitatively insight about the evolvement of the proposal toward embracing more opinions and improving through the deliberation process.

In terms of the technological aspects, PODS utilizes REST APIs for integrating with external user repositories, to provide secure user authentication. For data storage, PODS follows a hybrid approach, integrating a no-SQL database (mongo-DB) for storing material related to the proposals and the users' comments and a relational database for storing the system settings, the produced analytics, and other information related to the deployed deliberation setups.

The PODS evaluation process designed and implemented a model which is presented below. All phases are time constrained, meaning each phase is complete into a specified time frame, different for each theme deliberation.

- **Initialization**. The system administrator initiates the system by providing relevant information about the participants (name, location information, expertise, relevance, participation frequency, digital reputation, etc.). The above information is used by the system to ensure access to eligible participants (citizens of a specific local community, members of a social group, etc.) and by the authorities that have ordered the deliberation which need to set up the parameters and factors that will be used throughout the evaluation process to point out the "winning proposals."
- **Theme post**. The deliberation authority sets up a concept/problem/issue that is of interest to the local government. Any registered user can introduce a theme into the PODS system. The authority can issue from time-to-time calls inviting any interested individual to propose a theme for discussion. The theme is included in a list after approval of the deliberation authority. A list of active themes is always available to the users.
- **Proposal post**. In this phase, every participant can participate by introducing for each theme a proposal. All proposals with relevant documentation are displayed under a proposal list that is attached to each theme. For a large-scale deliberation scenario (e.g., a theme that concerns a country), the deliberation authority can choose to pose a pre-theme post phase. During this phase, the writers of the proposals shape their suggestion and invite other users to support it. A proposal passes to the next phase only if it gathers an adequate number of supporters

(the exact threshold is defined during the initialization phase by the deliberation authority). Following this workflow, only mature-enough proposals enter the evaluation phase.

- **Evaluation** ***phase***. This phase is the core of the system. It includes three major steps:
- Vote on a proposal. Each participant can cast a vote on a proposal. Although in certain electoral systems participants may be allowed to cast multiple votes, a citizen/participant can cast only a single vote. The total number of votes corresponds to the number of citizens who evaluate a given proposal, to any given proposal a positive, negative, or neutral vote, respectively. Each proposal can be evaluated based on the measure of acceptance where proposals having sufficient positive attraction are considered as candidates to qualify to the next phase. Participants can add comments, add relevant documentation supporting their vote, or suggest proposal merging.
- Proposal merging is a process where participants indicate proposals that seem similar in context and the relevant proposers are to encourage them to present a new proposal because of the comments of the participants. Proposers are notified if their proposal is recommended for "merging." The completely voting process follows the basic principles applied by all voting systems (Bouras, et al., 2003).
- Application of political analytics to be used for evaluating the proposal submitted either by the local government or by citizens. The general principles behind any realistic implementation of political analytics are presented in (Drakopoulos, et al., 2018) and are set either by the participants or the local government (based on political decisions). The decision criteria and factors should be transparent to users to reinforce a culture of openness and accountability.
- After the rating and acceptability phases, the remaining proposals are inserted back in the evaluation phase until a winning proposal pops up and is publicized.

• **Rating** ***phase***. The proposals are rated. The system displays at least the number of votes per proposal, the criteria and factors used to rank proposals, and the way the proposals were merged or rejected. The results are presented categorized by various criteria.
• **Acceptability phase**. This a repetitive phase which pops up the "winning proposals." At each round, the last ranked proposal is deleted, and a new round begins with the remaining proposals graded by the participants and is ranked according to their acceptance between individuals. The cycle ends when a *winning proposal* surfaces having the majority of acceptance between individuals.
• **Publication** ***phase***. At this phase, the outvoted proposal is presented. The proposal is accompanied by documented information that can be used by the local government in supporting the relevant proposal. Today, factors such as education, work experience, religious views, political alignment, exposure to social media, and family size play an important yet unconscious role in daily

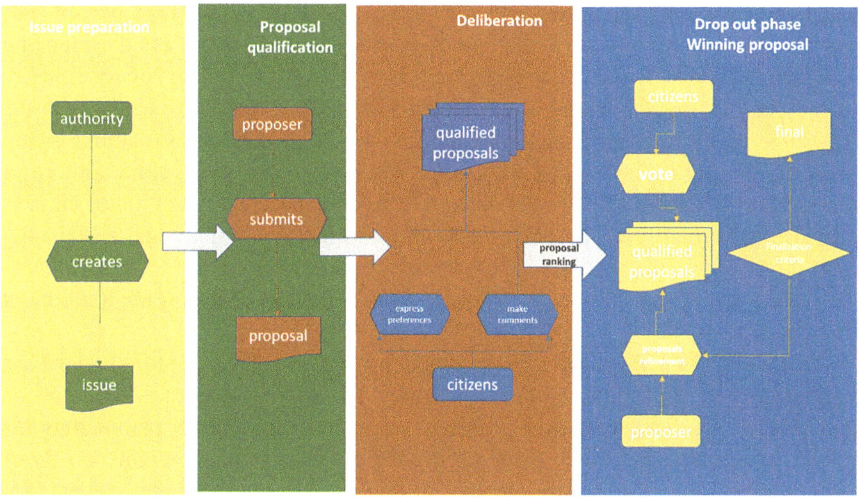

Fig. 3 PODS working flow

decision-making. The whole process is based on a repetitive schema, which is composed of consecutive rounds. Each round after evaluating and grading results in a set of acceptable proposals. The last proposal graded is deleted and a new round is introduced. The new round includes remaining proposals (undeleted) of new proposals (resulting from the merging process). After several rounds a final outvoted proposal surfaces and it is presented.

The final proposal is presented accompanied with all relevant documentation to the authority and the decision-makers. Decision-makers have access to the whole process data which can be used for further review. Figure 3 represents the process workflow which denotes the interaction between users and processes.

Comparing PODS architecture with the systems presented in the previous sections, it is important to mention that the design philosophy behind PODS does not target to sort out a "winning" proposal or opinion. On the other hand, PODS drives its users, through the merging phase, to cooperate, formulating entirely new proposals. Thus, the main objective of PODS is to create consensus among citizens, which at the end of the day is the outmost scope, for materializing the content of the deliberation outcome, and provide sustainability to smart cities' prosper living.

5 Conclusion

Digitally literate people constitute an important part of the society, as it is the part of the society which uses to a higher extent ICT in all aspects of life. These proportion of citizens form the society based on performance and information usage; thus, they usually are also the part of the society, which have higher participatory percentages to e-participation initiatives. Nevertheless, no citizen should be put aside from being a beneficiary of e-deliberation platforms. E-democracy demands smart cities to include in their policymaking and decision-making processes all citizens with the right to vote, by offering them equal and just access to public information, giving them the possibility to interact in a faster and easier way with public authorities and to have an impact on their activities. Higher and direct participation of citizens provides to a smart city the potential to become fair, supportive, and sustainably innovative. Thus, smart cities need to provide opportunities for increased awareness and access to e-participation platforms in such a manner that no individual be at disadvantage.

To conclude, an informed and participatory society means a strong society, a society able to contribute to the achievement of a true democracy. Through its participation, the society can help authorities to meet the public administration's general objective, namely, to respond to citizens' needs and requirements. By e-participating, citizens also offer support to public authorities and governments for them to carry out their duties in a more efficient manner, thus creating a closer relationship between all actors involved. This all results in a more uniform and thus more powerful state on a regional, local, and national level, a state in whose composition we can encounter smart cities. Finally, for the next generations of smart-cities, it is mandatory to educate "smart citizens" with skills not only for consuming smart services but also with the appropriate skills and state-of-mind to use technology to make their cities a better place to live. E-deliberation, especially through a "consensus-driven" platform, is a remarkable mean for enhancing participation of citizens in management and micromanagement of a smart city.

References

Anon. (2018). *Online consultation platform (GR0059)*. [Online] Retrieved October 28, 2021, from https://www.opengovpartnership.org/members/greece/commitments/GR0059/.

Behrens, J., Kistner, A., Nitsche, A., & Swierczek, B. (2014). *The principles of LiquidFeedback*. Interaktive Demokratie e. V.

Bouras, C., Katris, N., & Triantafillou, V. (2003). An e- voting service to support decision-making in local government. *Telematics and Informatics, 20*, 255–274.

Bryson, J. M., Quick, K. S., Slotterback, C. S., & Crosby, B. C. (2013). Designing public participation process. *Public Administration Review, 73*(1), 23–34.

Cardullo, P., Di Feliciantonio, C., & Kitchin, R. (Eds.). (2019). The right to the smart city. *Emerald Group Publishing*. ISBN: 978-1-78769-139-1.

Cardullo, P., & Kitchin, R. (2018). Being a 'citizen' in the smart city: Up and down the scaffold of smart citizen participation in Dublin, Ireland. *GeoJournal, 84*(1), 1–13.

Cardullo, P., & Kitchin, R. (2019). Smart urbanism and smart citizenship: The neoliberal logic of 'citizen-focused' smart cities in Europe. *Environment and Planning C: Politics and Space, 37*(5), 813–830.

CornelleRulemakingInitiative. (2017). *Cornell eRulemaking Initiative*, s.l.: s.n.

de Hoop, E., et al. (2019). Smart urbanism in Barcelona: A knowledge politics perspective. In J. Stissing Jensen, P. Späth, & M. Cashmore (Eds.), *The knowledge politics of urban sustainability transitions*. Routledge.
decidim. (2019). *decidim*. [Online]. Retrieved from https://decidim.org/
Drakopoulos, G., Triantafyllou, V. & Christodoulos, J. (2018). Political analytics for public electronic proposal deliberation at municipal level. *IADIS International Conference ICT*, Society, and Human Beings.
E-Democracia. (2016). *E-Democracia*, s.l.: s.n.
e-dialogos. (2012). *OnLinePolitics:*. [Online] Retrieved from https://onlinepolitics.wordpress.com/edialogos/
Fishkin, J. S., & Luskin, R. C. (2004). *Experimenting with a democratic ideal: Deliberative*.
Gillespie, T. (2010). The politics of platforms. *New Me, 12*(3), 347–364.
Goodin, H., & Stein, D. (2008). Deliberative discussion as an innovative teaching strategy. *Educational Innovations*.
Graeff, E. (2018). *Evaluating civic technology design for citizen empowerment*. Massachusetts Institute of Technology.
Habermas. (1987). *The theory of communicative action: Volumes 1&2*. Beacon.
Habermas. (1996). *Between facts and norms: Contributions to a discourse theory of law and democracy*. MIT Press.
Harrison, S., & Mort, M. (2003). Which champions, which people? Public and user involvement in health care as a Technology of Legitimation. *Social Policy & Administration*, 60–70.
Held, V. (2006). *The ethics of care: Personal, political, and global*. Oxford University. s.l.
Institute. (2004). *The consultation charter*. The Consultation Institute. s.l.
Joss, S., Cook, M., & Dayot, Y. (2017). Smart cities: Towards a new citizenship regime ? A discourse analysis of the British smart city standard. *Journal of Urban Technology, 24*(4), 29–49.
Kitchin, R., Cardullo, P., & Di Feliciantonio, C. (2019). The right to the smart city. *Journal of Urban Technology*.
Luskin, R. C., Fishkin, J. S., & Plane, D. L. (1999, November). Deliberative polling and policy outcomes: Electric utility issues in Texas. In *Annual meeting of the midwest political science association* (Vol. 4, p. 1999). Chicago.
March, H., & Ribera-Fumaz, R. (2014). Smart contradictions: The politics of making Barcelona a SelfSufficient City. *European Urban and Regional Studies, 23*(4), 816–830.
Marvin, S., & Luque-Ayala, A. (2017). Urban operating systems: Diagramming the city. *International Journal of Urban and Regional Research, 41*(1), 84–103.
Marvin, S., Luque-Ayala, A., & McFarlane, C. (2015). *Smart urbanism: Utopian vision or false Dawn?* Routledge, s.l.
Matei, A. I., & Savulescu, C. (2014). Enhancing the capacity for innovation of public administration. An exploratory study on e-governance ICT, knowledge management in Romania. *Theoretical and Applied Economics, 21*(11), 600.
Nogueira, R., Yang, W., Cho, K., & Lin, J. (2019). Multi-stage document ranking with bert. *arXiv preprint arXiv:1910.14424*.
Noveck, S. B. (2018). Crowdlaw: Collective intelligence and lawmaking. *Analyse & Kritik*, 359–380.
Päivärinta, T., & Sæbø, Ø. (2006). Models of E-democracy. *Communications of the Association for Information Systems, 17*(37).
Quinto, D., Landoli, L., & Deliddo, A. (2021). *Designing online collaboration for the individual and social good: A collective argumentation approach*. s.l.:s.n.
Rabari, C., & Storper, M. (2015). The digital skin of cities: Urban theory and research in the age of the sensored and metered city, ubiquitous computing and big data. *Cambridge Journal of Regions, Economy and Society, 8*(1), 27–42.
Rozenberg, O., & Viktorovitch, C. (2014). *Parlements et Citoyens*. [Online] Retrieved from https://parlement-et-citoyens.fr/

Senado, M. (2018). *Senado Colombia.* [Online] Retrieved from https://play.google.com/store/apps/details?id=gov.senado.app&hl=en_US

Sherriff, G. (2015). Voting on sustainable transport: Communication and governance challenges in greater Manchester's 'congestion charge'referendum. *Local Environment, 20*(12), 1507–1530.

Temma, S., Sugii, M., & Matsuno, H. (2019). *The document similarity index based on the Jaccard distance for mail filtering.* S.l., s.n., pp. 1–4.

Triantafyllou, V., Tsakanikas, V., Asimakopoulos, G., & Christodoulopoulos, J. (2019). Public on line deliberation system (PODS). *17th International Conference on e-Society 2019.*

Vanolo, A. (2016). Is there anybody out there? The place and role of citizens in Tomorrow's smart cities. *Futures, 82*, 26–36.

No-Code for Smart Cities

Ahmed ElBatanony and Giancarlo Succi

1 Introduction

In this chapter, we take a quick but deep dive into who the people living in smart cities are like, what they can and are expected to accomplish, how we can bridge the gap between human and machine using no-code, and lastly, we envision of a future in the making. The aim here is to introduce a vision of how no-code and modern software practices could change the way we develop future cities.

The second section discusses the reality of the citizens of smart cities and their relation to technology. The third section introduces the concept of no-code and the technologies and principles to be used to empower the citizens. The fourth section provides examples of how to apply no-code in the context of a smart city. The fifth section envisions a future powered by smart assistants. Lastly, the sixth section provides a summary of this chapter.

2 The Citizens

What is the difference between a citizen of a smart city and a non-smart city? This is an important question to ponder when designing methods of accessing and utilizing smart city systems. If we put aside some professions such as engineering and software development, there should be no inherent differences between smart and non-smart city citizens. Most of the smart city citizens in the near future will most likely not be born there. They might have recently relocated to the new city, and

A. ElBatanony (✉) · G. Succi
Innopolis University, Tatarstan, Russia
e-mail: a.elbatanony@innopolis.ru; g.succi@innopolis.ru

P. Fitsilis (ed.), *Building on Smart Cities Skills and Competences*, Internet of Things,
https://doi.org/10.1007/978-3-030-97818-1_15

there are no necessary unique attributes, at least on the surface, that differentiates them from others. Therefore, the difference between the two sorts of citizens is their new potential in the smart city. To maximize this potential, we should design systems and interfaces that empower the citizens.

Software developers and engineers welcome, or at least are not opposed to, communicating with machines in non-human languages or interfaces. What about the other citizens of the smart city? An important remark to be made is that most of the citizens of the smart city will not be engineers. Even if they start learning about the different ways they can use the city systems, they are not expected to reach an engineer's level of skill in the field. This is not elitism. I would not expect someone to become a doctor by providing them with "smart" scanning and examination equipment. Even if I do not need to trust them with my own health, I would not trust that they would be able to utilize the machines in an optimal way or get the best out of their use. Hence, it is fair to assume that to allow people to flexibly interact with and contribute to the city systems, we need to find a non-developer approach to the communication between humans and machines. Mobile applications do provide users with an array of options to interact with various platforms in a smart city, but they can be superficial, hard, or impossible to extend and customize, and they just do not provide the same level of control a developer has with an API (Sillitti et al., 2004; Corral et al., 2011, 2013, 2014, 2015).

It is indeed exciting to imagine a majority of a smart city's citizens able to develop, debug, deploy, and maintain software that significantly affects their lives and the lives of the ones around them. Unfortunately, we cannot expect this to happen today, in the near future, or possibly ever. Luckily, on the other hand, this is not a necessity anymore with the rise or revivification of the no-code development platforms. The no-code movement aims to provide non-developers with tools to enable them to create, manage, deploy, and maintain software without requiring any development or coding skills. This is usually done through a combination of clever abstractions and sophisticated background systems that the user is not directly exposed to.

Software engineering, at its essence, is an art form. An artist, or a developer, in this case, uses their paintbrushes (lines of code) to convey thoughts and meanings for others to experience (programs and interfaces for users). As the adage goes, the only constant is change. Software engineering is no exception. Painting, a "true" example of art, had its evolution over the ages. From painting on cave walls to pieces of wood, to temples, to canvases, using oil and other materials, to digital drawings, to graphic tablets, and now 3D painting in virtual reality. The process changed, but the artist and the art itself kept their essence, which is presenting the world with something creative, beautiful, and inspiring. A similar situation occurred and is still happening with software. It began with transistors, to punch cards, to the assembly language, to Fortran and the like, to C and the like, and to where we are today with modern high-level programming languages. Today, we are still seeing new eye-opening innovations in this fifty or more year-old industry. Examples include

GPT3 by OpenAI[1] that can construct simple user interfaces or websites from a text prompt, and the recently released GitHub Copilot[2] that combined the years of collective knowledge on Stack Overflow (a programming questions and answers website). This Copilot can implement full functions in your favorite programming language when provided a few words describing the functionality. It is also context-aware, and some may describe it as "magical," but time will be the judge on the revolutionary aspect of this innovation.

All of this is to say that software engineering did not and will not stop at C++, or Java, or Python. Software development still has ways to go and always will. We are here to explore one of such paths, which is the "absence" of code or using the more common term: no-code. Moreover, as many forms of creating arts, such as painting, music, and, of course, writing, have become more accessible to the general public, so will software development, eventually.

3 No-Code

The no-code movement aims to introduce software development tools that require no prior coding or development skills to the general population. Everyday users can now create software that only software engineers could create a few years ago. No-code is often associated with website builders, visual development tools, drag-and-drop, AI-assisted design, and similar tools.

The essence of no-code lies in empowering the user with the abilities of a developer without requiring the knowledge or years of training gained by an experienced developer. Additionally, no-code generally abstracts the software development process into a higher level, allowing the user to work directly with features and components, instead of lines of code. We explored the distribution of application features and how no-code can help developers become significantly more efficient in a previous work (ElBatanony and Succi, 2021a). Applying the no-code methodology and principles to smart cities would provide the citizen with much more control and influence over the city assets, whether it be personal assets or public ones.

In the smart cities of the future, citizens do not just interact with the city, but also with each other and with establishments. This is where the real possibilities lie. New tools for planning, governance, commerce, and collaboration open up when you provide the citizens with powerful and easy to use tools to shape their reality.

[1] https://openai.com/

[2] https://copilot.github.com/

3.1 City API

What if the city had an application programming interface (API)? Imagine if there is a library to import in a modern programming language that allows the developer to interact with the various systems in the city. This API would provide the ability to modify services' parameters and adjust the citizen's preferences. These functionalities would include the home utility systems, arranging transportation, home delivery preferences, etc.

Combined with no-code, such an API would be accessible to all the citizens through a modern and intuitive interface. Citizens would access all the resources without requiring the knowledge of using APIs, building software, or deploying it. This tool would additionally abstract away many of the development concerns, such as security and authentication, database management, version control, and so on. A significant portion of this effort would be providing a proper API, handling security concerns, and building an intuitive interface for the users.

Building such a platform requires thorough planning to ensure the reliability of the systems, as well as the security of the users' and businesses' data (Marino and Succi, 1989; Valerio et al., 1997; Vernazza et al., 2000; Musílek et al., 2002; Sillitti et al., 2002; Scotto et al., 2004; Pedrycz and Succi, 2005; Scotto et al., 2006; Moser et al., 2008a,b). This platform could start on a small scale including a handful of city systems interacting together. Then, using the principles of lean and agile development (Maurer et al., 1999; Kivi et al., 2000; Succi et al., 2001a, 2002; Clark et al., 2004; Ronchetti et al., 2006; Pedrycz et al., 2011; Sillitti et al., 2012; Pedrycz et al., 2012; Janes and Succi, 2014; Coman et al., 2014), the current systems could be expanded, and further systems could be developed. This platform is necessarily going to be open-source to provide a needed level of trust and oversight (Succi et al., 2001b; Kovács et al., 2004; Paulson et al., 2004; Rossi et al., 2010; Petrinja et al., 2010; Fitzgerald et al., 2011; Rossi et al., 2012; Di Bella et al., 2013). Thus, citizens who are developers would be able to extend and modify the platform.

3.2 Security and DevOps

A standard would be developed to guide the no-code platform developers and ensure the quality of the produced software. Tools having such a high impact on the city and the society would require special environments to be developed and deployed. Here is where DevOps engineers would come into play. Developing highly scalable, secure, and performant environments for building, testing, and deploying various forms of software is no easy feat. The no-code tools would range from mobile apps to web apps, to smart home applications and smart assistant bots. Furthermore, in addition to the applications the users interact with, there are the back ends that handle the applications logic, databases, storage, and the authentication and communication with the city API. The back-end components would include serverless functions, containers, web-hooks, and notification systems.

All of these no-code tools working together present a significant risk in terms of integration, security, and managing deployments. DevOps would allow multiple developers to edit their versions of the applications and deploy them without overwhelming hassle or conflict. The DevOps engineers would need to build a system that takes the no-code tool projects as input (as source code) and provide information on the build and deployment stages. An essential requirement would be ensuring the security of the applications through analyzing the code for vulnerabilities or exposed private keys for example. The endpoints would be deployed behind a global API gateway that handles authentication and load balancing.

4 Applications

The following are a sample of the sort of tools that could be offered to smart city citizens.

4.1 Public Records Database

Instead of relying on traditional methods of data entry and storage, a smart city could rely on newer, more reliable tools and methods of storing and manipulating data. A public ledger system could serve as the core of many of the city's systems. APIs could be developed to interact with this ledger, and tools could be developed to allow users to access this database and add records. Uses of such a tool could extend to the court system, trading platforms, public contracts, house leases, and so on.

Extending on the public ledger system, no-code tools could be developed to allow citizens to develop applications for managing businesses, buying and selling, renting, and employing. Businesses would be able to develop internal tools that interact with the city-wide systems such as driver-less cars or delivery systems.

4.2 Electronic Governments

Electronic governments are becoming more widespread. People are gaining more access and control over their interactions and relationships with their government through online services. Publicly funded services such as the police, the fire department, cleaning services, etc., incur a cost on the citizen in the form of taxes. Imagine if these services, which are connected to the city API, could inform the citizen of their impact, benefits, and costs. The citizen would additionally be able to calculate and submit all tax documents required by the city government and track any additional documents related to his residence in the city. Electronic voting has been a controversial topic due to the many security and fraud concerns. Nevertheless, if it is going to be used, it better comes from the developers who live in the city and can oversee and assess the security standards of the voting system. Citizens would then be able to engage with the government in a more granular fashion or delegate their voting rights to other citizens who act as a representative but on a smaller scale.

4.3 *Managing Public Transportation*

Business and public institutes can develop tools for managing the transportation of their employees using the various transportation systems found in the city. They would be able to interact with the busses, metros, trains, or taxis (manned or driverless). They could develop tools for citizens to adjust their preferences and adapt the transpiration to their schedules and needs. Such tools that interact with a city-wide system would allow the city to run more smoothly, reducing traffic jams, congested metros, and wasted commute time.

4.4 *Shopping and Services*

An ideal usage of the no-code tools would be in the form of tools for shops and service providers. Each shop would be able to build custom product pages with various customization options, but more importantly, it can communicate with city systems such as the delivery systems. Service providers would be able to create software that presents their services, the different packages they offer, create contracts (public or private), and integrate with the city in terms of tax calculations and providing utilities.

5 Smart Assistants

As with every innovative endeavor, it takes iterations to improve and optimize the process. In a previous work (ElBatanony and Succi, 2021b), we presented a vision of how no-code can lead to a new era of software development. Today, we suggested the use of no-code tools coupled with software engineering principles to facilitate the communication between citizens and the smart city. Tomorrow, we believe that this process could be abstracted even further and optimized in the form of a smart assistant that is capable of replacing these interfaces and APIs.

The smart assistants of today, such as the Google Assistant[3] or Apple's Siri[4], are already capable of significant non-trivial actions with complicated steps. Powerful APIs are developed for developers to harness the power of these smart systems to ease the experience of the user. Tools such as Dialogflow by Google make this technology more accessible to the average developer by abstracting away many of the implementation details and deployment hassles and by relying on machine learning at its core.

At the time of writing this book, Google is leading these innovative efforts through their breakthrough conversational AI technology. The product under development is called *LaMDA*, short for "Language Model for Dialogue Applications," and it is designed to engage the user in free-flow conversation in any domain. Although the technology is not yet released, the results are very promising. If Google follows through, we could expect to have an AI conversational system that can help our smart city residents to first understand the city systems, learn how to interact with them, and then issue commands to this AI to alter and customize the systems to their needs.

In the 1960s and 1970s, the general perception of the future in the 2000s was too optimistic. We thought that we would have many elements from what we still consider science fiction, such as flying cars and the like. Many of these ideas came into existence in a rather anticlimactic fashion. We do not have robotic butlers, but we have Roombas (robotic vacuum cleaners). The brightest example that comes to mind when thinking of smart cities and smart assistants would be the J.A.R.V.I.S AI, the household butler of Tony Stark (Iron Man), from the Marvel Cinematic Universe. It is not unlikely in the near future, to have some form of smart butler AI system that can help us with interacting with our homes and our cities. That is a future that we are excited for and looking forward to.

6 Summary

In this chapter, we explored our expectations of the citizens of the smart city, and how they can interact with the city systems. We discussed the no-code movement and the benefits it brings when applied in a smart city scenario. Example applications of the no-code platforms were provided. The technologies required to produce such a city-wide platform, such as DevOps, ledger systems, and lean and agile software development methodologies, were considered. Lastly, a future vision of smart assistants helping the citizens make the most out of a smart city was explored.

No-code could make a significant change in how we plan and construct our future cities. The way citizens interact with their environment is shaped and limited by

[3] https://assistant.google.com/

[4] https://www.apple.com/siri/

many aspects, and no-code could be a solution to many of the interactions between a citizen and city systems and infrastructures.

Acknowledgments This work was generously supported by Innopolis University.

References

Apple: Siri. Retrieved September 17, 2021, from https://apple.com/siri

Clark, J., Clarke, C., De Panfilis, S., Granatella, G., Predonzani, P., Sillitti, A., Succi, G., & Vernazza, T. (2004). Selecting components in large cots repositories. *Journal of Systems and Software*, *73*(2), 323–331.

Coman, I. D., Robillard, P. N., Sillitti, A., & Succi, G. (2014). Cooperation, collaboration and pair-programming: Field studies on backup behavior. *Journal of Systems and Software*, *91*, 124–134.

Corral, L., Georgiev, A. B., Sillitti, A., & Succi, G. (2013). A method for characterizing energy consumption in Android smartphones. In *Green and Sustainable Software (GREENS 2013), 2nd International Workshop on* (pp. 38–45). IEEE.

Corral, L., Georgiev, A. B., Sillitti, A., & Succi, G. (2014). Can execution time describe accurately the energy consumption of mobile apps? An experiment in Android. In *Proceedings of the 3rd International Workshop on Green and Sustainable Software* (pp. 31–37). ACM.

Corral, L., Sillitti, A., & Succi, G. (2015). Software assurance practices for mobile applications. *Computing*, **97**(10), 1001–1022.

Corral, L., Sillitti, A., Succi, G., Garibbo, A., & Ramella, P. (2011). Evolution of mobile software development from platform-specific to web-based multiplatform paradigm. In *Proceedings of the 10th SIGPLAN Symposium on New Ideas, New Paradigms, and Reflections on Programming and Software, Onward! 2011* (pp. 181–183). ACM.

Di Bella, E., Sillitti, A., & Succi, G. (2013). A multivariate classification of open source developers. *Information Sciences*, *221*, 72–83.

ElBatanony, A., & Succi, G. (2021a). The pareto distribution of software features and no-code. In *Proceedings of the 1st ACM SIGPLAN International Workshop on Beyond Code: No Code, BCNC 2021* (pp. 18–22). Association for Computing Machinery. https://doi.org/10.1145/3486949.3486964

ElBatanony, A., & Succi, G. (2021b). Towards the no-code era: A vision and plan for the future of software development. In *Proceedings of the 1st ACM SIGPLAN International Workshop on Beyond Code: No Code, BCNC 2021* (pp. 29–35). Association for Computing Machinery. https://doi.org/10.1145/3486949.3486965

Fitzgerald, B., Kesan, J. P., Russo, B., Shaikh, M., & Succi, G. (2011). *Adopting open source software: A practical guide*. The MIT Press.

GitHub: Github Copilot. Retrieved September 17, 2021, from https://copilot.github.com

Google: Dialogflow. Retrieved September 17, 2021, from https://cloud.google.com/dialogflow

Google: Google Assistant. Retrieved September 17, 2021, from https://assistant.google.com/

Janes, A., & Succi, G. (2014). *Lean software development in action*. Springer.

Kivi, J., Haydon, D., Hayes, J., Schneider, R., & Succi, G. (2000). Extreme programming: a university team design experience. In *2000 Canadian Conference on Electrical and Computer Engineering. Conference Proceedings. Navigating to a New Era (Cat. No.00TH8492)* (Vol. 2, pp. 816–820).

Kovács, G. L., Drozdik, S., Zuliani, P., & Succi, G. (2004). Open source software for the public administration. In *Proceedings of the 6th International Workshop on Computer Science and Information Technologies*.

Marino, G., & Succi, G. (1989). Data structures for parallel execution of functional languages. In *Proceedings of the Parallel Architectures and Languages Europe, Volume II: Parallel Languages, PARLE '89* (pp. 346–356). Springer-Verlag.

Maurer, F., Succi, G., Holz, H., Kötting, B., Goldmann, S., & Dellen, B. (1999). Software process support over the Internet. In *Proceedings of the 21st International Conference on Software Engineering, ICSE '99* (pp. 642–645). ACM.

Moser, R., Pedrycz, W., & Succi, G. (2008a). A comparative analysis of the efficiency of change metrics and static code attributes for defect prediction. In *Proceedings of the 30th International Conference on Software Engineering, ICSE 2008* (pp. 181–190). ACM.

Moser, R., Pedrycz, W., & Succi, G. (2008b). Analysis of the reliability of a subset of change metrics for defect prediction. In *Proceedings of the Second ACM-IEEE International Symposium on Empirical Software Engineering and Measurement, ESEM '08* (pp. 309–311). ACM.

Musílek, P., Pedrycz, W., Sun, N., & Succi, G. (2002). On the sensitivity of COCOMO II software cost estimation model. In *Proceedings of the 8th International Symposium on Software Metrics, METRICS '02* (pp. 13–20). IEEE Computer Society.

OpenAI: Gpt-3. Retrieved Sptember 17, 2021, from https://openai.com/blog/openai-api/

Paulson, J. W., Succi, G., & Eberlein, A. (2004). An empirical study of open-source and closed-source software products. *IEEE Transactions on Software Engineering*, *30*(4), 246–256.

Pedrycz, W., Russo, B., & Succi, G. (2011). A model of job satisfaction for collaborative development processes. *Journal of Systems and Software*, *84*(5), 739–752.

Pedrycz, W., Russo, B., & Succi, G. (2012). Knowledge transfer in system modeling and its realization through an optimal allocation of information granularity. *Applied Soft Computing*, *12*(8), 1985–1995.

Pedrycz, W., & Succi, G. (2005). Genetic granular classifiers in modeling software quality. *Journal of Systems and Software*, *76*(3), 277–285.

Petrinja, E., Sillitti, A., & Succi, G. (2010). Comparing OpenBRR, QSOS, and OMM assessment models. In *Open Source Software: New Horizons - Proceedings of the 6th International IFIP WG 2.13 Conference on Open Source Systems, OSS 2010* (pp. 224–238). Springer.

Prosus: Stack overflow. Retrieved September 17, 2021, from https://stackoverflow.com/

Ronchetti, M., Succi, G., Pedrycz, W., & Russo, B. (2006). Early estimation of software size in object oriented environments a case study in a CMM level 3 software firm. *Information Sciences*, *176*(5), 475–489.

Rossi, B., Russo, B., & Succi, G. (2010). Modelling failures occurrences of open source software with reliability growth. In *Open Source Software: New Horizons - Proceedings of the 6th International IFIP WG 2.13 Conference on Open Source Systems, OSS 2010* (pp. 268–280). Springer.

Rossi, B., Russo, B., & Succi, G. (2012). Adoption of free/libre open source software in public organizations: factors of impact. *Information Technology & People*, *25*(2), 156–187.

Scotto, M., Sillitti, A., Succi, G., & Vernazza, T. (2004). A relational approach to software metrics. In *Proceedings of the 2004 ACM Symposium on Applied Computing, SAC '04* (pp. 1536–1540). ACM.

Scotto, M., Sillitti, A., Succi, G., & Vernazza, T. (2006). A non-invasive approach to product metrics collection. *Journal of Systems Architecture*, *52*(11), 668–675.

Sillitti, A., Janes, A., Succi, G., & Vernazza, T. (2004). Measures for mobile users: an architecture. *Journal of Systems Architecture*, *50*(7), 393–405

Sillitti, A., Succi, G., & Vlasenko, J. (2012). Understanding the impact of pair programming on developers attention: A case study on a large industrial experimentation. In *Proceedings of the 34th International Conference on Software Engineering, ICSE '12* (pp. 1094–1101). IEEE Press.

Sillitti, A., Vernazza, T., & Succi, G. (2002). Service oriented programming: A new paradigm of software reuse. In *Proceedings of the 7th International Conference on Software Reuse* (pp. 269–280). Springer.

Succi, G., Benedicenti, L., & Vernazza, T. (2001a). Analysis of the effects of software reuse on customer satisfaction in an RPG environment. *IEEE Transactions on Software Engineering*, *27*(5), 473–479.

Succi, G., Paulson, J., & Eberlein, A. (2001b). Preliminary results from an empirical study on the growth of open source and commercial software products. In *EDSER-3 Workshop* (pp. 14–15).

Succi, G., Pedrycz, W., Marchesi, M., & Williams, L. (2002). Preliminary analysis of the effects of pair programming on job satisfaction. In *Proceedings of the 3rd International Conference on Extreme Programming (XP)* (pp. 212–215).

Valerio, A., Succi, G., & Fenaroli, M. (1997). Domain analysis and framework-based software development. ACM SIGAPP Applied Computing Review **5**(2), 4–15.

Vernazza, T., Granatella, G., Succi, G., Benedicenti, L., & Mintchev, M. (2000). Defining metrics for software components. In *Proceedings of the World Multiconference on Systemics, Cybernetics and Informatics* (Vol. XI, pp. 16–23).

Part IV
Smart Cities Innovative Technologies

A Big Data Analytics Conceptual Framework for a Smart City: A Case Study

Andreas F. Gkontzis, Dimitrios Kalles, Evgenia Paxinou, Rozita Tsoni, and Vassilios S. Verykios

1 Introduction

In modern cities, overpopulation causes huge and complicated issues such as housing crisis, congestion, unemployment, air pollution, poverty, drug abuse, and energy tension (Joshi et al., 2016). As a result, the so-called megacities are a hotbed for a plethora of problems related to the environment, health, society, etc. (Washburn et al., 2010). Our over-connected world spreads these problems around, transferring the most of them in every place on Earth.

A representative example of modern urban problems is certainly the recent COVID-19 pandemic situation. This big crisis was quickly magnified in big cities. The coexistence with our neighbors, our work colleagues, or our schoolmates who did not necessarily follow the hygiene precautions all the time seriously jeopardized our healthy lifestyle. Furthermore, during the pandemic days, several domains in our communities, such as education, transportation, and entertainment, have been strongly affected (Kunzmann, 2020). An additional and critical big city situation is also "The Tragedy of the Commons" (Garrett, 1968). According to this problem, the uncoordinated and selfish use of the common resources like water, atmosphere, ocean, public transportation, and roads could lead to the collapse of the open-access resource systems. The two above, but also many more serious problems, seem to be, most of the times, difficult to predict and even more difficult to control.

The need to come up with innovative solutions to all these problems is indisputable (Jalali et al., 2015). Smarting a city by using the traditional networks and

A. F. Gkontzis (✉)
Representative of the Innovation Network and Best Practices of Patras Municipality, Municipality of Patras, Patras, Greece
e-mail: andreas.gkontzis@ac.eap.gr

D. Kalles · E. Paxinou · R. Tsoni · V. S. Verykios
School of Science and Technology, Hellenic Open University, Patras, Greece

P. Fitsilis (ed.), *Building on Smart Cities Skills and Competences*, Internet of Things,
https://doi.org/10.1007/978-3-030-97818-1_16

services by making them more efficient with the use of digital solutions, for the benefit of its inhabitants, is a fashionable trend in the European urban development (Garagliu et al., 2011). Having in mind that the interaction of the citizens who live in a big or a small city is constant, there is a great opportunity to analyze all their digital traces and the big data that pile up, extracting this way knowledge that can be lifesaving for the management and the efficiency of the urban environment.

When seeking to explore the potentials of a city for smarting it, one needs to shed light on what benefits smart cities may bring about to the city. A smart city can be seen through three components, technological, human, and institutional, where the main sectors can be seen as energy, transportation, healthcare, safety, education, and environment (Nam & Pardo, 2011). Smart applications should be developed to improve the transport networks, to upgrade water supply and waste disposal facilities, and to efficiently light and heat the buildings to provide a more interactive and responsive city administration to offer safer public spaces that meet also the needs of an aging population (Hall et al., 2000).

At this point it is worth mentioning that nowadays, several diverse resources of data can be combined offering a great elevation in every scientific field. When talking about smart cities, big data analysis techniques can be used to process the data that are derived from all the smart applications, so as to improve the citizens' quality of life. In this project we present an effort to connect, through an interoperability center, five smart services already developed in the Municipality of Patras and analyze the different data sets through the use of new data management and processing techniques, known as data analytics pipelines.

2 Urban Problems and Smart Cities

Technology seems to play an important role in our lives. By surfing on the Internet, participating on the social networks, or using a self-driving car, our age is flooded with raw data that can provide useful information, if processed by the pundits of the big data analytics (Lotsari et al., 2014; Kagklis et al., 2016; Tsoni et al., 2019; Verykios et al., 2021). The digitalized services provided by the municipalities are also a rich source that contain an ever-increasing volume of important but hidden information. These data, if analyzed properly, could be of vital importance for the improvement of the citizens' daily interventions and for making decisions on creating new public services for them.

The ever-increasing production of the data coming from various sources in the municipalities hides trends that the city perceives only as a result, without having the opportunity to intervene. As the rate of technological development shows no signs of slowing down, sooner or later municipalities will be forced to adopt big data practices to improve the city inhabitants' lives. Smart services must be developed so that this rapid spread of the new technologies could be beneficial for the citizens. Data derived from these smart services could be transformed to necessary knowledge for taking short-term or long-term decisions.

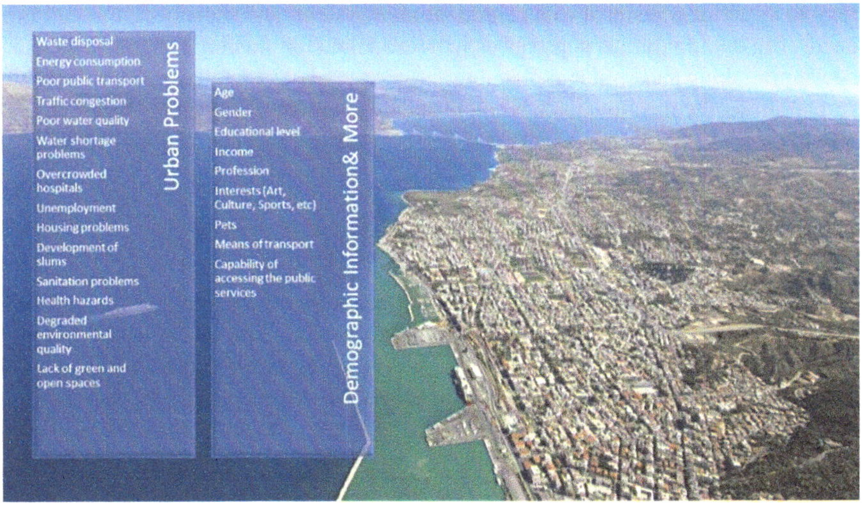

Fig. 1 Useful information for designing the ecosystem of a smart city

A smart city should be able to address the urban challenges, such as the waste shrinkage, the unemployment, the lack of green and open spaces, the accessibility to the public facilities, etc., in a direct and effective way (Su et al., 2011; Albino et al., 2015). Smart interventions, such as the exploitation of neglected industry buildings in theme parks or the upgrading and expansion of existing infrastructure, electricity networks, etc., could provide a chance to the city to be reborn. At the same time, dynamic and constant monitoring of the smart services, for any strategy adjustments, must occur to ensure their viability.

Smarting the city is a complicated procedure due to the complexity of the city itself (Reichental, 2020). The city is not an automated system that can be easily understood and predicted, but rather a living system that evolves every day, through variations and developments of its physical constructs, economic and political activities, social and cultural settings, and ecological systems (Yencken, 2013). The ecosystem of a smart city should address most of the urban problems based on the potentials of the city and on the citizens' needs, activities, and habits (Fig. 1), (Cocchia, 2014).

3 Our Methodology

In this study, we focus on big data analysis techniques to smart the city of Patras. In this section, we present our proposed methodology toward this goal. In more details, we propose an automated data analysis framework along with its accompanying infrastructure. Through this framework, we intend to collect data from different remote systems, transfer it to a central storage system, and transform it to suit our needs. Big data analytical approaches will be applied for drawing insights to support decision-making and strategic planning.

Our project will follow the next eight stages of the data science process:

3.1 The Project Understanding

This stage is crucial as it helps to clarify the goal of the project. Once the problem has been clearly stated, the data analyst can define more easily the approach that could be followed in the subsequent phases, in order to solve it (Kaufman et al., 2012). Possible questions that could be addressed in this stage are the following:

i. What exactly is the problem?
ii. What are the expected benefits?
iii. How would a solution look like?
iv. Which are the objectives of the project?
v. What is known about the domain?

3.2 The Data Understanding

In this stage, any knowledge about the data, about the needs that the data will satisfy, and about its content and location is gathered (Evans & Fisher, 2002). Possible questions that could be addressed in this stage are the following:

i. What data are available?
ii. Are the data sufficient in quality and quantity?

3.3 The Data Collection

The data collection is the third step in the data science process (Kohavi & Longbotham, 2007). Data can be extracted from different smart systems or applications that have been already developed in the target municipality. In our project, the data will be gathered from a web application where citizens can report problems occurring in their area. They can monitor the resolution through an application for the management of municipal lighting, an application for the registration of building infrastructure of the municipality, and an application for the monitoring of the municipality's vehicle fleet and the fuel ring.

3.4 The Data Storage

The data can be stored in a systematic and organized way for further exploitation. Systems that are used for data hosting are data warehouses, data lakes, etc. (Kohavi et al., 2000; Shearer, 2000). In this stage, the creation of a bridge between the different subsystems so as to collect their data in real time happens. This bridge is necessary as each one of the above applications is an independent subsystem with its own data access and retention policy, its own data generation flow, and its own interface.

3.5 The Data Preparation

Data collected from different systems and applications are very likely to have many peculiarities in terms of their structure and their formatting. Cleaning data and removing duplicates and detecting records that show some probabilistic similarity in nature are major challenges in this process, (Elmagarmid et al., 2006; Karapiperis & Verykios, 2014, 2016; Karapiperis et al., 2017). Furthermore, the integration of the data anonymization methods in the early stages of the data pre-processing is also essential (Verykios et al., 2004). The adherence to the GDPR law and similar requirements and the ethical aspect of research demand the protection of the anonymity and of the personal data. A very important part of the data preparation process is the selection of suitable features for the problem at hand, the reduction of high dimensionality, as well as the selection of suitable data samples, when the data themselves are very large in volume. Possible questions that could be addressed in this stage are the following:

i. Which data should we concentrate on?
ii. How can we increase the data quality?
iii. How could the data be best transformed for modeling?

3.6 The Modeling

The process of extracting knowledge from data, although it is more straightforward, requires a profound knowledge of the underlying characteristics of the data and the nature of the problem that needs to be solved (Hernández & Stolfo, 1995). The data themselves often guide a data scientist to adopt different analytical techniques, while most of the time, even simple improvements require very good mastering of the domain knowledge and of the intrinsic operation of the algorithms and their parameters that must be selected to achieve the best possible result. For the supervised techniques, data need to include a specific feature that characterizes them with respect to their resemblance with a certain category of the problem. Most of the time such a feature does not exist in the data, so other unsupervised techniques must be applied. Well-known supervised techniques comprise the classification and the regression, while unsupervised techniques include clustering, frequent itemset mining, and association rule discovery. Recent techniques applied to data of all kinds include text mining, social network analysis, time series analysis, outlier detection, and intrusion detection. Possible questions that could be addressed in this stage are the following:

i. What kind of model architecture suits the problem best?
ii. What is the best analytical technique to get the model?
iii. How good does the model perform technically?

3.7 The Evaluation

The evaluation is a very essential step in the whole process (Kohavi et al., 2009). In this stage, an investigation is occurred whether the proposed model fulfills the project requirements. In this stage, possible questions that could be addressed are the following:

i. How effective is the model in terms of project requirements?
ii. Have the objectives of the project been achieved?

3.8 The Deployment

In case the model passes the previous stage of evaluation, it can be deployed (Kohavi & Parekh, 2003). Possible questions that could be addressed in this stage are the following:

i. How can we best deploy the model?
ii. How do we know that the model is still valid?

In this project, we aim to automate all the tasks of the intelligent process, required in the above stages (Fig. 2), in order to discover new insights and knowledge from available data sets. These data sets are originating in different applications and services from the Municipality of Patras, a city located in Western Greece. The implementation of the process that will facilitate the analysis of the data sets will be accomplished through the use of new data management and processing techniques, known as data analytics pipelines. Data analytics pipelines are sets of

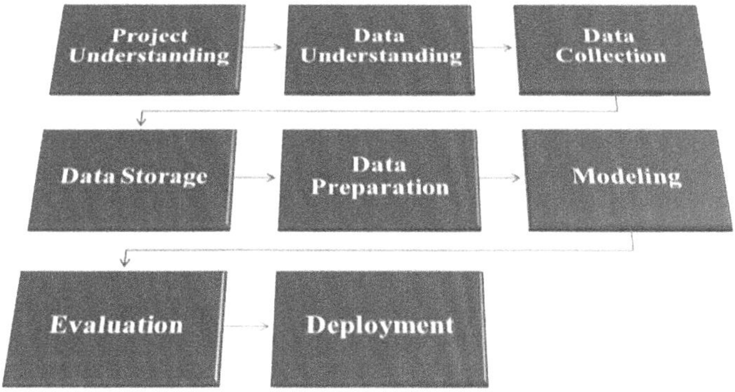

Fig. 2 The eight stages of the data science process followed in this project

well-organized, articulated processes that can support in a holistic way analytics, reporting, and machine learning capabilities (Densmore, 2021).

4 The City of Patras and Our Goal

Patras is the third largest city and the second largest port in Greece. It is located in the northwestern part of the Peloponnese peninsula, in the region of Western Greece, and it is the capital of Achaia. As of the 2011 census, the city of Patras has a population of 167,446 and the municipal unit has 170,896 inhabitants.

So far, the city of Patras has made timid steps toward its digital transformation and the adoption of smart services. The Municipality of Patras, facing most of the urbanization problems that are common in most of the cities in Southern Europe, had to guide smart city development through the introduction of smart services. Despite the sincere efforts of the technical teams in charge in the Municipality, many of the introduced smart services malfunction is mainly due to the lack of substantial support. The technical support staff in the Municipality is responsible mostly for the existing infrastructure which is not sophisticated and adequate enough to address the needs of the new smart services, a problem that is also highlighted by Cuquet et al. (2017). On the other hand, for the newly acquired staff even if they are educated on modern technologies, they cannot support, without further training, the multilevel management of such sophisticated systems, like the smart sensors and the smart platforms. As a result, in a short period of time, the control and the maintenance of all the already developed smart services will become of secondary importance in the Municipality agenda, and they will finally fade away.

The stakeholders who undertake the task of creating and maintaining a data analysis system to support smart city environments should obtain a set of relevant skills. These skills should be aligned with the demands of a KDD (Knowledge

Discovery from Databases) process. The first step that is data collection and preparation requires a combination of data understanding along with domain knowledge. Researchers should be able to spot problems, create research questions, and define the types of data available that can lead to the solution. It is of high importance to understand data within the context of a smart city environment. It is also essenand their potential use to resolve practical and everyday issues in order to improve the quality of citizens' lives. Initial visualization can be very helpful in this sense because they provide highly interpretable information. Thus, persons in charge of the analysis should be able to create and interpret graphs, making sure that the results would be understandable for the rest of the municipality staff involved.

Combining data from various sources is the key for a productive analysis. Therefore, some technical skills are required on record linkage techniques and managing large databases. Although programming skills provide a safe way to handle data preparation and also the following steps of the analysis, there is a shift of focus on platforms that use simple command-line tools in graphical interfaces. These tools allow people with no programming background, who, however, may have excellent domain knowledge in the field of interest, to create modern analytics workflows avoiding the complexity of a programming language. At the same time, they provide automation and reproducibility. Data management and curation can be achieved in a repeatable way.

Data modeling can also be addressed by the abovementioned platforms. However, numerical computing skills are required to ensure the knowledge of how algorithms work so that the process would be effectively superintended. Additionally, the understanding of the inner process of the modeling stage is important to draw valid inferences.

Especially, in a smart city application where human behavior is involved, it is crucial to be able to explain the results of the methods used in the analysis and to address issues of privacy, biases, and fairness. Machine learning techniques should be used to reap the benefits of artificial intelligence and provide solutions that respond to the true needs of the community and fit its particular characteristics, as long as their results can be explained and sufficiently justified.

While smarting a city, it is significant to present and evaluate the outcomes in a beneficial way. Graphic packages for data visualizations should be incorporated, and data storytelling skills will allow creating a final outcome that would be readable and understandable by nonexperts. The transformation of raw data into knowledge should be thoroughly designed and closely monitored so that this knowledge would be meaningful and actionable.

A solution regarding the management of the smart systems could have been to assign it to companies with which the Municipality of Patras already collaborates. Such a collaboration would have been ineffective since the contribution of these companies would have been limited to the recording of the flow data of the smart sensors and to the depiction of some statistical results derived from these big data. Anyway, the process of the management and the analysis of the big data that could give more descriptive and insightful suggestions for solving the urban problems are not included in any submitted proposal to the Municipality calls.

Our study reflects a framework for the interoperability of the various systems that manage smart sensors and services, in the Municipality of Patras. The integration of different systems is enriched by incorporating machine learning and visualization techniques for the calculation of evaluation indicators, the prediction of trends, and the visualization of results, in real time.

Our goal is to provide a central overview of the workflow between all current and future smart systems of the municipality in different groups of each department and each administration, in order to bridge the gap of immediate and collaborative support of the malfunctions of each system. At the same time, it will enlighten on a daily basis the decision-makers regarding the forecast trends and standards for each individual smart service and the correlations of the data of the smart systems. The interoperability center will integrate the data of all smart services into a central node, the attention of all services to the smooth operation of each system, and the advantage of all team decisions at each level of municipal administration to optimize timely decisions for the benefit of citizens. Figure 3 presents all the expected benefits after smarting the city of Patras.

More analytically, Fig. 4 presents the measurements that, if taken, could ensure the fulfillment of the aforementioned goals. The most important benefits are the public safety, the civil protection, the energy efficiency, the interaction with the citizens, the protection of the environment, and the accuracy of the transportation.

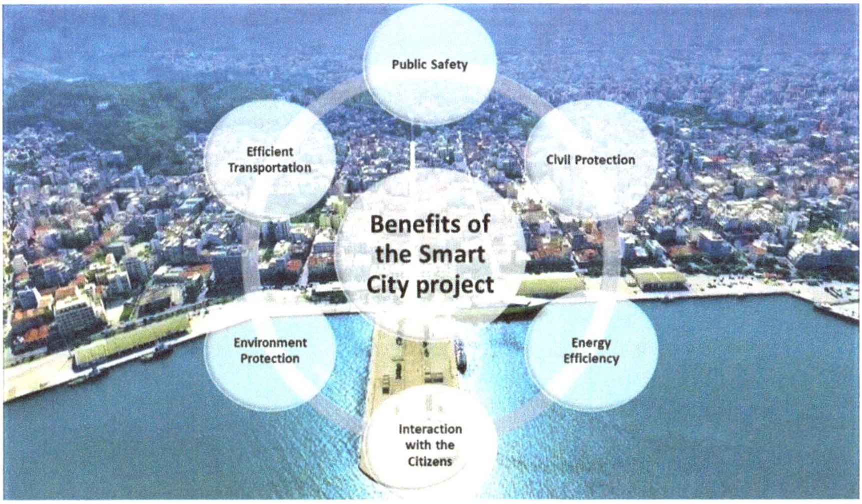

Fig. 3 The benefits of smarting the city of Patras

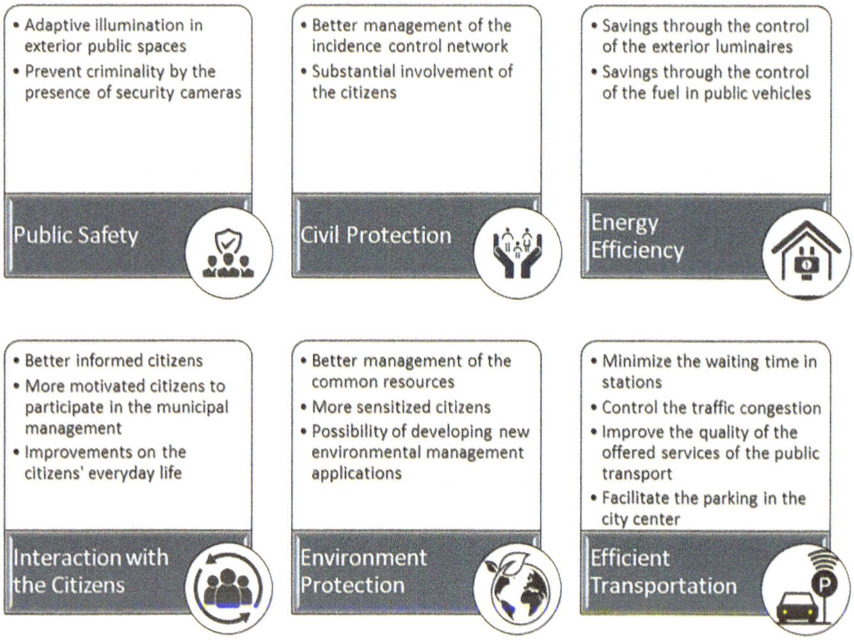

Fig. 4 The necessary measurements taken to fulfill the goals of smartening the city of Patras

5 The Developed Applications in the Municipality of Patras

Currently, systems and applications based on different technologies, forms, and schemes run in the Municipality of Patras acquiring the minimum possible support for their seamless operation, as a result, providing to all the interested parties the minimum of their potentials. The collected data are not used to draw conclusions, and no effort is made to correlate the data derived from these intelligent systems so as to make interpretations and extract new useful knowledge. Furthermore, these already developed systems do not include predictability models, an essential part of smart service personalization applications.

In more details, for the time being, the following smart systems are provided by the Municipality of Patras:

5.1 The Sense.City Application

The Sense.City application is a platform that enables the citizens to report and ask for help on various issues improving this way their daily city life. In details, the citizens can report and request from the municipality services, to remove a damaged bin or an oversized object, to cut some tree branches, to clean a specific street, to maintain a bus station, to fix a destroyed pavement, etc. This platform was released in January 1, 2018, and it is characterized by web and mobile applications adopting tools for activating citizens as city sensors. The citizens can download the Sense.City application on their smartphone, tablet, or PC, or they can visit the platform by clicking on the https://patras.sense.city. Thus, the citizens are actively involved in the processes of detecting and communicating not only to the services but also to their fellow citizens, a problem that appears in their neighborhood or in the wider surrounding area, aiming to immediate intervention and solution.

The mobile application was developed in order to make simpler and more efficient for the citizens to get in touch with the Municipality services and report directly a problem. The advantages of using this app are various. More analytically, the citizens can remotely report the problem by choosing to remain anonymous, or not. They can also get the certification of their report either via email or SMS, on their mobile. They can also get a detailed overview of the so far reported problems, divided in categories. On the other side, the Municipality is informed about the problems and directly assigns them to the appropriate department.

Figure 5 presents a diagram of the flow of the problems reported in the Sense.City platform from March 4 until April 4, 2021 (e.g., 04/03 stands for the fourth of March or 01/04 stands for the first of April).

5.2 The Municipal Lighting Management Application

Street lighting and mobility systems are often the first public services to come under scrutiny when a city heads for smartness. Based on that, both in Corinthou and Kanakari street, 232 new generation luminaires have been installed in combination with a smart communication controller that receives the luminaire data and transmits them over a wireless network to a cloud server with web platform management environment. Figure 6 presents a screenshot of the aforementioned platform.

Data that present the energy consumed by the luminaires, both individually and collectively during the day, are collected. The energy saving caption is presented as

Fig. 5 The flow of the problems reported per day, from March 4, 2021, to April 4, 2021

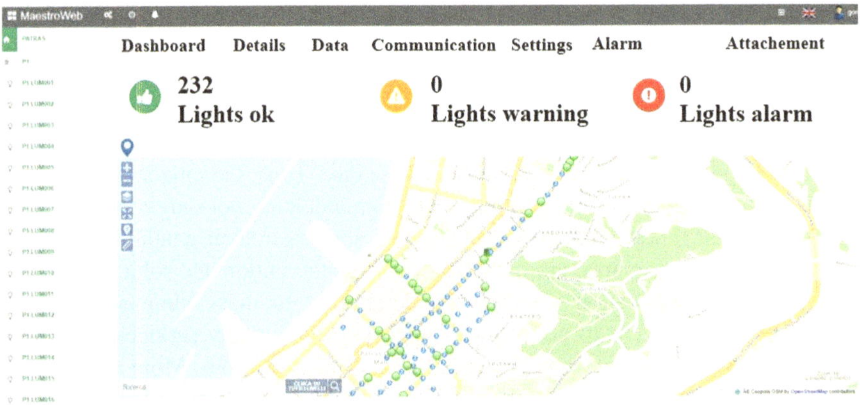

Fig. 6 A Screenshot from the lighting management application

the difference between the consumption of the new 70 W LED luminaires compared to the one of the previously used 250 W/150 W luminaires.

The platform has a built-in API and can be integrated into the interoperability center so as to connect data to other applications, such as environmental sensors, vehicle fleet monitoring applications, GIS, etc.

5.3 *The Computerized Maintenance Management System Application*

The computerized maintenance management system (GMMS) application contains information attributed to dashboards and diagrams for school and building complexes of the Municipality of Patras concerning the energy consumption per building, the fire safety studies and certificates, the defective lighting, the electromechanical equipment, the furniture, the building components, and the outdoor areas (Fig. 7). In addition, this application allows a multilevel management of logistics, maintenance and work scheduling information, and also an internal communication through the employees of different offices, in different departments in the Municipality.

The integration of this application with the interoperable center can be done through the API of the application, and the data can be correlated with those of the sense city platform.

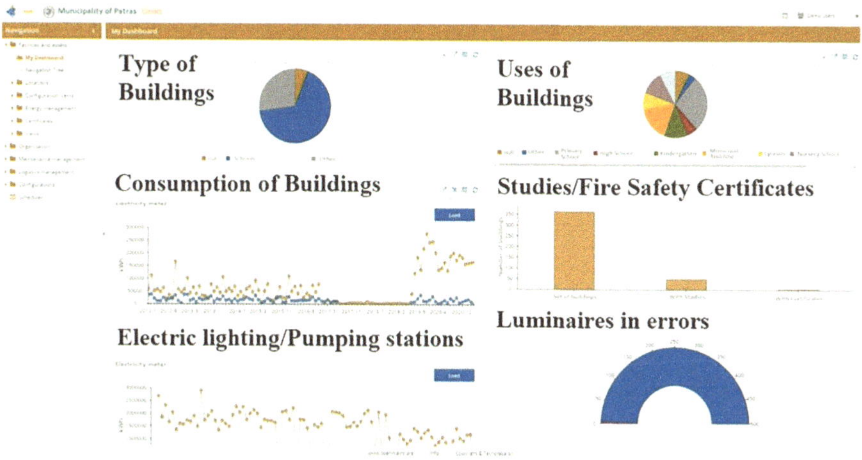

Fig. 7 A screenshot from the CMMS application

5.4 *The Geographic Information System Application*

The geographic information system (GIS) application is a system that can be described as an integrated system that combines hardware, software, and spatial data for the purpose of receiving, managing, analyzing, and displaying geographic information data. GISs track events (e.g., environmental disasters), activities (e.g., a construction), and objects (e.g., facilities, institutions, or natural resources) but also the areas that all the above occur and are located (Fig. 8). Currently, the Municipality of Patras reaps only a part of the capabilities of this system, and the database for geospatial data, plots, road names, settlement zones, road modifications, school units, and municipal plots is manually updated.

A very large volume of geospatial data information of the Municipality has been digitized, and it can be represented on a level map. As a first step, a spatial database could be created so that the GIS could provide useful information for both the Municipality and the citizens.

5.5 *The Vehicle Fleet and Fuel Ring Monitoring Application*

Currently, the vehicle fleet platform of this application manages 144 garbage trucks of the Municipality. As these trucks are equipped with a network card, the platform displays, in real time and through a web interface, the route followed by the vehicles, while at the same time, any deviations are also notified. Additionally, this application can distinguish, depending on the speed and the way of driving, which vehicle consumes more fuel than expected.

Fig. 8 A screenshot of the GIS application

Based on the traffic network updates, there is a constant change of the routes in the platform so as the collection of the wastes occurs faster and more efficiently.

The fuel rings are controlled by a separate platform related to the vehicle fleet monitoring application. Each vehicle has a fuel ring in its tank where the registration number of the vehicle, its refueling time, the pump that is used to fill in its tank, the current fuel price and quantity, and, finally, the receipt report are recorded.

A correlation of the vehicle fleet and the fuel ring monitoring application can be made through interoperability and user authentication. In this way, the citizens could be informed, in real time and through a dashboard, in which street and at what time a garbage truck is about to pass from their neighborhood. Thus, through this interaction, the citizens could be trained to throw their wastes in the nearest rubbish bin, in specific hours during the day, participating this way in the urban cleanliness.

Additionally, this application could be connected to the Sense.City, the CMMS, the GIS, and the Municipal Lighting Management platforms. Reported problems regarding the city streets, the municipal buildings and other constructions, the rubbish bins, and the recycling stations could be associated with the rescheduling of the vehicles routes and, as a result, with the alterations of the street lighting scenarios.

6 Interoperability: Needs and Benefits

Projects built on interoperability models have a higher success rate, since they provide flexibility and scalability and secure the possibility to integrate various and different applications on the same infrastructure, scaling up and adding more functionalities when and where needed. The interoperability offers a common

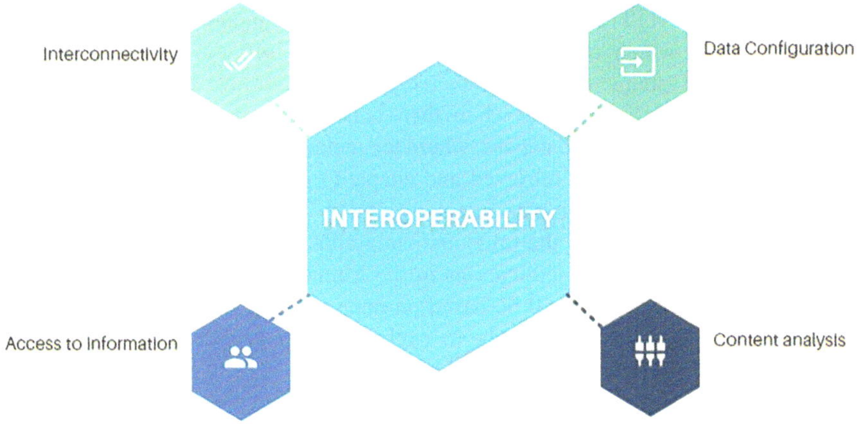

Fig. 9 The interoperability model

ground where different collections of data can be seamlessly integrated into an ecosystem where various independent applications can take advantage and deliver added value services to the citizens. In sequel, this creates an opportunity for scaling up different tasks that can be accomplished by a variety of users in an independent way, as well as for decentralizing the work load in dynamically changing environments.

In this ecosystem, appropriate data standards and data coordination guarantees are required to address the challenges of information exchanging as well as the smooth running of the interconnected systems. In this way, citizens in the Municipality, different departments, and other interested parties can be successfully served through a highly orchestrated centralized ecosystem, backing up the operations of the Municipality (Fig. 9).

The interoperability framework includes the institutional interoperability that legally ensures the collection, exchange, and management of the data and the organizational interoperability that refers to the objectives, procedures, collaborations, and the satisfaction of the involved parties. In addition, the interoperability framework includes the semantic interoperability that ensures the understanding of the concepts of the information exchanged by any application, standardizing the communication code in systems and services. This ensures an efficient processing of data coming from heterogeneous sources.

Finally, the technical interoperability is the ability to transfer and efficiently use the extracting information, in a homogeneous way between infrastructure and software of the IT systems of the Municipality, technically determining the storage, transfer, presentation, and security of each service. The technical interoperability is achieved by applying open communication and control standards and by providing implementations above the level of protocols provided specifically for each system, as well as by implementing a high-level Application Programming Interface

(API) in a commonly accepted practice, such as JSON packages with RESTful API services via HTTP. Thus, interoperability is implemented with peer-to-peer communication, which requires direct knowledge of service interfaces (service APIs) between systems and is characterized by collaborative immediacy as well as indirect collaboration, in cases where there is no API. In these cases, each system must have read access to its database, where the metadata and query-related services provided by the system are retrieved and adapted to the common communication language of the collaboration platform.

The system presented in the previous paragraphs will interconnect the already implemented and installed applications in the Municipality, as well as any future application. All techniques will be based on open standards, such as web services (XML via HTTP), with the ability to exchange data between any system and application in a practical and functional way.

All the intelligent systems that will be implemented by the Municipality in the future should, among other things, be harmonized with the documented implementation of web services – XML via HTTP – e.g., SOAP-, Interface-Programming Interface-APIs. Additionally, the collected data should be comprehensive, timely, and supported by metadata. All systems must provide access to historical data through the RESTful API, in order to avoid system queries either in real time or at a later time, resulting in time delays and depletion of computer resources and storage space. The systems should also be integrated to the interoperability platform according to the aforementioned objectives, adhering to the required standards and the trend monitoring systems for focused interventions.

A documented RESTful API is the system developer' responsibility. Therefore, in case an issue is identified during the integration with the interoperability platform, the system contractor should proceed to the necessary interventions in the forms and structures of the application base, to resolve issues related to the effectiveness of the integration of the application with the platform.

The basic components of the interoperability platform will include RESTful APIs of the connected systems, as well as other web services and Internet data sources (e.g., real-time weather data or traffic data) that can be combined and used in algorithmic models and conditions for highlighting and exploring new trends and categorizing them into classes. Figure 10 presents a simplified example of the resource tree structure of the interoperability platform. In details, APIs use *uri* to access the respective resources. The */oic* prefix defines the mechanism that leads to the organization of the resources of the interoperability ecosystem, */res* is used for resource discovery, and */p* allows for the discovery of information for a specific platform, */d* for the device properties, */con* the configuration of information for a specific device, and */mnt* for maintenance and diagnostics, while */sec* can be used for security.

The interoperability platform must be open, based on the cloud, which will allow the interconnection and visualized communication of the existing subsystems and the future applications of the Municipal departments. It will be an important control point, in terms of the level of management and the processing and analysis of all these data that will be collected by the individual systems. A first-level

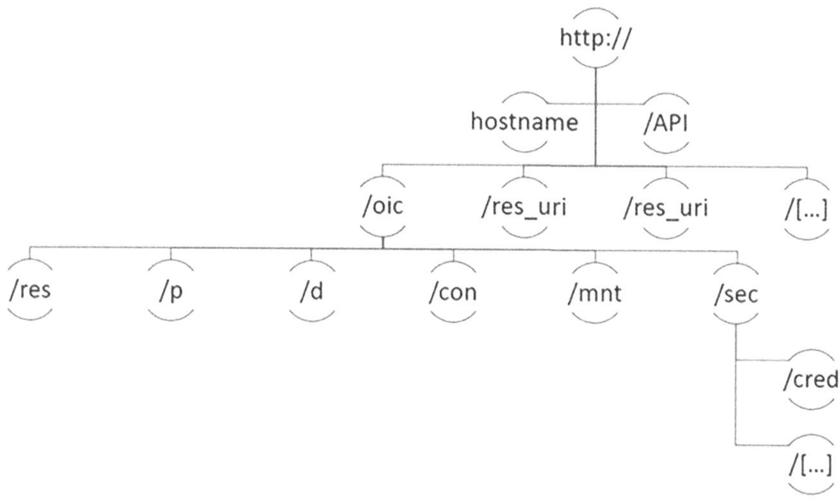

Fig. 10 A resource tree structure of the interoperability platform

control of the data coming from the individual systems will be carried out by the respective department that studied, auctioned, and organized the implementation of each system.

The interoperable platform will be able to securely manage and share data, to ensure that data generated by systems correlation are stored in third-party applications for additional analysis and synthesis and finally to display the data in dashboards. The main goal is the fast adaptability of the services and the immediate and effective interventions aiming to the improvement of the quality of life and the creation of a measurable benefit for the citizens and the enterprises, without additional administrative burden of the executives of the Municipality.

The platform will provide a control center of functions where the detailed tables and sub-tables of data and of the analysis and the results will be presented in screens giving this way the overall picture of the functions of the Municipality in a clear and understandable way. The goal is to minimize the operational complexity between the services of the Municipality and to strengthen the decision-making actions.

In the interoperable platform, the smart applications will be distributed in a separate dashboard, according to the levels of a digital map of the city which will be updated in real time as the data will flow from each application. Due to the multi-level architecture, accredited users will be able to be informed about the services provided in each area of the city with concise visualized tables of results and reports per hour, day, and month and with the possibility to correlate the results derived from the different applications (Fig. 11).

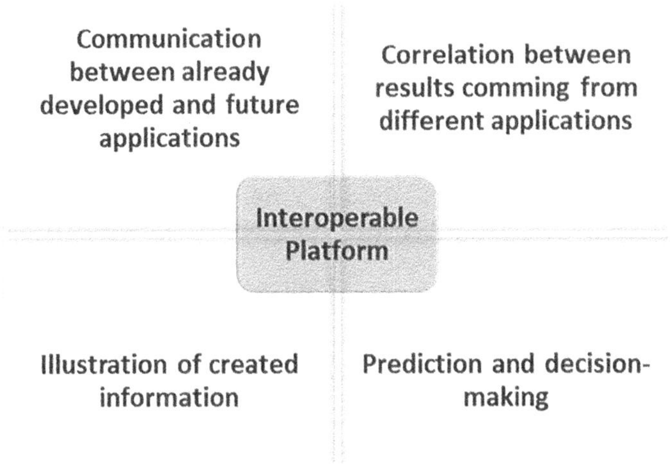

Fig. 11 Functions of the interoperable platform

7 User Access Levels

Figure 12 demonstrates the hierarchy in the Municipality of Patras. Apart from the Mayor and the Deputy Mayors, there are three directorates that supervise numerous different departments. In the following figure, only the departments that control the developed smart applications are presented.

Having in mind the hierarchy of the Municipality presented in Fig. 12, there are three user access levels in the interoperability system. In particular, these levels are the following:

A) Access in the Department Level

The department stakeholders control the operation of the application implemented by their department. They manage properly the workflow of their department and make the necessary interventions depending on the daily reports presented on the specific platform. The dashboard of the application offers a detailed monitoring of the daily data as well as ensures the data collection and analysis.

B) Access in the Directorate Level.

Each directorate supervises and manages all the applications controlled by the departments under it. For example, the Directorate of Projects and Environment manages four smart applications: the CMMS application, the Municipal Lighting Management application, the GIS, and the Vehicle Fleet and Fuel Ring Monitoring Application. The overall control of the workflow of these applications and the effectiveness of the interventions made by the specific departments are supervised by their directorate. It is worth mentioning that data analysis and statistics tables support the decisions made by the directorates.

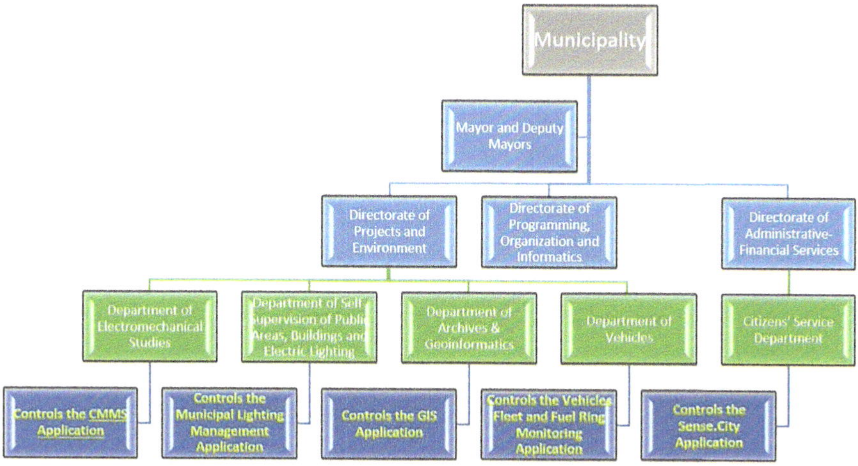

Fig. 12 A part of the structure of the Municipality of Patras

C) Access in the Interoperability Center Level.

The users of the interoperability center will have access and will manage all the data derived from all the applications from all the directorates. Multiple data sets will be accessible so as interactive control panels, maps, visualizations, correlations, and statistical analysis can be implemented.

8 Conclusions

This study sets the framework for the integration of the data, derived from already existing and any possible future smart systems in the Municipality of Patras, in an interoperability center. The rapid technological change that involves, among others, technologies like big data gives the Municipality the opportunity to organize all the available smart applications into this interoperability center in order to analyze useful information for the benefits of the citizens' everyday life. The interoperability center, in addition to having a supervisory role upon the smooth operation of the different smart applications, also focuses on any existing potentials to analyze, compare, and find the coherence between the different data sets. The analysis of these big data leads to wiser decisions and faster and more effective interventions regarding the reported problems.

Each individual smart application is a link in the process chain of the interoperability center that affects the overall workflow. Consequently, each department, responsible for the implementation and the operation of each application, ensures that the appropriate maintenance is made and that any malfunction is managed. Through this project we really do hope that the Municipality of Patras will successfully use the information and communication technologies and the already developed smart applications, with the aim to increase operational efficiency, share information with the public, and improve the quality not only of the government services but also its citizens' well-being.

References

Albino, V., Berardi, U., & Dangelico, R. M. (2015). Smart cities: Definitions, dimensions, performance, and initiatives. *Journal of Urban Technology, 22*(1), 3–21.

Cocchia, A. (2014). Smart and digital city: A systematic literature review. In *Smart City* (pp. 13–43). Springer.

Cuquet, M., Vega-Gorgojo, G., Lammerant, H., & Finn, R. (2017). *Societal impacts of big data: Challenges and opportunities in Europe*. arXiv preprint arXiv:1704.03361.

Densmore, J. (2021). *Data pipelines pocket reference*. O'Reilly Media.

Elmagarmid, A. K., Ipeirotis, P. G., & Verykios, V. S. (2006). Duplicate record detection: A survey. *IEEE Transactions on Knowledge and Data Engineering, 19*(1), 1–16.

Evans, R., & Fisher, D. (2002). Using decision tree induction to minimize process delays in the printing industry. In W. Klosgen & J. Zytkow (Eds.), *Handbook of data mining and knowledge discovery* (pp. 874–881). Oxford University Press.

Garagliu, A., Del Bo, C., & Nijkamp, D. (2011). Smart cities in Europe. *Journal of Urban Technology, 18*(2), 65–82.

Garrett, H. (1968). The tragedy of the commons. *Science, 162*(3859), 1243–1248.

Hall, R. E., Bowerman, B., Braverman, J., Taylor, J., Todosow, H., & Von Wimmersperg, U. (2000). *The vision of a smart city*. Retrieved from https://www.researchgate.net/publication/241977644_The_vision_of_a_smart_city

Hernández, M. A., & Stolfo, S. J. (1995). The merge/purge problem for large databases. *SIGMOD Record, 24*, 127–138.

Jalali, R., El-Khatib, K., & McGregor, C. (2015). Smart city architecture for community level services through the internet of things. In *Proceedings of the 8th international conference on intelligence in next generation networks* (pp. 108–113).

Joshi, S., Saxena, S., Godbole, T., & Shreya. (2016). Developing smart cities: An integrated framework. *Procedia Computer Science, 93*, 902–909.

Kagklis, V., Karatrantou, A., Tantoula, M., Panagiotakopoulos, C. T., & Verykios, V. S. (2016). A learning analytics methodology for detecting sentiment in student fora: A case study in distance education. *European Journal of Open, Distance and E-Learning, 18*(2), 74–94. https://doi.org/10.1515/eurodl-2015-0014

Karapiperis, D., Gkoulalas-Divanis, A., & Verykios, V. S. (2017). FEDERAL: A framework for distance-aware privacy-preserving record linkage. *IEEE Transactions on Knowledge and Data Engineering, 30*(2), 292–304.

Karapiperis, D., & Verykios, V. S. (2014). An LSH-based blocking approach with a homomorphic matching technique for privacy-preserving record linkage. *IEEE Transactions on Knowledge and Data Engineering, 27*(4), 909–921.

Karapiperis, D., & Verykios, V. S. (2016). A fast and efficient hamming LSH-based scheme for accurate linkage. *Knowledge and Information Systems, 49*(3), 861–884.

Kaufman, S., Rosset, S., Perlich, C., & Stitelman, O. (2012). Leakage in data mining: Formulation, detection, and avoidance. *ACM Transactions on Knowledge Discovery from Data (TKDD), 6*(4), 15.

Kohavi, R., Brodley, C., Frasca, B., Mason, L., & Zheng, Z. (2000). KDD-cup 2000 organizers' report: Peeling the onion. *ACM SIGKDD Explorations, 2*(2).

Kohavi, R., & Longbotham, R. (2007). Online experiments: Lessons learned. *Computer, 40*(9), 103–105.

Kohavi, R., Longbotham, R., Sommerfield, D., & Henne, R. (2009). Controlled experiments on the web: Survey and practical guide. *Data Mining and Knowledge Discovery, 18*(1), 140–181.
Kohavi, R., & Parekh, R. (2003). Ten supplementary analyses to improve e-commerce web sites. In *Proceedings of the Fifth WEBKDD workshop*.
Kunzmann, K. R. (2020). Smart cities after covid-19: Ten narratives. *disP, 56*, 20–31.
Lotsari, E., Verykios, V. S., Panagiotakopoulos, C., & Kalles, D. (2014). A learning analytics methodology for student profiling. In A. Likas, K. Blekas, & D. Kalles (Eds.), *Artificial intelligence: Methods and applications. SETN 2014* (Lecture notes in computer science) (Vol. 8445). Springer. https://doi.org/10.1007/978-3-319-07064-3_24
Nam, T.; & Pardo, T.A. (2011). Conceptualizing smart city with dimensions of technology, people, and institutions. In *ACM Int. Conf. Proc. Ser.* (pp. 282–291).
Reichental, J. (2020). *Smart cities for dummies*. Wiley.
Shearer, C. (2000). The CRISP-DM model: The new blueprint for data mining. *Journal of Data Warehousing, 5*(4), 13–22.
Su, K., Jie, L., & Hongbo, F. (2011). *Smart city and the applications*. 2011 International conference on electronics, communications and control (ICECC), Ningbo, China, IEEE.
Tsoni, R., Samaras, C., Paxinou, E., Panagiotakopoulos, C., & Verykios, V. S. (2019). From analytics to cognition: Expanding the reach of data in learning. In *The proceedings of the 11th international conference on computer supported education* (pp. 458–465). SCITEPRESS—Science and Technology Publications, Lda.
Verykios, S. V., Paxinou, E., Panagiotakopoulos, T. C., & Kalles, D. (2021). Latent space models for assessing dynamic student behaviour. In G. Ubachs (Ed.), *The envisioning report for empowering universities* (pp. 26–28). EADTU.
Verykios, V. S., Bertino, E., Fovino, I. N., Provenza, L. P., Saygin, Y., & Theodoridis, Y. (2004). State-of-the-art in privacy preserving data mining. *ACM SIGMOD Record, 33*(1), 50–57.
Washburn, D., Sindhu, U., Balaouras, S., Dines, R. A., Hayes, N. M., & Nelson, L. E. (2010). *Helping CIOs understand "Smart City" initiatives: Defining the smart city, its drivers, and the role of the CIO*. Forrester Research. Retrieved from http://public.dhe.ibm.com/partnerworld/pub/smb/smarterplanet/forr_help_cios_und_smart_city_initiatives.pdf
Yencken, D. (2013). Creative cities. In H. Sykes (Ed.), *Space place and culture*. Future Leaders.

RES-Q: Toward Semantic Interoperability for Risk and Disaster Management in Smart Cities

Omiros Iatrellis, Vasileios Kyriatzis, Nicholas Samaras, and Charalampos Dervenis

1 Introduction

A smart city is a well-defined geographical area (municipality), in which high technologies, logistics, and energy production cooperate to create benefits for citizens in terms of well-being, intelligent development, inclusion and participation, and environmental quality (Dameri, 2014). However, natural and manmade disasters and more recently the COVID-19 pandemic's unprecedented impact on every aspect of urban life have shown that smart city resilience is a major challenge for preserving a better quality of life, sustainable urban development, and improving environmental condition (Arafah et al., 2018). Smart city resilience can be defined as "the ability of smart cities to resist, absorb, and respond to the shock of disasters while maintaining and surging essential city services, and then to recover to its original state or adapt to a new one" (Meerow & Newell, 2019).

In the smart city domain, the disaster management operations require contributions and collaboration of different types of stakeholders and services with various functions, rules, protocols, and datasets, forming complex contexts in decision-making and event coordination. The collaboration among these various types of entities requires the establishment of a common semantic model, which will be widely utilized for the representation of any type of element or actor participating in the disaster/crisis management domain. The specific semantic model will also leverage the interoperability among the various IT systems and services since the exchanged datasets are heterogeneous and often proprietary. Consequently, the

O. Iatrellis (✉) · V. Kyriatzis · N. Samaras
Department of Digital Systems, University of Thessaly, Larissa, Greece
e-mail: iatrellis@uth.gr; kyriatzis@uth.gr; nsamaras@uth.gr

C. Dervenis
Department of Business Administration, University of Thessaly, Larissa, Greece
e-mail: dervenisch@uth.gr

P. Fitsilis (ed.), *Building on Smart Cities Skills and Competences*, Internet of Things,
https://doi.org/10.1007/978-3-030-97818-1_17

definition and establishment of a semantic model would be of major importance for the interoperation of the complete set of participating entities during the disaster/crisis management operations in a smart city domain, thus improving coordination and resilience.

In this paper, we present the cornerstone of the semantic infrastructure of RES-Q software environment, the RES-Q ontology, which realizes the appropriate semantics to allow the modeling of the multifaceted environment that is comprised in a disaster/crisis scenario, within a smart city. The RES-Q approach models the semantics across information at different levels of abstraction to allow data consolidation and harmonization techniques in a systematic way. Furthermore, the resulting solution will set the foundation for advanced reasoning and on-top analysis into the common knowledge base to facilitate the decision-making process and serialize the necessary recovery actions.

The current research work has been developed within the RES-Q Living Lab (LL) under the framework of the INVEST European Universities Erasmus+ program (INVEST, n.d.). European universities are ambitious transnational alliances of higher education institutions developing long-term structural and strategic cooperation. The INVEST collaboration envisions to support the establishment of living labs at each of the partner universities as platforms for collaboration for applied research and education. In this context, the RES-Q LL is conducted in Thessaly, Greece, by INVEST and targets to (a) improve the quality of the information on disaster risk and regional resilience and promote a culture of prevention, (b) improve the decision-making and management in the emergency phase, (c) promote a stronger collaboration and coordination of urban authorities with all stakeholders and resources involved in emergency management at all administrative levels (national, regional, local), and (d) encourage social awareness and involvement.

The remainder of this paper is organized as follows. The second section refers to our motivations, the RES-Q LL scope, and related work performed in this research area. Subsequently, we present the methodology we followed alongside with a discussion on conceptual issues arising. Section 4 analyzes the RES-Q semantic model, while Sect. 5 presents the proposed data harmonization and consolidation methodology. Finally, Sect. 6 concludes and proposes our thoughts for future work.

2 Motivations and Related Work

2.1 RES-Q Living Lab

As already mentioned, the proposed semantic model has been implemented within the RES-Q Living Lab. Primarily, the "RES-Q: knowledge for disaster risk reduction in integration to regional resilience" LL in Larissa, Greece, will deal with the situation and the response to two recent disasters affecting the region: the COVID-19 pandemic and a magnitude 6.3 earthquake, which occurred in 2021 and damaged

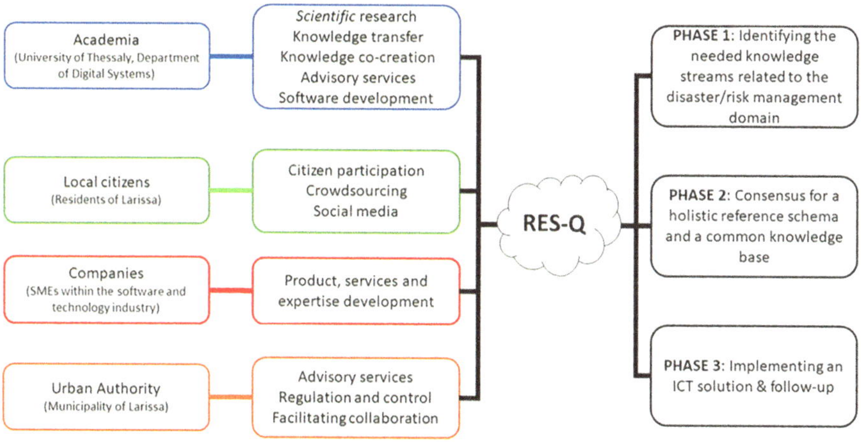

Fig. 1 RES-Q living lab

dozens of buildings in the region and many infrastructures including the water supply networks. Additionally, the LL will aim at contributing to the design of a flood risk management plan in order to reduce the flood risk. Today, despite the construction of flood protection works, floods remain a potentially perilous problem of the area.

The LL facilitates quadruple helix partnerships of research, education, companies/NGOs, and GO's collaboration with other stakeholders from the region of Larissa in Greece. However, the challenge of enabling "knowledge for disaster risk reduction in integration to regional resilience" requires firstly to consider what is hampering the use of information and the development of knowledge among different stakeholders so as to improve decision-making for risk reduction and enhanced resilience. The recent disasters that hit the region have shown that this goal is a major challenge for preserving a better quality of life, sustainable regional development, and improving environmental condition.

As Fig. 1 depicts, the RES-Q project is divided in three phases. During phase I, students and researchers in collaboration with the stakeholders have identified the needed domain knowledge streams semantically related to the crisis/disaster management domain. Initially, in order to define the different domains of discourse, which belong to the area of our interest, LL participants studied the available bibliography and conducted interviews with professionals related to stakeholders' awareness, perception, and preparedness. The research revealed the different domains of discourse that belong to the area of crisis/disaster management including disasters, smart city infrastructure and resources, population, stakeholder's roles, meteorology, etc.

In phase II, all the higher-level dimensions of the risk/disaster management domain that have been identified in phase I were combined to create an upper conceptual model (a) covering all the important concepts and (b) addressing the aforementioned interoperability and integration issues. In this phase, semantic web technologies were proposed as a promising solution for data harmonization, linkage, and consolidation and for achieving semantic interoperability between the various stakeholders. RES-Q ontology presented in this paper is the result of this phase.

Once the common semantic model is finalized, the RES-Q LL will design and implement an integrated ICT system for real-time knowledge sharing, decision-making, and orchestration of the processes during a crisis or emergency event for the region of Larissa (phase III).

2.2 Motivations

Alongside with the extensive study of the related work performed in the specific domain of our interest, the RES-Q LL conducted a set of interviews with professionals working in municipalities of smart cities in order to define the requirements of the semantic infrastructure. The interviews were conducted in person, where semi-structured interview methods were used. All participants were smart city professionals whose experience ranged from 5 to 14 years and whose professional tasks ranged from being solely an IT specialist to being a manager and department head. Six professionals were engaged in the procedure and discussed the tasks and open issues concerning the participating entities during the dynamic composition of resiliency actions. The whole procedure was executed in six (6) interviews and lasted for nine (9) hours in total. Once the procedure concluded, a concrete overview was reached concerning the open issues and challenges emerging in the domain of our interest:

- *Formalization of recovery actions*. The formalization of recovery actions constitutes one of the major challenges for the area (Kaufmann et al., 2020). The lack of their formal representation in a computer perceptible form creates a major need, since their translation into structured knowledge and their integration within an IT system could facilitate the establishment of a mechanism for the recommendation of the next appropriate step of a resilience plan in an automatic and consistent way.
- *Maintenance*. From municipality's perspective, resilience actions reflect on the available resources and infrastructure and on the multiple collaborative networks with various stakeholders and representatives from the emergency services, public and private companies, academic entities, media, citizens, and volunteer organizations. They must therefore be easily maintained so that the corresponding processes respond to the changes that may arise on the various involved stakeholders.
- *Semantic interoperability*. Smart city services, stakeholders, and their knowledge bases often use different terms, types, and values for data representation that does not support data interoperability. Without the provision of a uniform semantical

basis, data interoperability cannot be achieved, and this creates many difficulties during the integration and deployment of a resilience plan.

- *Knowledge discovery and evolution.* Current information feed to the resilience actions during execution mode is important, because the information collected could lead to major reconfigurations of the executed actions. Hence, the knowledge feedback is valuable, since it would be able to reconfigure the knowledge base together with the rules of the resilience actions.

2.3 Related Work

Resilience, with respect to the city or urban domain, is a research topic that has received great attention during the past decades. Numerous points of view and correlations with other scientific domains, technologies, and trends have created a plethora of information and knowledge. A key characteristic of the research conducted so far is that no single model/framework is established as a benchmark/standard for all to use. Even when we discuss the very definition of the word "resilience," we come up with many diverse versions, approaches, characteristics, and focal points [1–10]. Nowadays, more than ever, due to the COVID-19 pandemic, resilience is a topic of high importance and research interest, especially in terms of providing tangible results and solutions in the form of tools that will assist communities to assess, prepare, cope with, and recover from a crisis or disaster, be that manmade or natural.

Numerous frameworks are also reported in literature, all with their own focal points, interdependencies, impacts assessed, assessment techniques used, attributes, or even geographic applicability constraints. NIST (McAllister, 2016) provides an extensive deliberation and evaluation of various resilience frameworks available (viz., SPUR, PEOPLES, FEMA Hazus, NOAA CRI, Rockfeller CFR & CRI, BRIC, CART, CARRI CRS, UNISDR Scorecard, Oregon Res. Plan).

Among those that stand out, in terms of comprehensiveness and applicability in various situations (i.e., not focusing only in specific topics like buildings or infrastructure resilience), is the City Resilience Framework, developed by ARUP (Arup, 2014) (City Resilience Framework Research Report—Volume 2: Fieldwork Data Analysis 2014). It is defining resilience as the capacity of cities to function, so that the people living and working in cities survive and thrive no matter what stresses or shocks they encounter. A list of eight city functions that are critical to resilience was drafted, proposing that a resilient city delivers basic needs; safeguards human life; protects, maintains, and enhances assets; facilitates human relationships and identity; promotes knowledge; defends the rule of law, justice, and equity; supports livelihoods; and stimulates economic prosperity. Moreover, based on extensive field research conducted, the City Resilience Index, a tool to assess a city's resilience, was formulated, comprising of 4 categories, 12 indicators, more than 50 sub-indicators, and 150 variables/metrics.

Moreover, (Sharifi, 2016) discerns the various characteristics of the reviewed tools with respect to the type of assessment used, namely, "formative" and "summative" assessment; the former method involves ex ante evaluation and continuous monitoring of the conditions from the early stages of the planning process, making it suitable for addressing dynamic evolving issues and accounting for future uncertainties, while the latter method focuses on ex post measurement of the effectiveness of interventions, allowing thus communities to diagnose their current resilience status and hence provide the required evidence needed for inducing appropriate strategies. The assessment format used (index, toolkit, scorecard), the type of data required (primary, secondary), and the assessment criteria used (assessment against baseline conditions, thresholds reflecting program objectives, principles of good resilience, peers-benchmarking, and based on the speed of recovery) are also discussed. In order to evaluate the performance using the aforementioned methods, the tools under review make use of both quantitative methods using numerical data and also qualitative methods based on public perceptions and expert opinions/judgment, illustrating the different approaches used and hence the potential subjectability of the outcome. Finally, for evaluating community resilience, toolkits, indices, models, and scorecards are the main assessment forms used, with toolkits recognized as the most appropriate one to use as they provide directions on issues like the timeline of the assessment, the required stakeholders involved, and interventions required based on the assessment results. A key finding of this study is that community resilience tools have failed to sufficiently address the dynamic nature of resilience. Moreover, in order to be able to cope with the frequently changing conditions and dynamics, these tools need to employ modeling and projection techniques apart from using historical trends and baseline conditions.

Taking in consideration the multitude and magnitude of the various dimensions reflected in these frameworks, one can easily appreciate that assessing a city's resilience is a nontrivial task. This becomes even more challenging when the aspect of timely and accurate provision of suitable data is recognized as a cornerstone. This challenge is clearly reflected in the fact that despite the numerous frameworks available, there is a profound lack of relevant software tools that transform these frameworks to tools that produce tangible, actionable results, past the resilience assessment and toward a functional instrument that can be of assistance to decision-making and action orchestration, as is the proposed RES-Q semantic model aiming at. This situation though is in a transition, since during the past few years, a fair number of semantic models have been proposed in the literature.

For example, in (Nicola et al., 2019), a framework to support creative design of emergency management scenarios is presented, with key points being that it gathers and organizes knowledge about emergency management situations by automatically generating conceptual models related to fragments of emergency scenarios, using semantics-based techniques to enable the application of a computational creativity approach. This Smart Cities and Emergency Management Ontology includes 284 concepts and 117 relationships.

In (Ni et al., 2019), a knowledge model and respective ontology for emergency response based on contingency planning system of China are presented, illustrating the need for the use of models/ontologies for providing stakeholders with helpful information.

Another upper ontology model is the Dynamic Flood Ontology (DFO) (Kurte et al., 2019), which is formulated to represent/model the spatiotemporal changes that occur in a flood disaster situation. The approach followed is presented, and its applicability and effectiveness, in the form of situational awareness indicators, are assessed with promising results.

Empathi (Gaur et al., 2019) is an another interesting ontology for emergency management hazard situational awareness and events during emergency scenarios. It imports numerous external vocabularies and ontologies, containing 423 classes and 338 relations. Its goal is to capture and integrate information from numerous sources like satellite images, local sensors, and social media content created by people while including concepts like impact, affected population, human prayer, volunteer support criminal activity, etc.

In (Iatrellis et al., 2020), a synergistic approach is proposed, which draws from the strengths of both semantic-driven and cloud-based approaches for integrating ubiquitous competence-based management and learning pathways for smart cities. The semantic infrastructure consists of an ontological framework, which leverages interconnections with well-distinguished ontology modules, thus enclosing the required knowledge streams for the holistic representation of the specific domain in an integrated way as well as for the implementation of a semantic rule repository. The skill gap analysis is built in a distinct architectural layer on top of the semantic model and is enhanced through the utilization of the well-known rule-based expert system approach.

In (Wang et al., 2020), emergency decision-making model of environmental emergencies based on case-based reasoning is proposed, the scenario evolution mechanism is discussed, various algorithms for different types of factors (i.e., accurate numerical data, fuzzy semantic data, and symbolic data) are presented in view of solving the problem of incomplete data and diverse data types, and finally the model is validated against a real environmental emergency.

The POLARISC ontology (Elmhadhbi et al., 2019) is a modular ontology developed in order to explore the knowledge and semantics of the various emergency respondents involved in a disaster response process. Its goals are to provide a basis for semantic interoperability and alleviate the semantic heterogeneity among different ERs. It comprises of seven modules related to knowledge shared by the various crisis actors like firefighters, healthcare units, police, public authorities, and messages exchanged. It includes 669 classes and 177 relations and is available at (POLARISCO, n.d.).

In (Chehade et al., 2020) Rescue MODES, a communication system focused on supporting situation awareness among French emergency actors in rescue operations is discussed, with focus placed on the ResOnt application ontology. Sharing common characteristics with the POLARISC ontology, it diversifies by representing apart from the abstract entities (like tasks and phases) material entities too (like hazardous materials) involved in emergency situations.

Finally, in (Benaben et al., 2020), a model-based AI framework for describing collaborative situations between organizations during crises is presented. COSIMMA (Collaborative Situation Metamodel) is structured around a core layer which describes concepts and relations of any collaborative situation and a layer composed with four packages describing crisis management domains, namely, context, partners, objectives, and behavior.

3 Methodology

The approach used to develop this conceptual model follows the Ontology 101 development process proposed by (Noy & Mcguinness, 2000):

1. *Determine the domain and scope of the ontology*. A core concern in the process of defining the RES-Q ontology was to propose a common model for **sematic interoperability** between the participating stakeholders as well as to provide a semantically enhanced repository for **data consolidation** and **integration** from different sources during disaster risk management. The cornerstone of the RES-Q semantic model is a top-level ontology, which captures in a uniform manner the inherent semantics of this complex domain in order to facilitate the transformation of the original raw data derived from heterogeneous sources into standard RDF (Resource Description Framework), which will be easily queried and utilized for advanced analysis and decision-making. Thus, in contrast with existing research works that present large and monolithic ontology models, the RES-Q approach proposes a two-dimensional semantic model which interlinks a higher-level conceptual backbone with vertical domain-specific ontological components (Fig. 2). The upper model encodes the concepts that define crosscutting characteristics of the disaster risk management domain, while the vertical components formalize a specific knowledge with concepts from the domain ontologies extending the high-level concepts of the top-level ontology.
2. *Consider reusing existing ontologies*. RES-Q reuses existing ontologies and classifications to propose a holistic conceptualization of the participating entities in the disaster risk management process. Some ontologies have served as an inspiration for the modeling of the proposed semantic model.
3. *List the relevant terms of the domain*. Terms that are important in describing the domain have been captured in the preparatory phase (literature review and interviews) during which we defined the main set of concepts that needed to be covered. This set of terms includes disaster, impact, cause, response, agent, risk, process, and infrastructure.
4. *Define the classes and the class hierarchy*. From the list of terms that were selected in step 3, a set of superclasses has been defined for the upper model. These superclasses are *disaster*, *risk*, *agent*, *process*, *infrastructure*, and *quality assurance* and are associated with other classes to represent the relationships where appropriate.

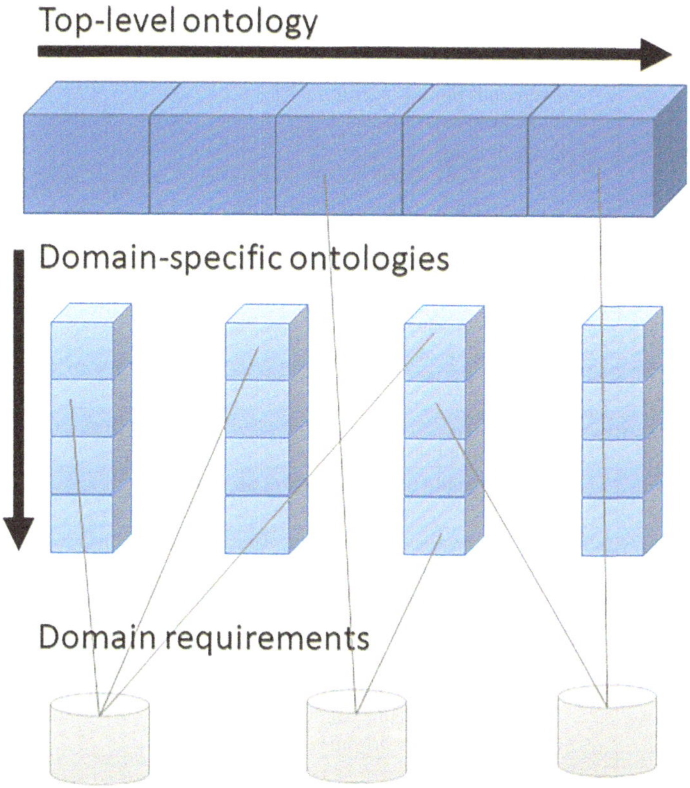

Fig. 2 RES-Q multidisciplinary knowledge: domain requirements are described by specific domain ontologies and top-level ontology

5. *Define the properties of classes.* Object properties are added to describe the way classes are related. Examples of object properties are shown in Table 1.
6. *Define the datatype properties.* The datatype property relates an individual to a data value, numeric information, as opposed to an object property which relates an individual to another individual. Examples of data properties are shown in Table 2.
7. *Create instances.* The final step in the development of an ontology is to create actual instances from the abstract representation. In our case, individuals will be obtained by mapping data from SQL-dump operations and other IT systems to RDF according to the RES-Q consolidation and harmonization methodology described later.

Table 1 RES-Q object properties

Object property	Description logic
Responds	$\exists$ responds thing $\subseteq$ agent T $\subseteq \forall$ responds disaster
Causes	$\exists$ causes thing $\subseteq$ disaster T $\subseteq \forall$ causes disaster
Utilizes	$\exists$ utilizes thing $\subseteq$ process T $\subseteq \forall$ utilizes infrastructure
Executes	$\exists$ executes thing $\subseteq$ agent T $\subseteq \forall$ executes process
hasImpact	$\exists$ hasImpact thing $\subseteq$ disaster T $\subseteq \forall$ hasImpact agent

Table 2 RES-Q datatype properties

Data property	Description logic
AgentId	$\exists$ AgentId datatype literal $\subseteq$ agent T $\subseteq \forall$ AgentId datatype int
InfrastructureTypeName	$\exists$ InfrastructureTypeName datatype literal $\subseteq$ infrastructure T $\subseteq \forall$ InfrastructureTypeName datatype string
AgentRoleDescription	$\exists$ AgentRoleDescription datatype literal $\subseteq$ agent T $\subseteq \forall$ AgentRoleDescription datatype string
ProcessId	$\exists$ ProcessId datatype literal $\subseteq$ process T $\subseteq \forall$ ProcessId datatype int
IndicatorName	$\exists$ IndicatorName datatype literal $\subseteq$ KPI T $\subseteq \forall$ IndicatorName datatype string

4 RES-Q Semantic Model

The cornerstone of the RES-Q semantic model is a top-level ontology, which consists of the very general terms (including objects, properties, and relations) that are common across all associated domains. An important function of the RES-Q upper model is to support broad semantic interoperability among several domain-specific ontologies related to smart city risk/disaster management by providing a common starting point for the formulation of the main definitions. Terms in the domain ontologies are ranked under the terms in the RES-Q top-level ontology, e.g., the upper ontology classes will be superclasses or supersets of all the classes in the domain ontologies.

Figure 3 provides an abstract view of the RES-Q upper ontology that presents the main conceptualization principles formalized by means of OWL-DL description logic language. These classes represent the following knowledge:

1. *Disaster*. The concept refers to the disasters that take place and require response. It includes a classification of natural and manmade disasters. Disaster acts as a general concept by adopting the Emergency Disasters Database (EM-DAT),

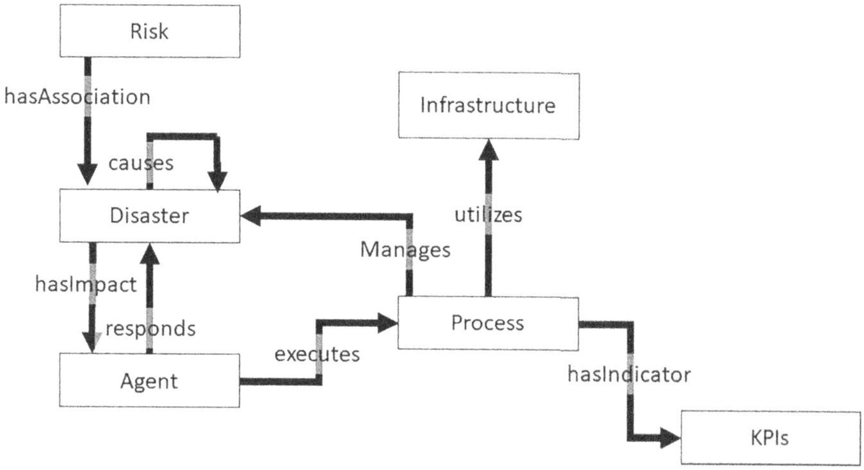

Fig. 3 RES-Q upper model

which provides a precise definition of concepts and furthermore a categorization of disturbance-related events.

2. *Risk.* This concept represents risks associated with disasters and facilitates policy making in prevention, preparedness, response, and reconstruction activities. Codenamed DOAM, the Description of a Model risk model is a domain ontology that aims to represent and categorize knowledge about risk management concepts using semantic web information technologies.
3. *Agent as a person with a role.* This concept represents the human individual. It contains the basic categories that describe a smart city agent who undertakes specific roles in the context of a disaster management response event. RES-Q uses friend of a friend (FOAF) to describe people, relations, and respective roles. In order to identify an agent during a disaster, a datatype property associates an agent with a unique anonymous ID. Examples of this category are the smart city professions that were identified during the Smart DevOps research project (SmartDevOps, 2018):

   (a) *Smart city IT* manager can be defined as an ICT consultant with responsibilities that include setting objectives and strategies for the IT department, through the implementation of suitable technological solutions, in order to support all internal operations while being responsible of designing and customizing new smart city's systems.
   (b) *Smart city planner* can be a high-level official that is able to bridge the needs that arise from cities' traditional development and operational needs; smart and sustainable cities' frameworks, best practices, standards, and technologies; and strategic priorities of the city's political leadership.
   (c) *Smart city IT officer* is an IT technical expert that should be able to analyze smart city's organizational data; determine information system requirements

Table 3 Ontologies/classifications mapping to different dimensions of the RES-Q model

Domain	Ontologies/vocabularies
Agents	FOAF[a]
Risk	DOAM risk function ontology[b]
Infrastructure	SAREF,[c] s4city[d]
Physical location	a-loc[e]
Weather phenomena and exterior conditions	BIMERR[f]
Roles	Smart DevOps role profiles[g]
Disasters	EM-DAT[h]

[a] http://xmlns.com/foaf/spec/
[b] https://www.openriskmanual.org/ns/doam/index-en.html
[c] https://saref.etsi.org/core/v3.1.1/
[d] https://saref.etsi.org/saref4city/v1.1.2/
[e] https://lov.linkeddata.es/dataset/lov/vocabs/a-loc
[f] https://bimerr.iot.linkeddata.es/def/weather/
[g] https://smartdevops.eu/
[h] http://www.emdat.be/Glossary

and define project objectives; apply software development process, with appropriate tools and techniques; make recommendations for necessary IT systems; design, implement, deploy, and operate new IT services; and provide support and training to various types of users.

4. *Process*. This concept represents a process that is executed by a smart city agent in order to mitigate the impact of a disaster. During the design mode, a recovery process repository will be created in a centralized database that will allow an agent to retrieve the appropriate recovery plans to use during an emergency.
5. *Infrastructure*. The concept represents the set of fundamental facilities, resources, and systems that can be used during the response to a disaster. RES-Q ontology utilizes SAREF4CITY, an extension of SAREF, in order to create a common core of general concepts for smart city data oriented to material equipment, resources, IoT devices, and infrastructure.
6. Key performance indicators. The specific module that is modeled and interfaced as part of the RES-Q ontology is that of key performance indicators in an arbitrary municipality by covering a number of quality assurance indicators. It supports a set of key performance indicators (KPIs) derived from domain experts' experience and perceptions.

Table 3 presents the vertical domain-specific ontologies and the state-of-the-art vocabularies that improve the prowess of the RES-Q top-level ontology.

As the above table shows, the upper model acts as a semantic bridge supporting very broad semantic interoperability between various relevant domains, while at the same time it can be extended to cover more dimensions if necessary. In this way, it (a) facilitates knowledge reuse across various smart city systems and applications, (b) enables distributed engineering of ontology modules over different locations and different areas of expertise, (c) supports effective management and browsing of modules, and (d) offers easier implementation, maintenance, and update.

5 Data Harmonization and Consolidation Methodology

Once the ontology model is finalized, the RES-Q data consolidation methodology will be applied to allow the integration of the different data sources originated from various services and stakeholders, according to this model. Figure 4 shows a general overview of this methodology, where the conceptual layer (CL) defines the vocabulary with concepts and relationships in the domain of smart city disaster risk management. Within CL, the RES-Q ontology is implemented in OWL format according to which concepts and relationships are represented by classes and data or object properties, respectively. This upper model allows the interconnection with other vertical domain-specific ontologies oriented to different aspects as described in Table 3.

At a different layer, the RES-Q data layer (RDL) realizes all the instances in the knowledge domain involving the smart city risk-/disaster-related data. These instances are stored in RDF triple format in a Virtuoso repository with efficient

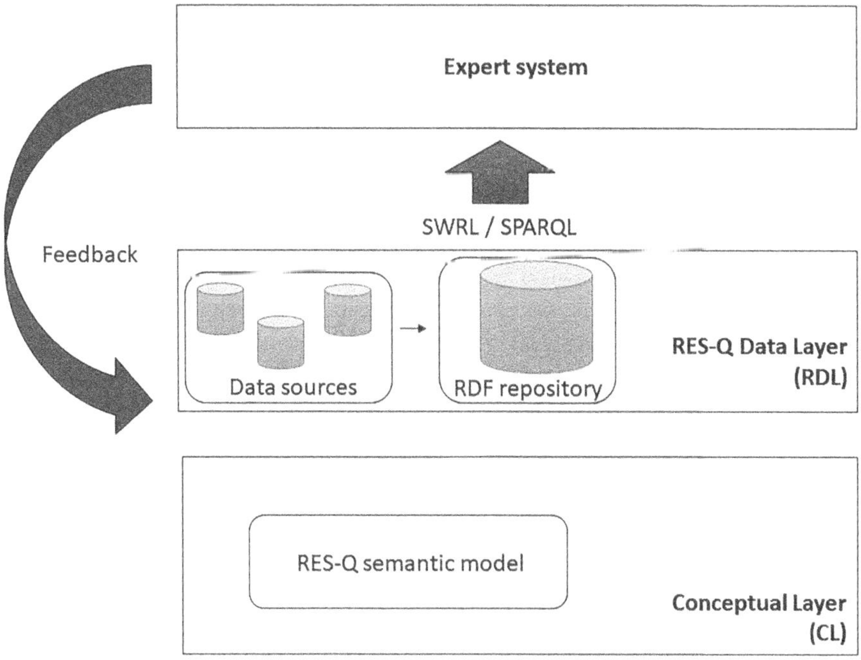

Fig. 4 General overview of the RES-Q approach

reasoning capabilities. To do so, a set of mapping functions have been implemented and tested to transform the data originating from the different sources into RDF according to the RES-Q scheme. For example, in the case of weather data, they are mapped from a set of SQL-dump statements on a relational database regarding a national weather database.[1] A second data source that was tested was integrated from the European Integrated Data Archive (EIDA),[2] which is a federated European data center that archives and provides access to seismic waveforms and their related metadata including station inventory and seismic waveforms' quality parameters from the European research infrastructures.

At the upper layer, having the data consolidated in the common RDF repository, it is now possible to perform complex reasoning operations based on semantic web rule language (SWRL) or query them using SPARQL, independently of the source of the data, their structure or the syntax of the original format. In this way, the knowledge that is modeled as part of the common repository can be fed in the expert system to perform analysis, produce recommendations, and further orchestrate the disaster recovery processes. By combining the reasoning power of RES-Q holistic semantic model, SWRL and SPARQL, our methodology can be utilized by an integrated software environment for the recommendation and execution of the appropriate next process in order to minimize the impact of a potential risk or disaster.

Another important characteristic to the RES-Q approach is that in any execution cycle, new knowledge can be created that will be utilized in the future reasoning processes that belong to a particular disaster category. This attribute ensures the constant update of the knowledge that is stored inside the CL layer together with the RDL repository.

6 Conclusions

RES-Q provides an ontological representation of the disaster risk management domain for smart cities for the semantic integration of multiple data sources to form the foundation for efficient reasoning and querying independently (or in a transparent way) of the origin. In contrast with other research developments, we proposed a top-level semantic model extended with vertical ontological components that facilitate the capture of domain knowledge at different levels of granularity in order to leverage semantic interoperability between various smart city agents and data consolidation from heterogeneous sources. The proposed approach promotes the production of extensive linked knowledge to be further utilized in the future by an expert system for decision-making and for the dynamic composition of disaster recovery processes. Under the RES-Q pilot, the areas of the LL—as micro versions

[1] http://www.hnms.gr

[2] https://www.orfeus-eu.org/data/eida/

of larger territories—can be seen as appropriate sites for validating and spreading new resilience solutions serving as showcases at macro level. Thus, scalability and replicability to other regions is strongly embedded in the RES-Q project.

Our future work in the RES-Q LL will be devoted to implement the expert system following the data harmonization and consolidation methodology so as eventually to provide an integrated software environment for disaster risk management. Once the expert system is completed, extensive elaborations will be conducted with heterogeneous datasets to provide valuable results concerning the performance of the system and the usability and further enhancement of the implemented semantic model. Furthermore, our intentions for further work focus on semantic enhancements concerning organizational and quality assurance modeling issues and health emergency knowledge representation.

Acknowledgments Supported by the INVEST4EXCELLENCE project under the H2020-IBA-SwafS-Support-2-2020 program (project no.: 101035815, www.invest4excellence.eu). Special thanks to the other project partners.

References

Arafah, Y., Winarso, H., & Suroso, D. S. A. (2018). Towards smart and resilient city: A conceptual model. *IOP Conference Series: Earth and Environmental Science, 158*(1).

Arup. (2014). *City resilience framework research report—Volume 2: Fieldwork data analysis* (Vol. 2, April). Retrieved from https://www.arup.com/-/media/arup/files/pdf-downloads/cri_research_report_vol2_comp.pdf.

Benaben, F., et al. (2020). An AI framework and a METAMODEL for collaborative situations: Application to crisis management contexts. *Journal of Contingencies and Crisis Management, 28*(3), 291–306.

Chehade, S., Matta, N., & Remi, J.-B. P. (2020). Handling effective communication to support awareness in rescue operations. *Journal of Contingencies and Crisis Management, 28*(3), 307–323.

Dameri, P. P. (2014). Searching for Smart City definition: A comprehensive proposal. *International Journal of Computers & Technology, 10*(1), 2146–2161.

Elmhadhbi, L., Karray, M. H., & Archimède, B. (2019). A modular ontology for semantically enhanced interoperability in operational disaster response. In *Proceedings of the international ISCRAM conference* 2019-May (May) (pp. 1021–1029).

Gaur, M., Shekarpour, S., Gyrard, A., & Sheth, A. (2019). Empathi: An ontology for emergency managing and planning about hazard crisis, *2019 IEEE 13th International Conference on Semantic Computing (ICSC)*, pp. 396–403. https://doi.org/10.1109/ICOSC.2019.8665539

Iatrellis, O., et al. (2020). Cloud computing and semantic web technologies for ubiquitous management of smart cities-related competences. *Education and Information Technologies, 26*(2), 1–22.

INVEST. (n.d.). *INVEST*. Retrieved July 19, 2021, from https://www.invest-alliance.eu/

Kaufmann, H. R., et al. (2020). DevOps competences for Smart City administrators. In *CORP* 2020 (pp. 213–223).

Kurte, K., Potnis, A., & Durbha, S. (2019). Semantics-enabled Spatio-temporal modeling of earth observation data: An application to flood monitoring. In *Proceedings of the 2nd ACM SIGSPATIAL international workshop on advances in resilient and intelligent cities, ARIC 2019* (pp. 41–50).

McAllister, T. (2016). Community resilience planning guide for buildings and infrastructure systems. *National Institute of Standards and Technology, 1*, 126. https://doi.org/10.6028/NIST.SP.1190v1

Meerow, S., & Newell, J. P. (2019). Urban resilience for whom, what, when, where, and why? *Urban Geography, 40*(3), 309–329. https://doi.org/10.1080/02723638.2016.1206395

Ni, Z., Rong, L., Wang, N., & Cao, S. (2019). Knowledge model for emergency response based on contingency planning system of China. *International Journal of Information Management, 46*(November 2018), 10–22. https://doi.org/10.1016/j.ijinfomgt.2018.10.021

Nicola, D., Antonio, M. M., & Villani, M. L. (2019). Creative Design of Emergency Management Scenarios Driven by semantics: An application to smart cities. *Information Systems, 81*, 21–48.

Noy, N. F., & Mcguinness, D. L. (2000). Ontology development 101: A guide to creating your first ontology (pp. 1–25).

POLARISCO. (n.d.). POLARISCO. Retrieved July 19, 2021, from https://github.com/LindaElmhadhbi/POLARISCO

Sharifi, A. (2016). A critical review of selected tools for assessing community resilience. *Ecological Indicators, 69*, 629–647.

SmartDevOps. (2018). *Smart DevOps*. Retrieved January 14, 2020, from http://devops.uth.gr/

Wang, D., Wan, K., & Ma, W. (2020). Emergency decision-making model of environmental emergencies based on case-based reasoning method. *Journal of Environmental Management, 262*(April 2019), 110382. https://doi.org/10.1016/j.jenvman.2020.110382

Blockchain for Smart Cities: Findings from a Systematic Literature Review

Ifigenia Georgiou, Juan Geoffrey Nell, and Angelika I. Kokkinaki

1 Introduction

Rapid urbanization and developments in information and communications technology (ICT) lead to the origin of the concept of smart city (Denyer & Tranfield, 2009; Giffinger et al., 2007). Massive efforts are still needed to enable the various services of a smart city (see Gori et al., 2015) to connect to one another and to gain real value through such connectivity (Batty et al., 2012). The challenges include integration and interoperability of systems in the application layer; infrastructure and information security concerns including cyberattacks, privacy, and confidentiality; and the exponential growth of data that needs to be managed (Silva et al., 2018). New forms of database design that can be distributed at a city-wide scale to accommodate data collection that can rely on crowdsourcing to elicit the preferences of citizens and enable the city to engage in social experimentation around key urban problems are required (Batty et al., 2012). This is where blockchain becomes relevant. In fact, the United Nations, in their 2018 Revision, explicitly points out under "Goal 11" that "blockchain provides an opportunity to collaborate in a transparent and secure way across the many components of smart cities, ensuring sustainability and accountability" (United Nations, n.d.). Blockchain is a specific type of database that is defined as a decentralized, distributed digital ledger distributed on a network of computers (nodes). The ledger is like a chain that consists of individual blocks of data linked together using cryptography. When new data is added to the network, a new block is created and added to the chain, and then all nodes update their version of the blockchain ledger, so every node has an identical version. How these new blocks are created is key to why blockchain is considered highly secure.

I. Georgiou (✉) · J. G. Nell · A. I. Kokkinaki
School of Business, University of Nicosia, Nicosia, Cyprus
e-mail: georgiou.i@unic.ac.cy; nell.j@unic.ac.cy; kokkinaki.a@unic.ac.cy

P. Fitsilis (ed.), *Building on Smart Cities Skills and Competences*, Internet of Things,
https://doi.org/10.1007/978-3-030-97818-1_18

The majority of nodes must verify and confirm the legitimacy of the new data before a new block can be added to the ledger. This ensures the immutability and irreversibility attributes of blockchain.

Smart cities and blockchains have been studied extensively yet in a decoupled manner in previous works, as Xie et al. (2019) point out while they provide a survey of the state-of-the-art blockchain technology that can be applied in smart cities. To the best of our knowledge, Xie et al. (2019) is the first and only attempt to look into the field of blockchain applications to smart cities. We extend this work into two important ways: first, we focus on the reasons that blockchain was used as the solution for smart cities, i.e., what issues it was called to address. Moreover, our study differs in scope as we employ a systematic literature review methodology that derives data from journals that meet predefined search criteria.

In this chapter we provide the results of our systematic review of the literature on the application of blockchain technology for smart cities answering the following research questions: (i) Why was blockchain chosen as the solution? (ii) What blockchain-based applications are being researched for smart cities? Further, we highlight the skills that would be useful for implementing blockchain technologies for smart cities based on the framework suggested by Fitsilis and Kokkinaki (2021). The results of our study are useful to researchers in the fields of blockchain and smart cities; they could serve also as a guide to practitioners.

The remaining of this chapter is structured as follows: Sect. 2 outlines the methodology followed for the systematic literature review, and Sect. 3 presents our findings, while Sect. 4 concludes.

2 Methodology

Having formulated the research questions above, the next step in the systematic literature review is to identify appropriate keywords and formulate the search strings. The main keywords used were "blockchain," "smart city," and variations of them, namely, "distributed ledger" or "DLT" for "blockchain" and "smart district," "digital city," and "smart towns" for "smart city." We searched the following widely recognized repositories: (a) Scopus, (b) ScienceDirect, and (c) IEEE Explore digital library. The inclusion criteria used are (a) English language papers only, and (b) the search terms had to be present in the title or keywords or abstract to ensure that the results were narrowly focused on the topic. The search was conducted across all years. The exclusion criteria specified that only papers published in peer-reviewed academic journals are to be included. This process provided 641 results. In Step 5, these 641 papers were then further evaluated and screened. Duplicated papers, those missing full text, or papers that focus solely on improving the blockchain technology were removed, resulting in 60 unique, relevant papers, selected based on a review of the title, keywords, abstract, and document type. The last screening required reading the full paper to ensure that the paper did provide insight on the research questions. Literature reviews were also eliminated from our sample at this stage. This left us

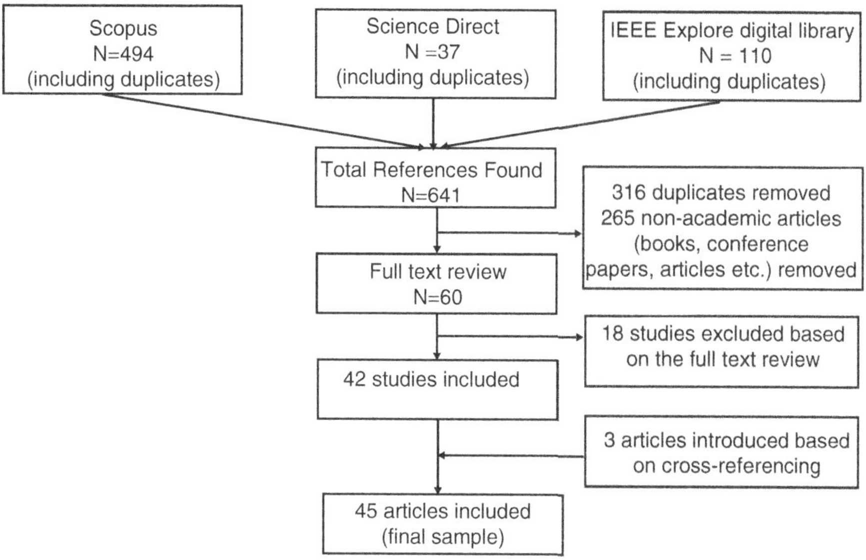

Fig. 1 Data mapping process

with 42 studies. Next, the references of these papers were checked, and three papers that satisfy our criteria were introduced based on this cross-referencing process. Data from the resulted 45 papers were populated into a data collection table to facilitate a content analysis focusing on answering the research questions. Figure 1 provides a visual representation of the steps 1–5 described above.

3 Results

The final steps of the systematic review process consist of the descriptive review of the literature, followed by the thematic analysis. As the concept of smart cities is new, few papers on this topic were published prior to 2018. Through these years, there has been a gradual evolvement of the focus of research from smart contracts toward a more thorough examination of blockchain applications relevant to smart cities' use cases. Geographically—based on the first author's location—most studies originate from China, followed by South Korea and India. We proceed to address the research questions.

3.1 Why Was Blockchain Chosen as the Solution?

Table 1 highlights the key reasons that blockchain was proposed as a potential solution including increased security and its trustworthiness due to immutability that emanates from the decentralized nature of the blockchain; it is nearly impossible to hack a blockchain—to alter the database, the chain, a hacker would need to take control of more than half of all the computers in the same distributed ledger, an event that is highly unlikely. Moreover, a blockchain is a trustless system, that is, the need for participating parties to trust each other or a third party for the system to function is eliminated or minimized, as no single entity has authority over the system, but instead entries on the chain are verified through consensus, i.e., majority agreement in a decentralized manner. Privacy was also cited that can be achieved on a blockchain despite transparency of transactions.

Blockchain for increased security. Security was the top reason cited for the use of blockchain. Due to the massive volume of data, the stakeholders involved, and necessary controls, smart cities require high levels of security. Blockchain would

Table 1 Reasons for implementing blockchain applications

Reason for blockchain	No. of papers	Cited by
Increased security	32	Pieroni et al. (2018), Altulyan et al. (2020), Banerjee et al. (2018), Chaudhary et al. (2019), Dagher et al. (2018), Dwivedi et al. (2019), Ferraro et al. (2018), Fernandez-Carames and Fraga-Lamas (2019), Gong et al. (2019), Hadi et al. (2019), Hammi et al. (2018), Jo et al. (2019), Kamel Boulos et al. (2018), Khan and Salah (2018), Kumar et al. (2019), Liao and Wang (2018), Liu et al. (2019), Mohanta et al. (2019), Park et al. (2018), Rahman et al. (2019), Rathore et al. (2019), Sa and Umamakeswari (2018), Sharma et al. (2017), Sharma et al. (2019), Shen et al. (2019), Singh et al. (2019), Sun and Zhang (2016), Wang et al. (2019), Yin et al. (2019), Zhang et al. (2019), Zhou et al. (2019)
Privacy	16	Dagher et al. (2018), Dwivedi et al. (2019), Fernandez-Carames and Fraga-Lamas (2019), Ferraro et al. (2018), Jo et al. (2019), Khan et al. (2019), Kumar et al. (2019), Liao and Wang (2018), Marsal-Llacuna (2018), Mohanta et al. (2019), Rahman et al. (2019), Shen et al. (2019), Singh et al. (2019), Zhao et al. (2019), Zhou et al. (2019), Zou et al. (2019)
Trust	14	Bruneo et al. (2019), Chen et al. (2019), Fernandez-Carames and Fraga-Lamas (2019), Gori et al. (2015), Jo et al. (2019), Khan et al. (2019), Khare et al. (2020), Liu et al. (2019), Marsal-Llacuna (2018), Park et al. (2018), (Scekic et al. (2018), Sun and Zhang (2016), Yu et al. (2018), Zou et al. (2019)

improve security across Internet of Things (IoT) in a smart city and sharing economy (Banerjee et al., 2018; Ferraro et al., 2018; Dwivedi et al., 2019; Altulyan et al., 2020; Singh et al., 2019; Rahman et al., 2019; Shen et al., 2019; Jo et al., 2019; Sa & Umamakeswari, 2018; Khan & Salah, 2018; Mohanta et al., 2019; Liu et al., 2019).

Rathore et al. (2019) provide a framework to be used across the IoT ecosystem to detect potential attacks by monitoring and analyzing all the traffic data based on blockchain—a costly solution, as blockchain requires significant computer power (Marsal-Llacuna, 2018). Hammi et al. (2018) propose a "bubble of trust" where secure virtual zones are created to enable communication of IoT devices. Gong et al. (2019) focus on secure IoT device management to ensure integrity, data availability, and confidentiality.

Blockchain can be used to build a secure, autonomous transport system for autonomous vehicles (Sharma et al., 2017; Yin et al., 2019; Chen et al., 2019). Zhang et al. (2019) focus on solving the security issues of vehicular ad hoc networks (VANETs), whereas Wang et al. (2019) discusses the use of blockchain to implement secure energy delivery services for electric vehicles and energy nodes involving both the transport and energy industry. Chaudhary et al. (2019) and Zhou et al. (2019) discuss secure energy trading for electric vehicles.

Moreover, in the energy industry, Pieroni et al. (2018) and Park et al. (2018) discuss a secure blockchain-based electrical energy trading system. In the healthcare industry, blockchain can enable secure sharing of patient medical records (Dagher et al., 2018). Other studies focus on blockchain for securing information in the casino and entertainment industry (Liao & Wang, 2018), smart campus cybersecurity (Ferraro et al., 2018), and supply chain security (Sharma et al., 2019).

Blockchain for improved privacy. Marsal-Llacuna (2018) discusses the questionable privacy practices used by central authorities. Ferraro et al. (2018) discusses the social contract enforcement of identity privacy. Identity and access management could be securely controlled through blockchain (Khan et al., 2019). Dagher et al. (2018) define a privacy-preserving framework for access control and interoperability, and Kumar et al. (2019) presents a privacy preservation technique to protect user identity; Kamel Boulos et al. (2018) specifically focus on securing patient and healthcare provider identities, and Dwivedi et al. (2019) proposes a privacy-preserving blockchain for the analysis of big data in healthcare. Moreover, an interesting use case is "Reportcoin," a blockchain-based incentive anonymous reporting system to preserve privacy when reporting law violations (Zou et al., 2019).

The use of blockchain to tackle IoT-related privacy issues is proposed by Jo et al. (2019), Singh et al. (2019), Mohanta et al. (2019), Rahman et al. (2019), and Sharma and Park (2018). Shen et al. (2019) explores the Privacy-Preserving Support Vector Machine training over blockchain-based encrypted IoT data in smart cities, while Zhao et al. (2019) explores machine learning-based privacy-preserving fair data trading. Moreover, Zhou et al. (2019) discusses the privacy challenges of Internet of Vehicles (IoV).

Increased trust. User trust in the entire system is critical (Bruneo et al., 2019) and can be viewed from two different perspectives: first, from the inherent blockchain attributes of transparency, immutability, and auditability that ensure data integrity (Park et al., 2018; Khan et al., 2019; Liu et al., 2019; Zou et al., 2019; Chen et al., 2019) and second from the decentralized nature of blockchain technology that eliminates the need to trust centralized third parties (Khan et al., 2019; Scekic et al., 2018; Khare et al., 2020; Sun & Zhang, 2016; Fernandez-Carames & Fraga-Lamas, 2019; Yu et al., 2018; Marsal-Llacuna, 2018; Hadi et al., 2019; Jo et al., 2019). As explained by Zou et al. (2019) "blockchain converts trust in people or institutions into trust in the system."

Other reasons. Blockchain technology provides a traceable and irreversible mechanism for incentives in the context of loyalty and reward programs (Chen et al., 2019; Bruneo et al., 2019; Nam et al., 2019). Applications involved motivating citizens to participate in the collection and sharing of data (Chen et al., 2019; Yin et al., 2019), the reporting of issues (Zou et al., 2019), and validation of data (Kamel Boulos et al., 2018).

Other reasons cited involved blockchain's peer to peer (P2P) nature (Fernandez-Carames & Fraga-Lamas, 2019; Kamel Boulos et al., 2018; Nam et al., 2019). Notably, Sun and Zhang (2016) suggests a P2P approach that prevents units of unilaterally taking actions on behalf of the community

Another important reason cited for using blockchain as a solution was efficiency (see Khan et al., 2019; Sharma et al., 2019, and Kamel Boulos et al., 2018) and the capability of currency/token management (Yin et al., 2019; Liao & Wang, 2018; Nam et al., 2019; Ferraro et al., 2018; Wang et al., 2019).

3.2 What Blockchain-Based Applications Are Being Researched for Smart Cities?

We use the taxonomy framework proposed by Silva et al. (2018) to classify the applications and use cases. Sometimes those would fit into more than one category; for example, Liao and Wang (2018) discusses applications for integrated resorts which include healthcare, transport, logistics, supply chains, hospitality, etc. (Fig. 2).

Below is a detailed breakdown of the top 6 major categories or themes, comprising 80% of the use cases identified in the research.

Smart transportation/mobility. Smart transportation/mobility has the highest amount of use cases researched for smart cities. Table 2 lists the use cases and the papers referring to them.

The concept of "traffic lights" and "car pooling" is examined in Bruneo et al. (2019). Sharma and Park (2018) focuses on vehicular network architecture in smart cities and investigate vehicle resource discovery and sharing, intelligent transport systems that communicate with the home, and incorporating ride sharing

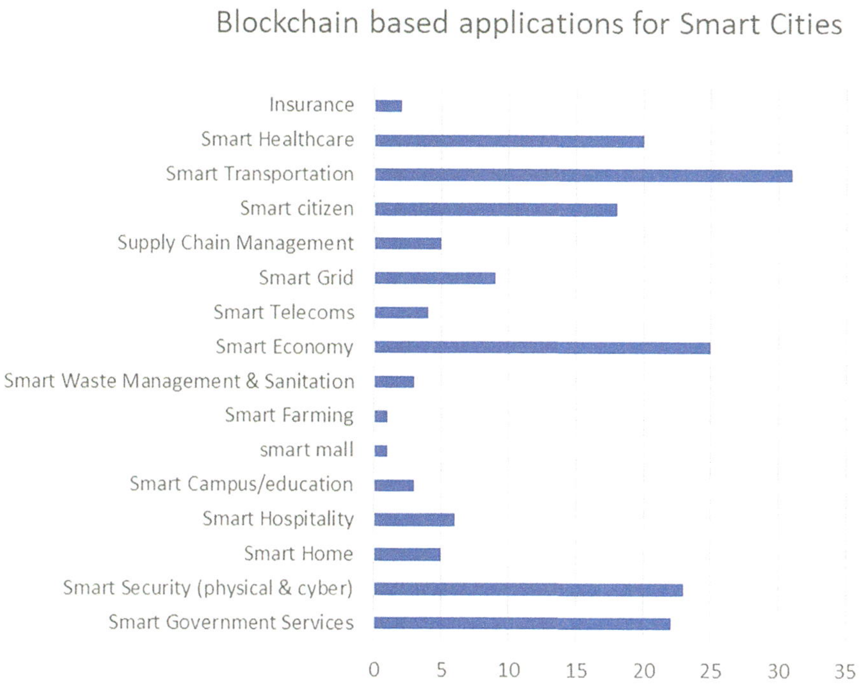

Fig. 2 Blockchain-based applications for smart cities

and scheduling into the home and vehicular network, as part of a smart home. Wang et al. (2019) investigates integrating electric vehicles into smart homes by examining data management application of the vehicular energy network (VEN) and renewable energy (RE) transportation.

Bruneo et al. (2019) investigates the possibilities of environmental data collection using taxi data (speed, GPS, etc.). Zhang et al. (2019) highlights a use case for data gathering on the communication between smartphones and vehicles. Hadi et al. (2019) investigates data forensics in IoT and Hammi et al. (2018) focus on IoT authentication.

"Smart transportation" supporting logistics is the overlap between this category and the "logistics and supply chain" classification. Both Liao and Wang (2018) and Yin et al. (2019) investigate mobile crowdsensing and relevant incentives. Karale and Ranaware (2019) and Liao and Wang (2018) research into mobility logistics in the pharmaceutical and gaming industries, respectively.

Smart economy. Applications for the smart economy were the second most researched topic, with 25 use cases by different authors. Those are shown in Table 3.

Cryptocurrencies are viewed as trading tokens (Zhou et al., 2019), as incentives, as rewards tokens (Bruneo et al., 2019; Nam et al., 2019; Zou et al., 2019), or as e-money across the ecosystem (Liao & Wang, 2018).

Table 2 Transportation/mobility applications

Application	Citation
Smart transportation/mobility	Hadi et al. (2019), Liao and Wang (2018), Rahman et al. (2019), Sharma and Park (2018), Sharma et al. (2019), Sharma et al. (2017), Khare et al. (2020), Fernandez-Carames and Fraga-Lamas (2019)
Electric vehicles	Chaudhary et al. (2019), Wang et al. (2019)
Automotive loyalty/incentive programs	Sharma et al. (2019), Wang et al. (2019), Chen et al. (2019)
Parking	Bruneo et al. (2019), Sharma and Park (2018)
Car pooling	(Bruneo et al., 2019)
Logistics	Karale and Ranaware (2019), Liao and Wang (2018)
Ride sharing/bike sharing	Scekic et al. (2018), Sharma et al. (2017)
Data sharing	Singh et al. (2019)
Drones	Kamel Boulos et al. (2018)
Environment data using taxi infrastructure	Bruneo et al. (2019)
Traffic data	Bruneo et al. (2019)
Automotive automated payments	Sharma et al. (2019)
Mobile crowdsensing	Yin et al. (2019)
Vehicle registration and maintenance services	Sharma et al. (2019)
Vehicle and smart phone communications	Zhang et al. (2019)
Vehicular network	Sharma et al. (2017)
Smart roads	Hammi et al. (2018)

Table 3 Smart economy applications

Application	Citation
Virtual currencies	Bruneo et al. (2019), Liao and Wang (2018), Zhou et al. (2019), Nam et al. (2019), Zou et al. (2019)
Payments	Bruneo et al. (2019), Nam et al. (2019), Sharma et al. (2019), Sharma et al. (2017)
Loyalty and rewards	Kamel Boulos et al. (2018), Wang et al. (2019), Chen et al. (2019)
Marketplace	Chaudhary et al. (2019), Scekic et al. (2018), Zhao et al. (2019)
Investing	Kumar et al. (2019), Marsal-Llacuna (2018)
Financial transactions and trades	Hadi et al. (2019)
Sharing economy	Sun and Zhang (2016)
Econometric models	Marsal-Llacuna (2018)
Clinical research and data monetization	Kamel Boulos et al. (2018)
Smart economy	Fernandez-Carames and Fraga-Lamas (2019)

Sharma et al. (2019) looks into the automated payment process in the automotive industry and, as part of their research on an IoT service ecosystem, Bruneo et al.

Table 4 Smart security (physical and cyber) applications

Application	Citation
Secured information/data management and exchange	Yin et al. (2019), Sharma et al. (2017), Wang et al. (2019), Khan and Salah (2018), Zhao et al. (2019), Shen et al. (2019), Sa and Umamakeswari (2018), Liu et al. (2019)
Identity management	Rahman et al. (2019), Nam et al. (2019), Gong et al. (2019), Singh et al. (2019), Hammi et al. (2018), Kamel Boulos et al. (2018)
Detect and mitigate security attacks in IoT	Rathore et al. (2019), Khare et al. (2020), Mohanta et al. (2019)
Firmware detection, updates, and self-healing IoT	Banerjee et al. (2018), Gong et al. (2019)
Data forensics	Hadi et al. (2019)
Big data auditing	Yu et al. (2018)
Secure communication in a distributed environment	Jo et al. (2019)
Anonymous reporting	Zou et al. (2019)

(2019) looks into parking and university services payments and Liao and Wang (2018) into blockchain as a platform for the integrated casino and entertainment industry.

Investing use cases focus on real estate investing (Kumar et al., 2019) and smart malls (Sharma & Park, 2018). Loyalty and rewards incentives are a critical element of influencing participation and engagement with stakeholders in a smart economy and are discussed by Karale and Ranaware (2019). Wang et al. (2019) discusses a financial framework for energy delivery for vehicular networks. Chen et al. (2019) look into a quality-driven auction-based incentive mechanism.

Finance is one of the topics tackled by Hadi et al. (2019) while looking into blockchain technology and its integration with IoT. Zhao et al. (2019) explored data trading in the big data market, with a specific focus on machine learning and preserving financial privacy. Pieroni et al. (2018) research a smart energy grid based on blockchain technology, whereby energy providers and private citizens can freely exchange energy both as consumers and prosumers.

Smart security (physical and cyber). Smart security is a critical element of a smart city as it is at high risk of cyberattacks (Table 4).

The most prominent use cases for smart security concern securing information and data. Eight papers focus on secure information management and exchange with Sharma et al. (2017) describing a model allowing vehicles within a network of vehicles to discover and share their resources and data securely.

Wang et al. (2019) examine secure energy delivery services for electric vehicles and energy nodes. Yin et al. (2019), Zhao et al. (2019), Shen et al. (2019), Sa and Umamakeswari (2018), and Liu et al. (2019) all focus on securing IoT data, whereas Khan and Salah (2018) look into ensuring the security of the collection and trading of data on the network. Marsal-Llacuna (2018) explored blockchain as the next enabling network, illustrating how blockchain technology can be used to connect data processing technologies securely for IoT.

Identity management is the topic of five papers. Hammi et al. (2018) presents a use case for a decentralized blockchain-based authentication system for identity and access management to be applied to IoT. The topic of identity management networks to integrate decentralized identity management and distributed credential storage is investigated by Hadi et al. (2019) and by Nam et al. (2019) in a smart tourism context.

Identity management is also highlighted as a potential use case within logistics and supply chain management by Hadi et al. (2019), where integration of blockchain technology with IoT is discussed and use cases of decentralized identity management and distributed credential storage are presented.

Another subtopic under secure identity management is the sharing economy. Rahman et al. (2019) presents a blockchain- and IoT-based cognitive edge framework for sharing economic services in a smart city. Banerjee et al. (2018) with a focus on healthcare looks into the data management of firmware detection and self-healing through IoT.

The remaining use cases under "security" include a blockchain-based secure device management framework for an IoT network (Gong et al., 2019). Khare et al. (2020) investigates the design of a trustless smart city system that involves the acquisition, storage, and consumption of sensor data. Jo et al. (2019) focuses on sustainable smart city network security, looking into secure communication in a distributed environment. Wang et al. (2019) presents a blockchain-based secure incentive scheme for energy delivery in a vehicular energy network, to be integrated into the IoT through the management of devices at home, vehicle, and city.

Smart government. With blockchain technology, the interaction with governmental services and administration can be trustless, secure, and more transparent and encourage citizen participation (see Table 5). Karale and Ranaware (2019) examines loyalty and rewards platforms, birth and death registries, court case filings, property registration, local business registration, and voting platforms.

Both Marsal-Llacuna (2018) and Bruneo et al. (2019) highlighted the need of policies, rules, laws, regulations, and standards to govern smart cities and the role of blockchain. Citizen engagement and participation within smart cities and government is the focus of Marsal-Llacuna (2018) that discusses use cases for citizens to submit their urban needs onto the blockchain encouraging involvement in policy decisions. Marsal-Llacuna (2018) also focuses on geographical information systems, econometric models, mayors' dashboards, and statistical projections. Finally, Khan and Salah (2018) discusses an immutable log of events and management of access control to government data.

Smart healthcare. Smart healthcare has also garnered a lot of interest (Table 6). Altulyan et al. (2020) and Dagher et al. (2018) refer to opportunities in healthcare at a high level. Hadi et al. (2019) delves deeper into the topic in combination with IoT and describes financial, transactional, and trade aspects of intelligent healthcare networks. Hammi et al. (2018) researches the concept of secure virtual zones

Table 5 Smart government applications

Application	Citation
Administration including court case files and building information	Karale and Ranaware (2019), Marsal-Llacuna (2018)
Policies, rules, laws, regulations, standards	Marsal-Llacuna (2018), Bruneo et al. (2019)
Crowdsensing and crowdsourcing	Kamel Boulos et al. (2018), Bruneo et al. (2019)
Voting platforms	Karale and Ranaware (2019), Bruneo et al. (2019)
Infrastructure and an ecosystem of services	Bruneo et al. (2019)
Incentive mechanisms	Bruneo et al. (2019)
Registrations (birth and death, property, local business)	Karale and Ranaware (2019)
Citizen data (sharing)	Khan et al. (2019)
District area monitoring and safety	Khare et al. (2020)
Urban budgeting	Khare et al. (2020)
Social compliance	Ferraro et al. (2018)
Log of events and management	Khan and Salah (2018)
Geographic information system	Marsal-Llacuna (2018)
Mayors dashboards	Marsal-Llacuna (2018)
Statistical projections	Marsal-Llacuna (2018)
Crisis mapping and recovery	Kamel Boulos et al. (2018)

Table 6 Smart healthcare applications

Application	Citation
Smart healthcare	Altulyan et al. (2020), Hadi et al. (2019), Dagher et al. (2018), Liao and Wang (2018), Rahman et al. (2019), Hammi et al. (2018)
Collection, processing, and storage of healthcare data	Kamel Boulos et al. (2018)
Transparent pharmaceutical and medical device supply chains	Karale and Ranaware (2019), Kamel Boulos et al. (2018)
Secure sharing of data	Singh et al. (2019), Kamel Boulos et al. (2018)
Remote patient monitoring	Dwivedi et al. (2019)
Single electronic health record for the citizens	Karale and Ranaware (2019)
Smart hospital	Sharma and Park (2018)
Clinical research and data monetization, medical fraud detection, public health surveillance, wearables	Kamel Boulos et al. (2018)
Healthcare insurance claims	Mohanta et al. (2019)

(bubbles) where things can identify and trust each other in the context of hospitals, healthcare, and medical use cases.

Table 7 Smart citizenship applications

Application	Citation
Citizen participation and engagement	Khan et al. (2019), Scekic et al. (2018), Khare et al. (2020), Marsal-Llacuna (2018)
Energy trading and marketplace	Altulyan et al. (2020), Chaudhary et al. (2019), Zhou et al. (2019), Pieroni et al. (2018)
Crowdsensing (problem reporting)	Bruneo et al. (2019), Zou et al. (2019)
Incentives and rewards	Bruneo et al. (2019), Karale and Ranaware (2019)
Sharing economy	Sun and Zhang (2016), Scekic et al. (2018)
Remote patient monitoring	Dwivedi et al. (2019)
Smart living	Hadi et al. (2019)
Smart people (education)	Fernandez-Carames and Fraga-Lamas (2019)

Rahman et al. (2019) focuses on a sharing economy including smart health services and Liao and Wang (2018) includes smart health in their study of use cases for integrated casino and entertainment. Kamel Boulos et al. (2018) researches aspects including medical fraud detection and mechanisms for validating, crediting, and rewarding crowdsourced geotagged data, public health surveillance, and wearables.

Smart citizenship. This pertains to the social element of smart cities where citizen participation and engagement play a major role (Table 7). Scekic et al. (2018) explores citizen cocreation, both at the neighborhood scale and bidirectional and city-wide. Scekic et al. (2018) introduces the social aspects of exchanging societal value and cocreation through the "WeValue" smart city platform. Marsal-Llacuna (2018) investigates the "people's layer" for urban technologies and geographic information systems.

Bruneo et al. (2019) and Khan et al. (2019) specifically look into "citizen participation" in terms of listing issues, data sharing, and participation in governmental decision-making. Ferraro et al. (2018) focuses on education falling into the "smart people" subcategory.

Other classifications. Energy grid application for "energy exchange" is a popular use case for blockchains. Pieroni et al. (2018) and Park et al. (2018) explore citizens as prosumers, trading energy between providers, prosumers, and consumers of smart homes. Scekic et al. (2018) note that a smart city platform for societal value exchange may be utilized for energy-savings. Fernandez-Carames and Fraga-Lamas (2019) research smart governance, smart living, and smart economy of energy management and Pieroni et al. (2018) look into the data associated with energy trading and a sustainable electrical energy transaction ecosystem between prosumers and consumers of smart homes. Wang et al. (2019) proposes a blockchain-based incentive scheme for energy delivery in vehicular energy network creating a case for integrated wireless power transfer technology into the smart home and building in incentive schemes through intelligent reporting. Karale and Ranaware (2019) and Bruneo et al. (2019) find notable use cases for the use of blockchain in renewable energy.

Another topic of interest is waste management. Hammi et al. (2018) and Khare et al. (2020) focus on waste management and sanitation. Bruneo et al. (2019) suggest a blockchain-based system for payments for waste management. Khare et al. (2020) focuses on the mobility of waste management services.

Smart home is another interesting topic. Hadi et al. (2019) explores the topic of IoT for smart lives and smart homes. Sharma et al. (2019) had an architectural approach to smart homes within the smart city.

Supply chain management is discussed in association with applications of blockchain technology to logistics management in integrated casinos and entertainment (Liao & Wang, 2018), the automotive industry (Sharma et al., 2019), and pharmaceutical supply chains (Karale & Ranaware, 2019).

4 Discussion and Conclusions

The intersection of blockchain and smart cities is a complex topic that spans across multiple disciplines. In this chapter through a systematic literature review methodology, we analyze 45 academic papers published in 33 different journals to answer two important research questions.

The first question is why blockchain was chosen as the solution, that is, what it has to offer to smart cities. The main reasons cited in the literature were first security, followed by privacy and trust. Due to the vast volume of data, the stakeholders involved, and the necessary controls, smart cities require a very high level of security. Privacy issues involve identity privacy, healthcare privacy, privacy when reporting crimes, and privacy issues of the Internet of Vehicles. Trust, another reason cited, is enhanced by transparency, immutability and auditability, and trustless systems enabled by blockchain technology.

The second question focuses on the applications of blockchain proposed for smart cities thematically taxonomized using the (Silva et al., 2018) framework. Our findings show that smart cities' blockchain applications focus on smart transportation, smart economy, smart security, smart government, smart healthcare, and smart citizenship applications, creating important links between the social, economic, and industry elements of smart cities.

Interestingly, in reviewing the literature, we observed that studies often tend to focus on smaller subsets of a smart city; so, a "smart city" is a conglomeration word for "smart places." These include smart campuses, smart suburbs, smart hospitals, integrated casinos, smart malls, etc.; the concept, however, can also expand to include a whole country. The "place" becomes the centerpiece of the model, and there are six smart themes, facilitated through different blockchains (Fig. 3). Each blockchain-based smart city application falls within one or more of the six themes of this taxonomy. The six themes are economy, which associated the applications of sharing, investing, and marketplace; environment, associated with mobility, energy, and telecoms; governance, with its associated laws and regulations and record keeping; services, associated with education, healthcare, insurance,

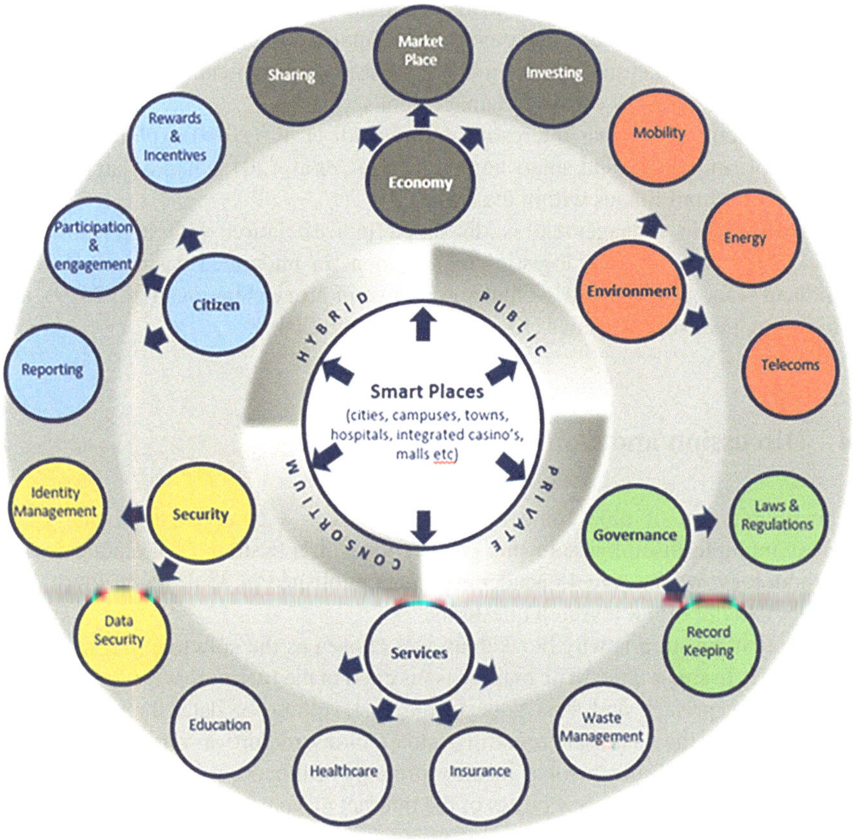

Fig. 3 Graphical view of blockchain-based applications

and waste management; security and its associated identity management and data security; and, citizenship, associated with rewards and incentives, participation and engagement, and reporting.

Blockchain solutions need to be carefully thought through and implemented as the very attributes that make blockchain desirable for smart cities could also create challenges that need to be overcome, including the need for notaries, quality data input, regulatory privacy, and the right to be forgotten. On a more technical note, additionally future research is needed around blockchain scalability, standards, and interoperability across multiple blockchains to enable a holistic ecosystem that brings together all the elements of a smart city.

Planning and building a conventional brick and mortar city requires highly skilled planners, designers, and builders, as implementing smart cities requires specific skills from smart cities' professionals and participants in general. We follow the framework from Fitsilis and Kokkinaki (2021) regarding the learning objectives in developing skills for smart cities to determine the skills needed for implementing blockchain-based solutions for smart cities; we observe that all the skills listed under the four learning objectives are all important. Specifically, LO1: "Development of transversal skills"; LO2: "Building an adequate IT knowledge background"; LO3: "Developing advances for software development and operation skills (DevOps)"; and LO4: "Developing smart city management skills" capture the whole range of skills required for utilizing blockchains in smart cities. The skills described under LO1, for example, will be valuable throughout the whole design and implementation phases. The group of skills described under L2 should be cultivated not only among professionals but also among smart citizens.

Blockchain-based applications within the smart city context are steadily gaining more momentum; however, it must be noted that they are still in an evolution state. Significant developments and change management will be required to position blockchain-based applications for smart cities as an opportunity for transparency, trust, and participative citizenry, like this is described in L4, 8 "Citizen Driven/Citizen Orientation/User Experience Design" which is about learning how to engage citizens.

Based on the assumption that skills under LO1 and LO2 are already present in smart cities' professionals, the skills under LO3 and LO4 are the relevant ones for blockchain applications. Specifically, the skills under LO3 will ensure that developing blockchain-based apps for smart cities will take place efficiently and meet high standards of quality. Furthermore, developing blockchain services in the context of smart cities will specifically require skills that pertain to smart cities' business models and financial management (skill 2 under LO4), to smart services and operating procedures (skill 3 under LO4), to the legal issues and standards (skill 5 under LO4) emerging from the use of blockchain for smart cities' applications, and to understanding the concept of digital twins (skill 10 under LO4).

Skills are needed to change the current mindset of centralized control and trusted third parties to a more participative engagement model across smart cities.

Acknowledgments This work was supported by the Erasmus+ KA2 under the project "DevOps competences for Smart Cities" (Project No.: 601015-EPP-1-2018-1-EL-EPPKA2-SSA). This **book chapter** reflects the views only of the authors, and the Commission cannot be held responsible for any use which may be made of the information contained therein.

References

Altulyan, M., Yao, L., Kanhere, S. S., Wang, X., & Huang, C. (2020). A unified framework for data integrity protection in people-centric smart cities. *Multimedia Tools and Applications, 79*(7), 4989–5002.

Banerjee, M., Lee, J., & Choo, K.-K. R. (2018). A blockchain future for internet of things security: a position paper. *Digital Communications and Networks, 4*(3), 149–160.

Batty, M., Axhausen, K. W., Giannotti, F., Pozdnoukhov, A., Bazzani, A., Wachowicz, M., ... Portugali, Y. (2012). Smart cities of the future. *The European Physical Journal Special Topics, 214*(1), 481–518.

Bruneo, D., Distefano, S., Giacobbe, M., Minnolo, A. L., Longo, F., Merlino, G., … Tapas, N. (2019). An iot service ecosystem for smart cities: The# smartme project. *Internet of Things, 5*, 12–33.

Chaudhary, R., Jindal, A., Aujla, G. S., Aggarwal, S., Kumar, N., & Choo, K. K. R. (2019). BEST: Blockchain-based secure energy trading in SDN-enabled intelligent transportation system. *Computers & Security, 85*, 288–299.

Chen, W., Chen, Y., Chen, X., & Zheng, Z. (2019). Toward secure data sharing for the IoV: A quality-driven incentive mechanism with on-chain and off-chain guarantees. *IEEE Internet of Things Journal*, 1–1.

Dagher, G. G., Mohler, J., Milojkovic, M., & Marella, P. B. (2018). Ancile: Privacy-preserving framework for access control and interoperability of electronic health records using blockchain technology. *Sustainable Cities and Society, 39*, 283–297.

Denyer, D., & Tranfield, D. (2009). Producing a systematic review. In D. A. Buchanan & A. Bryman (Eds.), *The Sage handbook of organizational research methods* (pp. 671–689). Sage Publications Ltd.

Dwivedi, A. D., Gautam, S., Dhar, S., & Singh, R. (2019). A decentralized privacy-preserving healthcare blockchain for IoT. *Sensors, 19*(2), 326.

Fernandez-Carames, T. M., & Fraga-Lamas, P. (2019). Towards next generation teaching, learning, and context-aware applications for higher education: A review on blockchain, IoT, Fog and edge computing enabled smart campuses and universities. *Applied Sciences, 9*(21), 4479.

Ferraro, P., King, C., & Shorten, R. (2018). Distributed ledger technology for smart cities, the sharing economy, and social compliance. *IEEE Access, 6*, 62,728–62,746.

Fitsilis, P., & Kokkinaki, A. (2021). Smart cities body of knowledge. Retrieved from https://smartdevops.eu/scbok

Giffinger, R., Fertner, C., Kramar, H., & Meijers, E. (2007). *City-ranking of European mediumsized cities* (vol. 9, 1–12). Vienna, UT: Centre for Regional Science.

Gong, S., Tcydenova, E., Jo, J., Lee, Y., & Park, J. H. (2019). Blockchain-based secure device management framework for an internet of things network in a smart city. *Sustainability, 11*(14), 3889.

Gori, P., Parcu, P. L. & Stasi, M., (2015). *Smart cities and sharing economy.* s.l.:s.n.

Hadi, F. A., Hussein, A. R., Rashed, J. R., Saad, N., & Alseelawi, H. A. (2019). A vision of blockchain technology and its integration with IOT: Applications, challenges, and opportunities; from the authentication perspective. *Journal of Theoretical and Applied Information Technology, 97*(15), 4048.

Hammi, M. T., Hammi, B., Bellot, P., & Serhrouchni, A. (2018). Bubbles of trust: A decentralized blockchain-based authentication system for IoT. *Computers & Security, 78*, 126–142.

Jo, J. H., Sharma, P. K., Sicato, J. C. S., & Park, J. H. (2019). Emerging technologies for sustainable smart city network security: Issues, challenges, and countermeasures. *Journal of Information Processing Systems, 15*(4), 765–784.

Kamel Boulos, M. N., Wilson, J. T., & Clauson, K. A. (2018). Geospatial blockchain: Promises, challenges, and scenarios in health and healthcare. *International Journal of Health Geographics, 17*(25).

Karale, S., & Ranaware, V. (2019). Applications of blockchain technology in smart city development: A research. *International Journal of Innovative Technology and Exploring Engineering, 8*(11S).

Khan, M. A., & Salah, K. (2018). IoT security: Review, blockchain solutions, and open challenges. *Future Generation Computer Systems, 82*, 395–411.

Khan, Z., Abbasi, A. G., & Pervez, Z. (2019). Blockchain and edge computing-based architecture for participatory smart city applications. *Concurrency and Computation Practice and Experience, 32*(12).

Khare, A., Merlino, G., Longo, F., Puliafito, A., & Vyas, O. P. (2020). Design of a trustless smart city system: The# SmartME experiment. *Internet of Things, 10*, 100126.

Kumar, S. E., Talasila, V., & Pasumarthy, R. (2019). A novel architecture to identify locations for Real Estate Investment. *International Journal of Information Management, 56*, 102012.

Liao, D.-Y., & Wang, X. (2018). Applications of blockchain technology to logistics management in integrated casinos and entertainment. *Informatics, 5*(4), 44.

Liu, Y., Wang, K., Qian, K., Du, M., & Guo, S. (2019). Tornado: Enabling blockchain in heterogeneous internet of things through a space-structured approach. *IEEE Internet of Things Journal, 7*(2), 1273–1286.

Marsal-Llacuna, M.-L. (2018). Future living framework: Is blockchain the next enabling network? *Technological Forecasting & Social Change, 128*, 226–234.

Mohanta, B. K., Jena, D., & Satapathy, U. (2019). Trust management in IOT enable healthcare system using ethereum based smart contract. *International Journal of Scientific & Technology Research, 8*(9).

Nam, K., Dutt, C. S., Chathoth, P., & Sajid Khan, M. (2019). Blockchain technology for smart city and smart tourism: Latest trends and challenges. *Asia Pacific Journal of Tourism Research, 26*, 454–468.

Park, L. W., Lee, S., & Chang, H. (2018). A sustainable home energy prosumer-chain methodology with energy tags over the blockchain. *Sustainability, 10*, 658.

Pieroni, A., Scarpato, N., Di Nunzio, L., Fallucchi, F., & Raso, M. (2018). Smarter city: Smart energy grid based on blockchain technology. *International Journal on Advanced Science, Engineering and Information Technology, 8*(1), 298–306.

Rahman, M. A., Rashid, M. M., Hossain, M. S., Hassanain, E., Alhamid, M. F., & Guizani, M. (2019). Blockchain and IoT-based cognitive edge framework for sharing economy services in a smart city. *IEEE Access, 7*, 18611–18621.

Rathore, S., Kwon, B. W., & Park, J. H. (2019). BlockSecIoTNet: Blockchain-based decentralized security architecture for IoT network. *Journal of Network and Computer Applications, 143*, 167–177.

Sa, B., & Umamakeswari, A. (2018). Role of blockchain in the Internet-of-Things (IoT). *International Journal of Engineering & Technology, 7*(2.24), 109–112.

Scekic, O., Nastic, S., & Dustdar, S. (2018). Blockchain-supported smart city platform for social value co-creation and exchange. *IEEE Internet Computing, 23*(1), 19–28.

Sharma, P. K., Kumar, N., & Park, J. H. (2019). Blockchain-based distributed framework for automotive industry in a smart city. *IEEE Transactions on Industrial Informatics, 15*(7), 4197–4205.

Sharma, P. K., Moon, S. Y., & Park, J. H. (2017). Block-VN: A distributed blockchain based vehicular network architecture in smart city. *Journal of Information Processing Systems, 13*(1), 184–195.

Sharma, P. K., & Park, J. H. (2018). Blockchain based hybrid network architecture for the smart city. *Future Generation Computer Systems, 86*, 650–655.

Shen, M., Tang, X., Zhu, L., Du, X., & Guizani, M. (2019). Privacy-preserving support vector machine training over blockchain-based encrypted IoT data in smart cities. *IEEE Internet of Things Journal, 6*(5), 7702–7712.

Silva, B. N., Khan, M., & Han, K. (2018). Towards sustainable smart cities: A review of trends, architectures, components, and open challenges in smart cities. *Sustainable Cities and Society, 38*, 697–713.

Singh, S. K., Rathore, S. & Park, J. H., 2019. BlockIoTIntelligence: A blockchain-enabled intelligent IoT architecture with artificial intelligence. Future Generation Computer Systems. 110:721-743

Sun, J., & Zhang, K. (2016). Blockchain-based sharing services: What blockchain technology can contribute to smart cities. *Financial Innovation, 2*(26).

United Nations (n.d.). Department of Economic and Social Affairs, Population Division, 2019. *World urbanisation prospects: The 2018 revision (ST/ESA/SER.A/420)*, New York: United Nations.

Wang, Y., Su, Z., & Zhang, N. (2019). BSIS: Blockchain-based secure incentive scheme for energy delivery in vehicular energy network. *IEEE Transactions on Industrial Informatics, 15*(6), 3620–3631.

Xie, J., Tang, H., Huang, T., Yu, F. R., Xie, R., Liu, J., & Liu, Y. (2019). A survey of blockchain technology applied to smart cities: Research issues and challenges. *IEEE Communications Surveys & Tutorials, 21*(3), 2794–2830.

Yin, B., Wu, Y., Hu, T., Dong, J., & Jiang, Z. (2019). An efficient collaboration and incentive mechanism for internet of vehicles (iov) with secured information exchange based on blockchains. *IEEE Internet of Things Journal, 7*(3), 1582–1593.

Yu, H., Yang, Z., & Sinnott, R. O. (2018). Decentralized big data auditing for smart city environments leveraging blockchain technology. *IEEE Access, 7*, 6288–6296.

Zhang, D., Yu, F. R., & Yang, R. (2019). Blockchain-based distributed software-defined vehicular networks: A dueling deep Q-learning approach. *IEEE Transactions on Cognitive Communications and Networking, 5*(4), 1086–1100.

Zhao, Y., Yu, Y., Li, Y., Han, G., & Du, X. (2019). Machine learning based privacy-preserving fair data trading in big data market. *Information Sciences, 478*, 449–460.

Zhou, Z., Wang, B., Guo, Y., & Zhang, Y. (2019). Blockchain and computational intelligence inspired incentive-compatible demand response in internet of electric vehicles. *IEEE Transactions on Emerging Topics in Computational Intelligence, 3*(3).

Zou, S., Xi, J., Wang, S., Lu, Y., & Xu, G. (2019). Reportcoin: A novel blockchain-based incentive anonymous reporting system. *IEEE Access, 7*, 65544–65559.

Artificial Intelligence, Big Data Analytics, and Smart Cities

Yiannis Kiouvrekis, Theodor Panagiotakopoulos, Iakovos Ouranos, and Ioannis Filippopoulos

1 Introduction

The rapid increase in population in urban areas has made it imperative to provide services and infrastructure to meet the needs of the inhabitants of these areas. As a result, the demand for integrated systems with sensors, actuators, and smartphone has increased, leading to important solutions in the age of the Internet of Things (IoT), in which all devices are able to connect and communicate with each other via the Internet. Future systems will include smart home sensors, vehicle networking, weather and water sensors, smart parking sensors, etc. (Luechaphonthara & Vijay-alakshmi, 2019). It is evident that we are dealing with systems that will manage a huge amount of data and processes. Big data and AI are two fields which are of interest to us due to the fact that data analysis and mining from big data sets are no longer dealt with in traditional ways (Favaretto et al., 2020). The major challenges

Y. Kiouvrekis (✉)
Department of Public and One Health, University of Thessaly, Thessaly, Greece

School of Business, University of Nicosia, Nicosia, Cyprus
e-mail: kiouvrekis.y@uth.gr

T. Panagiotakopoulos
School of Technology and Science, Hellenic Open University, Patras, Greece
e-mail: panagiotakopoulos@eap.gr

I. Ouranos
Hellenic Civil Aviation Authority, Heraklion, Greece

School of Applied Mathematical and Physical Sciences National Technical University of Athens, Athens, Greece
e-mail: iouranos@central.ntua.gr

I. Filippopoulos
Shipping and Operation, Cyprus International Institute of Management, Nicosia, Cyprus

P. Fitsilis (ed.), *Building on Smart Cities Skills and Competences*, Internet of Things,
https://doi.org/10.1007/978-3-030-97818-1_19

of data analytics include data mining, data storage, data analysis, search, sharing, transportation, visualization, information, information privacy, and data sources (Favaretto et al., 2020). The previous concepts are quite complex and refer to a very wide spectrum of scientific backgrounds.

Furthermore, it is well known that sampling as a part of data analysis presents several challenges, such as multidimensional data integration (Diego et al., 2016). The size and number of available data sets has increased exponentially as data are collected from devices such as mobile devices, Internet information detectors, antennas, cameras, microphones, radio frequency (RFID) readers, and wireless sensor networks.

Data analysis and applications of artificial intelligence find applications in almost all areas of human activity. The Internet of Things (IoT) combined with the rapid development of artificial intelligence play an important role in this process. In the field of smart cities, the widespread use of IoT has led to the creation of huge volumes of data, with the use of platforms that allow collection, analysis, and distribution of data that are applied in different areas of life, such as medicine, education, environmental health, etc. (Rjab & Mellouli, 2018). In the age of big data, understanding and interpreting scientific results from the fields of artificial intelligence, artificial learning, and statistics is becoming more and more pressing. At the same time, smart environments are growing rapidly, whether they involve macro-infrastructures like smart cities or micro-infrastructures like those in smart health.

The infrastructure of smart environments is equipped with wireless sensor networks that collect, analyze, and communicate autonomous data, and this infrastructure is referred to as "smart monitoring" (Vázquez et al., 2014). AI algorithms provide capabilities for analyzing big data and detecting patterns and features that could not be detected using traditional approaches. But regardless of the scope, smart environments require large-scale data analysis and decision-making tools. This fact has naturally created the introduction of increasingly specialized tools of artificial intelligence and big data analysis.

2 Artificial Intelligence and Big Data Analytics

AI constitutes a meeting point for many methodologies and tools from separate fields of mathematics and computer science (Haenlein & Kaplan, 2019), namely:

- Cognitive sciences, such as philosophy, logic, and psychology.
- Mathematical science and in particular mathematical logic, mathematical analysis, and topology.
- The wider scientific fields of informatics, as the main product of artificial intelligence, are intelligent information systems, expert systems, proofs of theorems, etc.

- Automation and control theory in collaboration with biotechnology, which propose the development of intelligent systems at the hardware level.

AI has several subfields, problem-solving being one of them. When we try to solve a problem, we must first define the initial conditions and then design an algorithm that will solve it. In order to give a problem as input to an algorithm, it must first have been clearly formally formulated and then defined in an appropriate way to be an input to an algorithm (Pearl, 1984).

Furthermore, complex process design as a subfield of AI intrinsically includes programming problem-solving, which is inherent in almost all industrial applications. Programming problem-solving in industrial applications is crucial and is based on appropriate AI algorithms (Pearl, 1984).

In addition, another area is the area of **autonomous roboting**. Autonomous robots have the ability to act on their own without the need for outside control (Ingrand & Ghallab, 2017). This type is programmed in such a way that it can respond to external stimuli. In fact, it is equipped with a sensor which allows it to detect obstacles, and every time it comes across an obstacle, it changes direction, according to its programming. Of course, the most advanced robots use stereovision to see the world around them, and in fact their software enables them to sort various objects as well as to calculate how far they are. Today robots can navigate effectively in various environments (Ingrand & Ghallab, 2017).

An experienced system is a system or simply a tool that is designed to solve difficult decision-making problems based on knowledge gathered by experts. Thus an experienced system is expected to act similarly to how an expert would act. On the other hand, a knowledge-based system is a software system that shows intelligent behavior in a specific problem or a specific function that represents and uses knowledge in a formal way.

An intelligent system essentially processes information in order to do something purposeful. Examples of intelligent systems are biological systems, computers, and robots. One type of intelligent system that is of particular interest to scientists is that of the intelligent agent. Intelligent agents are artificial agents, and they have the ability to act on their experience and the information they perceive (Wooldridge & Jennings, 1995). They are also looking for the best action plan for the situation they are called upon to resolve in order to achieve the best possible result. It is worth noting that intelligent agents find applications in areas such as networking, medicine, and e-commerce.

Siri and Alexa, for example, are smart agents because they use sensors such as microphones and other inputs to sense a request and draw on their collective experience and knowledge through supercomputers and databases around the world to make a decision.

What is required from this kind of tools is not only making a decision but also the user to understand of why and how the system reached the specific conclusion. The universal application of artificial intelligence algorithms has created a sense of distrust expressed by engineers, scientists, and citizens, resulting in the need for transparency in the way decisions are made. This is where XAI emerges as a new branch of AI. At XAI what we are called to do is explain how our system reached the final conclusion. Why did an autonomous car turn right and not left? Why did the algorithm decide to manage the resources of a system in this way? Why did the agent decide to act this way? Why did its algorithm decide that this is a better option than another? Does the system meet the specifications? The above are questions that the user can and does ask, but here is the first fundamental goal: the answer needs to cover as wide a range of users as possible, and this is where the XAI classification comes in.

3 Two Approaches in AI and Data Analysis

There are two fundamental approaches to artificial intelligence, the symbolic logic- and rule-based approach and the machine learning approach.

3.1 Symbolic AI and Formal Methods for Smart Cities

Mathematics and artificial intelligence have always had a symbiotic relationship. Every aspect of artificial intelligence has mathematical roots, and there have been bilateral developments. Efforts to improve computational logic have led to new results in mathematical logic. With the help of artificial intelligence, we can solve problems and prove theorems. Proof of theorems specifically includes logical and analogical reasoning.

Rule-based systems are the best available tool for coding problem-solving know-how by simulating the human expert (Hayes-Roth, 1985). Formal methods are techniques, languages, and tools that provide a rigorous basis for specifying and verifying software and hardware systems. Based on mathematical logics, they can offer high levels of certainty and consistency in the analysis of models, design, and programs (Woodcock et al., 2009). Several languages and tools supporting formal analysis and verification have been developed, and one should consider their advantages and disadvantages and the characteristics of the problem under study to choose the most appropriate formal method. Formal methods can be useful during the requirements phase of a system's development process. Their tools can provide automated support for checking completeness, traceability, verifiability, and reusability and for supporting requirements evolution, diverse viewpoints, and inconsistency management (George & Vaughn, 2003; Ghose, 2000). They are also used in specifying software, i.e., developing a precise description of what the software does while avoiding implementation details. Popular methods include ASM (Borger & Stark, 2003), B (Abrial, 1996a), and VDM (Jones, 1990) and algebraic specification languages such as CafeOBJ (Diaconescu & Futatsugi, 1998) and Maude (Clavel et al., 2007). Design by contract (Meyer, 1991) is another formal method technique which divides a complex specification into sub-specifications, each describing a subcomponent of the system. In addition, during the design of the system, formal methods can be used for data refinement which involves state machine specification, abstraction functions, and simulation proofs. Code verification is another application of formal methods, where every program-specification pair implicitly asserts a correctness theorem under conditions, and code verification is an attempt to prove this theorem or to establish why this theorem fails (Woodcock et al., 2009). There are several formal verification approaches supported by tools. In (Urban & Mine, 2021), the authors mention the fundamental undecidability of correctness properties of programs as a consequence of Rice's theorem, which states that it is impossible to design a tool that can decide precisely for every program whether it is correct or not. Tools must give up either full automation, generality, or completeness. Well-known approaches to formal verification include (Woodcock et al., 2009) deductive verification (Filliatre, 2016; Cuoq et al., 2012), design by refinement (Abrial, 1996b), interactive theorem provers-proof assistants (Diaconescu & Futatsugi, 1998; Bertot & Casteran, 2004; Isabelle, n.d.; HOL Interactive Theorem Prover, n.d.), model checking (Clavel et al., 2007; Jackson, 2012; Holzmann, 2003; TLA+, n.d.; UPPAAL, n.d.; mCRL2, n.d.; NuSMV, n.d.; JAVA PATHFINDER, n.d.), semantic static analysis, etc.

It is important to ensure that the system will work as intended. Especially when dealing with industrial safety critical software and hardware systems, a malfunction can be very dangerous for people's lives and can have environmental consequences and/or loss or severe damage to equipment/property, as well as financial consequences. In recent years, several companies such as Intel, Boeing, Siemens, Rolls Royce, Microsoft, Amazon, Facebook, NASA, IBM, SAP, BAE Systems, etc. use formal method techniques to conduct research on real-life development projects.

Formal methods can be used to model and verify systems that are implementing smart city applications. These systems are complex engineered systems with a large number of heterogeneous components and capable of multiple complex functions, leading to the ubiquitous cyber-physical systems (CPSs) (Baras, 2019). IoT together with CPSs has enabled the emergence of smart cities around the world, where a vast amount of sensing data and smart services are utilized to improve citizens' safety, wellness, and quality of life. But while significant research efforts have been spent toward building smarter services, sensors, and infrastructure in cities, the research challenge of how to ensure that a city's real-time operations satisfy safety and performance requirements has received only scant attention. Failure to check such requirements can lead to conflicts among smart services, with catastrophic consequences (Ma et al., 2021). The two key issues that the authors of (Ma et al., 2021) deal with are about monitoring whether city states satisfy a wide range of city requirements at runtime and how someone can predict a city's future states and check if the prediction application will satisfy city requirements. They propose a novel special aggregation signal temporal logic (SaSTL) (Meiyi et al., 2021) which extends STL (Maler & Nickovic, 2004) with logical operators for spatial aggregation and counting. Other extensions of STL include signal spatiotemporal logic (SSTL), spatial temporal logic, and spatial temporal reach and escape logic (STREL) (Bartocci et al., 2018). In (Keerthi et al., n.d.) the authors apply symbolic modeling checking techniques such as binary decision diagrams (BDD) (Bryant, 1986) and SAT-based bounded model checking (Clarke et al., 2001) with the CMBC tool, which is a C-based model checker (Clarke et al., 2004), to verify security issues in IoT devices, such as functional correctness of implementations, programming bugs, side-channel analysis, and hardware Trojans. In (Hofer-Schmitz & Stojanovic, 2020), a review of formal methods for an extensive variety of protocols used in the IoT environment is presented, together with detailed descriptions of the considered properties and the applied methods. The authors distinguish four application fields, namely, functional checks, checks for security properties, suggestions for enhanced schemes including a priori security property checks, and implementation checks of protocols.

In (Krichen, 2019), the authors suggest several techniques to apply formal verification and model-based testing to IoT and smart city systems. They have also presented a simple case study of a temperature measuring system comprising one collector and four sensors. Another paper (Roig et al., 2020) presents a formal algebraic specification of an IoT/Fog environment, where users may be moving around and their associated computing assets are meant to migrate among hosts, in order to follow their respective users so as to be as close as possible to them. They use the algebra of communicating processes (ACP) (Padua, 2011), which is a type of process algebra.

Machine learning and artificial intelligence algorithms are also used in systems supporting smart cities. In (Urban & Mine, 2021), the authors present a review of formal methods applied to machine learning, the large majority of which verify trained neural networks and employ either SMT, optimization, or abstract interpretation techniques. Verified artificial intelligence as a goal of designing AI-based systems with strong, ideally provable assurances of correctness with respect to mathematically specified requirements is presented in (Seshia et al., 2020). The authors present how formal methods can be applied to AI systems and identify five main challenges. They have developed two open-source tools, VerifAI (Anonymous, n.d.-a) and Scenic (Anonymous, n.d.-b), that have been applied to industrial-scale systems in the autonomous driving and aerospace domains.

3.2 Machine Learning

Machine learning is the creation of models or templates from a set of data derived from a computer system. Depending on the type of problem, various machine learning techniques have been developed. We will distinguish two types of machine learning: supervised learning, where the system is called to learn from a set of data, and unsupervised learning, where the system must learn from the creation of standards.

Supervised learning is the process where the algorithm constructs a function that represents given inputs (the training set) in well-known outputs, with the goal of generalizing this function for inputs with unknown output as well. It is used for the following:

- Classification
- Prediction
- Interpretation

Supervised learning involves use of data set, an input value set, and an output value set, so that the artificial intelligence (AI) network is trained to find the appropriate function that displays the input data at the output. Classic methods in supervised learning are regression and classification. The most common supervised learning methodologies are linear regression, support vector machine, and random forest. Classification is the process of determining the category of an observation. Unlike clustering, here the object categories are already known for training and testing purposes. Once a classification algorithm completes training for a given classification task, it can then assign a recently observed object to a category. Machine learning classifiers are also useful for classifying multiple classes.

Unsupervised learning involves the algorithm constructing a model for a set of inputs in the form of observations without knowing them desired outputs. It is used for the following:

- Association analysis
- Clustering

In unsupervised learning, there is no such thing as guidance; the AI model is trained to find hidden patterns, and one of the most popular methodologies in this case is k-means algorithm.

Clustering is used to group sets of objects based on their similarities in a multidimensional space. Objects in the same cluster are more alike than those in different clusters. Clustering is considered unsupervised machine learning because the types of object categories are not known in advance. In clustering, objects are grouped within a space, and the similarity between two objects is measured by a function of distance similarity. Clustering analysis reveals team patterns, provides information on key effectiveness factors, and identifies best practices for business activities. Reinforcement learning, the algorithm learning, involves a strategy of action through direct interaction with the environment. It is used mainly in planning problems, such as traffic control robots and optimization of work in factory areas.

Machine learning is used to identify patterns in data as well as to predict future events. An important difference between classification and forecasting is that classification is used to derive a rule or equation related to the current situation, while forecasting is about predicting what will happen in a new state. For example, predictive machine learning can check data to detect signals that will affect the future performance of a system.

The goal of a mark of decision process is to find solutions to problems modeled on successive decision problems. MDP is a model of a nondeterministic stochastic process and more specifically can represent the interaction of an agent with the environment or system.

The application of the MDP in the smart cities framework is remarkably big as asset. In (Yousefi et al., 2018) the MDP is used to create mobile agents for route planning in IoT smart environment. In (Turitsyn et al., 2011), the authors propose a device-based Markov decision process (MDP) to model individual devices which are expected to participate in future demand-response markets on distribution grids.

Even more interesting are the universal approaches as presented in (Mohammad, 2019), where the authors have modeled the security service infrastructure of smart city using an abstraction of MDPs which can be used for studying the security of various smart city infrastructures, deployment configurations, and attack vectors.

Continuing with the applications of MDP, we can see that in the field of robotics there are just as many and different applications. In (Simmons & Koenig, 1995), the methodology is typically used for tracking a robots' location in office environments. In (Beynier & Mouaddib, 2011), the authors used MDPs for decentralizing the control of multi-robot systems.

The Naive Bayes algorithm belongs to the supervised machine learning algorithms. It is a simple probabilistic classifier based on Bayes' theorem. The Naive Bayes algorithm is one of the basic classification techniques, and despite its simplicity and the independence assumptions it makes, it performs well on many problems. The k-nearest neighbors (KNN) algorithm is a well-known and widely used classification technique based on the use of distance-based measures. The central idea is that the value of the target function for a new snapshot is based solely on the corresponding values of the k "closest" training snapshots, which are its "neighbors." Decision trees are the best known supervised induction algorithm learning and have been successfully implemented in many areas where classification is required. The DT algorithm leads to creation of a tree form whose leaves represent classes. This tree form can also be read as a set of rules called "rules classification." In (Aloqaily et al., 2019) the authors used DT methods to create a detection system for connected vehicles in smart cities, and in (Zekić-Sušac et al., 2021) we get a system for managing energy efficiency of the public sector for smart cities.

Data mining is effectively applied in the field of smart health. There are many applications for creating and discovering models for the diagnosis of diabetes, coronary heart disease, and cancer among the available data. In (Jackins et al., 2021), the authors proposed AI-based smart prediction of clinical disease using Naive Bayes. The previous methodologies are also used in the analysis of the main types of crimes that occurred in the city. In (Pradhan et al., 2019), we read that the authors aim to observe the trend over the years and to determine how different characteristics contribute to specific crimes. More specifically, the model predicts the type of crime that will occur in each district of the city.

In the age of smart cities, we cannot exclude applications that include smart homes, smart kitchens, etc. In the context of the concept of smart nursery, we have the need for monitoring and ensuring the safety of infants. In (Mahmoud et al., 2020), we find an intelligent crib system in a smart children's room that automates cradle functions based on baby sounds.

As mentioned previously, the field of urban planning deals with the solution of smart city problems. AI and big data analysis have become an important part of urban planning in order to build smart cities. For example, in (Nallaperuma et al., 2019) the authors have proposed an expansive smart traffic management platform (STMP) based on unsupervised online incremental machine learning, deep learning, and deep reinforcement learning. Furthermore, we can find similar applications in smart parking. The number of vehicles on the roads is increasing rapidly, and the difficulty of finding a vacant parking spot is increasing in turn, making it a major problem for cities. In (Bhavani & Ghalib, 2018), the intelligent parking system helps to find the nearest free parking spot. More specifically, the k-nearest neighbor algorithm is used for this purpose, and the exact results are obtained for each test case.

4 Conclusions

The European Commission's strategy is to address the digital skills gap and to promote projects and strategies to improve the level of digital skills. All citizens need digital skills to study, work, communicate, access online public services, and find reliable information. The previous sections show the wide scientific spectrum from which the concepts of big data and artificial intelligence with applications in smart cities originate. In addition, it is important to note that the Digital Economy and Society Index (DESI) shows that 40% of adults in Europe lack basic digital skills. The European Commission has set goals in the European Skills Agenda and Action Plan for Digital Education to ensure that 70% of adults will have basic digital skills by 2025.

Based on the analysis of the two previous chapters, we end up with the following list of skills in relation to artificial intelligence and smart cities: (a) data integration, (b) data analysis, (c) predictive analysis, (d) data visualization, (e) modeling in formal systems, and (f) problem-solving.

Digital integration and the elimination of digital illiteracy in AI are not a sprint but a marathon that requires strategic planning and coordinated action. Supporting people unfamiliar with AI is a self-evident obligation of benevolent societies. Digital technologies—having invaded all fields of smart cities—are radically shaping the way of life, work, and education. They can thus be an ideal ally for tackling everyday pressing challenges. Understanding the issue and coordinated action through lifelong learning can build more inclusive, more equitable, and more sustainable societies, where everyone will be able to benefit to the fullest in this new digital age from its potential and skills for a more sustainable development. That is why it is important to create even more and more innovative curricula for the citizens of smart cities.

References

Abrial, J. R. (1996b). The B book: Assigning programs to meanings.

Abrial, J.-R. (1996a). *The B-book: Assigning programs to meanings*. Cambridge University Press.

Aloqaily, M., Otoum, S., Al Ridhawi, I., & Jararweh, Y. (2019). An intrusion detection system for connected vehicles in smart cities. *Ad Hoc Networks, 90*, 101,842, ISSN 1570-8705. https://doi.org/10.1016/j.adhoc.2019.02.001

Anonymous (n.d.-a). *VerifAI: A toolkit for design and verification of AI-based systems*. Retrieved from http://github.com/BerkeleyLearnVerify/VerifAI.

Anonymous (n.d.-b). *Scenic environment modeling and scenario description language*. Retrieved from http://github.com/BerkeleyLearnVerify/Scenic.

Baras, J. (2019). Formal methods and tool suites for CSP security, safety and verification. In A. Pretschner et al. (Eds.), *Engineering secure and dependable software systems* (pp. 1–7). IOS Press. https://doi.org/10.3233/978-1-61499-977-5-1

Bartocci, E., Deshmukh, J., Donzé, A., Fainekos, G., Maler, O., Ničković, D., & Sankaranarayanan, S. (2018). Specification-based monitoring of cyber-physical systems: A survey on theory, tools and applications. In E. Bartocci & Y. Falcone (Eds.), *Lectures on runtime verification* (pp. 135–175). Springer.

Bertot, Y., & Casteran, P. (2004). Interactive theorem proving and program development.

Beynier, A., & Mouaddib, A.-I. (2011). Applications of DEC-MDPs in multi-robot systems. In L. E. Sucar et al. (Eds.), *Decision theory models for applications in artificial intelligence: concepts and solutions*, SCOPUS.

Bhavani, D., & Ghalib, M. (2018). Internet of things based smart car parking system using K-nearest neighbor algorithm to find the nearest slot. *Journal of Computational and Theoretical Nanoscience, 15*, 2040–2045. https://doi.org/10.1166/jctn.2018.7403

Borger, E., & Stark, R. (2003). *Abstract state machines: A method for high-level system design and analysis*. Springer-Verlag.

Bryant, R. (1986). Graph-based algorithms for Boolean-function manipulation. *IEEE Transactions on Computers, 35*(8), 677–691.

Clarke, E. M., Biere, A., Raimi, R., & Zhu, Y. (2001). Bounded model checking using satisfiability solving. *Journal of Formal Methods in System Design, 19*(1), 7–34.

Clarke, E.M., Kroening, D., & Lerda, F. (2004). A tool for checking ANSI-C programs. In *Tools and algorithms for the construction and analysis of systems, 10th international conference TACAS 2004*, Held as part of the joint European conferences on theory and practice of software, ETAPS 2004, Barcelona, Spain, 29 March–2 April 2004, Proceedings (pp. 168–176).

Clavel, M., Durán, F., Eker, S., Lincoln, P., Martí-Oliet, N., Meseguer, J., & Talcott, C. (2007). *All about Maude—A high-performance logical framework, how to specify, program and verify systems in rewriting logic*. LNCS 4350. Springer.

Cuoq, P., Kirchner, F., Kosmatov, N., Prevosto, V., Signoles, J., & Yakobowski, B. (2012). Frama-C: A software analysis perspective. In *SEFM*.

Diaconescu, R., & Futatsugi, K. (1998). Cafeobj Report. *World Scientific*. https://doi.org/10.1142/3831

Diego, G., Loyola, R., Pedergnana, M., & García, S. G. (2016). Smart sampling and incremental function learning for very large high dimensional data. *Neural Networks, 78*, 75–87, ISSN 0893-6080. https://doi.org/10.1016/j.neunet.2015.09.001

Favaretto, M., De Clercq, E., Schneble, C. O., & Elger, B. S. (2020). What is your definition of big data? Researchers' understanding of the phenomenon of the decade. *PLoS One, 15*(2), e0228987. https://doi.org/10.1371/journal.pone.0228987

Filliatre, J. C. (2016). Deductive program verification with Why3—A tutorial.

George, V., & Vaughn, R. (2003). Application of lightweight formal methods in requirement engineering. *Crosstalk: The Journal of Defense Software Engineering*.

Ghose, A. (2000). Formal methods for requirements engineering. In *2000 International Symposium on Multimedia Software Engineering (ISMSE 2000), Taipei, Taiwan, 11–13 Dec. 2000* (pp. 13–16). IEEE Computer Society.

Haenlein, M., & Kaplan, A. (2019). A brief history of artificial intelligence: On the past, present, and future of artificial intelligence. *California Management Review, 61*(4), 5–14. https://doi.org/10.1177/0008125619864925

Hayes-Roth, F. (1985). Rule-based systems. *Communications of the ACM, 28*(9), 921–932. https://doi.org/10.1145/4284.4286

K.Hofer-Schmitz,B.Stojanovic,Towards formal verification of IoT protocols: A review, Computer Networks 174 (2020), Elsevier, DOI: doi: https://doi.org/10.1016/j.comnet.2020.107233.

HOL Interactive Theorem Prover. (n.d.). Retrieved from https://hol-theorem-prover.org/#doc

Holzmann, G. J. (2003). *The spin model checker: Primer and reference manual*. Addison Wesley.

Ingrand, F., & Ghallab, M. (2017). Deliberation for autonomous robots: A survey. *Artificial Intelligence, 247*, 10–44, ISSN 0004-3702. https://doi.org/10.1016/j.artint.2014.11.003

Isabelle. (n.d.). Retrieved from http://isabelle.in.tum.de/

Jackins, V., Vimal, S., Kaliappan, M., et al. (2021). AI-based smart prediction of clinical disease using random forest classifier and naive Bayes. *The Journal of Supercomputing, 77*, 5198–5219. https://doi.org/10.1007/s11227-020-03481-x

Jackson, D. (2012). *Software abstractions* (2nd ed.). MIT Press.

JAVA PATHFINDER. (n.d.). Retrieved from https://github.com/javapathfinder

Jones, C. B. (1990). *Systematic software development using VDM* (2nd ed.). Prentice-Hall International.

Keerthi, K., Roy, I., Hazra, A., & Rebeiro, C. (n.d.). Internet of Things. In R. S. Chakraborty et al. (Eds.), *Security and fault tolerance in internet of things*. Springer. https://doi.org/10.1007/978-3-030-02807-7_9

Krichen, M. (2019). *Improving formal verification and testing techniques for internet of things and smart cities, Mobile networks and applications*. Springer. https://doi.org/10.1007/s11036-019-01369-6

Luechaphonthara, K., & Vijayalakshmi, A., IOT based application for monitoring electricity power consumption in home appliances, Journal of Electrical & Computer Engineering, vol. 9, pp. 2088–8708, Feb 2019.

Ma, M., Stankovic, J. A., & Feng, L. (2021). Toward formal method for smart cities. *IEEE Computer*, 39–48. https://doi.org/10.1109/MC.2021.3082991

Mahmoud, A. A., Alawadh, I. N. A., Latif, G., & Alghazo, J. (2020). Smart nursery for smart cities: Infant sound classification based on novel features and support vector classifier, In *2020 7th international conference on electrical and electronics engineering (ICEEE)* (pp. 47–52), doi: https://doi.org/10.1109/ICEEE49618.2020.9102574.

Maler, O., & Nickovic, D. (2004). Monitoring temporal properties of continuous signals. In Y. Lakhnech & S. Yovine (Eds.), *Formats/FTRTFT 2004. LNCS* (Vol. 3253, pp. 152–166). Springer. https://doi.org/10.1007/978-3-540-30206-3_12

mCRL2. (n.d.). Retrieved from https://www.mcrl2.org/web/user_manual/introduction.html

Meiyi, M., Bartocci, E., Lifland, E., Stankovic, J., & Feng, L. (2021, early access). A novel spatial-temporal specification-based monitoring system for smart cities. *IEEE Internet of Things Journal*. https://doi.org/10.1109/JIOT.2021.3069943

Meyer, B. (1991). Advances in object-oriented software engineering. In D. Mandrioli & B. Meyer (Eds.), *Chapter design by contract* (pp. 1–50). Prentice Hall.

Mohammad, N. (2019). A multi-tiered defense model for the security analysis of critical facilities in smart cities. In *IEEE Access* (Vol. 7, pp. 152,585–152,598). doi: https://doi.org/10.1109/ACCESS.2019.2947638.

Nallaperuma, D., et al. (2019). Online incremental machine learning platform for big data-driven smart traffic management. *IEEE Transactions on Intelligent Transportation Systems, 20*(12), 4679–4690.

NuSMV. (n.d.). Retrieved from https://nusmv.fbk.eu/

Padua, D. A. (2011). *Encyclopedia of parallel computing*. Springer.

Pearl, J. (1984). *Heuristics: Intelligent search strategies for computer problem solving*. Addison-Wesley Publishing Company, Web.

Pradhan, I., Potika, K., Eirinaki, M., & Potikas, P. (2019). Exploratory data analysis and crime prediction for smart cities. In *Proceedings of the 23rd International Database Applications & Engineering Symposium (IDEAS '19)* (pp. 1–9). Association for Computing Machinery., , Article 4. https://doi.org/10.1145/3331076.3331114

Rjab, A. B., & Mellouli, S. (2018). Smart cities in the era of artificial intelligence and internet of things: Literature review from 1990 to 2017. In *Proceedings of the 19th annual international conference on digital government research: Governance in the data age (dg.O '18)* (pp. 1–10). Association for Computing Machinery., , article 81. https://doi.org/10.1145/3209281.3209380

Roig, P. J., Alcaraz, S., Gilly, K., Filiposka, S., & Aknin, N. (2020). Formal algebraic specification of an IoT/fog data centre for fat Tree or leaf and spine architectures. In *Proc. of 2nd ICECCE*. IEEE.

Seshia, S., Sadigh, D., & Sastry, S. (2020). Towards verified artificial intelligence, arXiv:1606.08514v4[cs.AI].

Simmons, R., & Koenig, S. (1995). Probabilistic robot navigation in partially observable environments. *IJCAI*.

TLA+. (n.d.). Retrieved from https://lamport.azurewebsites.net/tla/tla.html

Turitsyn, K., Backhaus, S., Ananyev, M., & Chertkov, M. (2011). Smart finite state devices: A modeling framework for demand response technologies. In *Proceedings of the 50th IEEE conference on decision and control and European control conference, Orlando, FL, USA, 12–15 December 2011* (pp. 7–14).

UPPAAL. (n.d.). Retrieved from https://uppaal.org/

Urban, C., & Mine, A. (2021). A review of formal methods applied to machine learning. arXiv 2104.02466v2 [cs.PL] 21April 2021.

Vázquez, J., et al. (2014). Making smart cities smarter using artificial intelligence techniques for smarter mobility. In *A: International Conference on Smart Grids and Green IT Systems. "SMARTGREENS 2014: proceedings of the 3rd International Conference on Smart Grids and Green IT Systems"*. SciTePress, p. IS7-IS11.

Woodcock, J., Larsen, P. G., Bicarregui, J., & Fitzgerald, J. (2009). Formal methods: Practice and experience. *ACM Computing Surveys, 41*(4), 1–36. https://doi.org/10.1145/1592434.1592436

Wooldridge, M., & Jennings, N. (1995). Intelligent agents: Theory and practice. *The Knowledge Engineering Review, 10*(2), 115–152. https://doi.org/10.1017/S0269888900008122

Yousefi, S., Derakhshan, F., & Bokani, A. (2018). Mobile agents for route planning in internet of things using Markov decision process. In *2018 IEEE international conference on smart energy grid engineering (SEGE)* (pp. 303–307). doi: https://doi.org/10.1109/SEGE.2018.8499517.

Zekić-Sušac, M., Mitrović, S., & Has, A. (2021). Machine learning based system for managing energy efficiency of public sector as an approach towards smart cities, 102074, ISSN 0268-4012. *International Journal of Information Management, 58*. https://doi.org/10.1016/j.ijinfomgt.2020.102074

Index

A
Ability to survive, 142
ADDIE process
  analysis, 58, 59
  design, 58, 59
  development, 58, 59
  evaluation, 58, 59
  implementation, 58, 59
Administration and participation, 188
Agglomeration effects, 21, 33
Algebra of communicating processes (ACP), 320
Alternation schemes, 224
Apple's Siri, 253
Application programming interface (API), 250
Artificial intelligence (AI), 134–135
  algorithms, 317
  data analysis, 316
  digital skills, 323
  experienced system, 317
  intelligent systems, 317
  machine learning, 320
  methodologies and tools, 316
  smart environments, 316
  symbolic SI/formal methods
    code verification, 319
    definition, 319
    formal verification and model-based testing, 320
    industrial safety critical software, 319
    proof of theorems, 318
    rule-based systems, 318
    verified, 320

B
Bibliometric, 39
Bibliometric analysis, 23, 156, 189, 197
Bibliometric approach, 156
  citizen-centered approach, 165
  co-occurrence network, 164
  country scientific production, 161
  data used, 159
  literature review
    citizen satisfaction, 156–157
    ranking systems, 158
  methodology, 157–159
  "motor" themes, 160
  semi-structured interviews, 165
  sustainability journal (MDPI), 159
  thematic evolution, 162, 163
Bibliostatistic analysis, 156
Big data, 124, 191
Big data analytics
  automated system, 261
  data science process
    data collection, 262
    data preparation, 263
    data storage, 263
    deployment model, 264–265
    evaluation, 264
    modeling, 263–264
    project understanding, 262
  digitalized services, 260
  smart applications, 260
  urban challenges, 261
Binary decision diagrams (BDD), 320
Blockchain, 134
  applications, 300, 302, 310, 311

P. Fitsilis (ed.), *Building on Smart Cities Skills and Competences*, Internet of Things,
https://doi.org/10.1007/978-3-030-97818-1

Blockchain (*cont.*)
    attributes, 302
    data mapping process, 299
    decentralized nature, 300
    governance, 306
    privacy, 301
    repositories, 298
    security, 300, 301
    systematic review, 298
Bounce-back ability, 142
Business sector, 31

C
Circular cities, 111
Circular economy strategies, 110–111
Cities in Motion Index (CIMI), 10
Citizen, 247–249
    agency, 189
    participation, 190, 191
City API, 250
City management, 145–148
    managing and planning, 146
    proper strategic plan, 146
City of Patras
    benefits, 267
    data modeling, 266
    digital transformation, 265
    KDD process, 266
    measurements, 268
Civic engagement, 188
    adoption, 197
    applications, 193
    bibliometric analysis, 189
    corresponding author's country, 190, 193
    education/knowledge, 188
    in Greek, 193–197
    levels, 188
    MCP, 191
    methodology, 189–190
    network map, 192, 195
    SCP, 191
    three-field plot, 191, 194
    world growth per year, 190, 192
Civic paternalism, 189
Civilizational competences, 39
Civil society, 30–31
Cloud computing, 124
Clustering, 321
Competence, 37
    smart cities, 38–39
    in statistics, 133
Copilot, 249
Cornell E-Rulemaking Initiative (CeRI), 233
COSIMMA (Collaborative Situation Metamodel), 288
Covariance-based-structural equation modelling (CB-SEM), 80
COVID-19, 26, 102, 259, 281
Creativity, 7–8
Cryptocurrencies, 303
Customer-focused approach, 188
Cyberattacks, 135
Cyber-physical systems (CPSs), 319
Cyber-security risks, 123
Cyberthreats, 134

D
Data analysis and mining, 315
Data analytics
    challenges, 316
    definition, 316
    pipelines, 260, 264–265
Data integration, 316
Data mining, 322
Data [illegible], 40
Decision trees, 322
Deliberation authority, 239
Democracy, 188
Design of People-Centered Smart Cities, 226
Design requirements (DRs), 56–58
    ADDIE process, 58–59
    design cycles and major decisions, 59
    didactical concept development, 57–58
    digital transformation module, 60
    problem situations, 61–64
DevOps, 250–251
DevOps skills, 11, 14, 15
Digital Agenda 2020, 187
Digital city, 22, 23, 190
Digital Economy and Society Index (DESI), 323
Digitalization, 99–100
Digitalization of Cities and Traffic, 54
Digital participatory platforms, 188
Digital pedagogy, 119–120
Digital transformation (DX), 11–12, 54, 76, 103–104
Direct democracy, 229
Disaster/crisis management domain, 281
Disruptive technologies, 97, 100
Distance-learning
    challenges, 54
Distributed ledger (DLT), 298
Dynamic Flood Ontology (DFO), 287

E

Eco-centric approach, 110–112
Ecole 24 model, 33
Economic engine, 95
E-deliberation, 230
  benefits of, 238
  CeRI, 233
  CrowdLaw, 232
  DebateHub, 234
  Decidim, 234–236
  E-Democracia, 233
  e-dialogos, 232
  LiquidFeedback, 234
  MiSenado, 233
  online consultation platform, 234
  ParlementetCitoyens, 233
  participatory democracy, 238
  PODS models, 239–243
  smart citizens, 236–238
  system categories, 230
    deliberative poll, 231
    discussion method, 231
    e-petition, 232
    online forums, 230–231
    questionnaires, 231
    voting, 231
E-dialogos, 232
e-government, 190, 191
Electronic governments, 251-252
Electronic voting, 252
Energy grid application, 308
Engage citizens
  civic engagement (*see* Civic engagement)
  civic paternalism, 189
  in decision-making processes/cycles, 188
  democracy, 188
  economy, 188
  neoliberal conception, 189
  participatory applications, 187
  stewardship, 189
  technocracy, 188
Entrepreneurship, 8
E-participation, 230
European Innovation Partnership (EIP), 76
European Innovation Partnership on Smart Cities and Communities (EIP), 187
European Integrated Data Archive (EIDA), 294

F

Formal educational system, 31
Friend of a friend (FOAF), 291

G

Google Assistant, 253
Governance, 191
GPT3, 249
Greece, 190, 193–197
Greek municipalities, 73, 82, 83

H

Heterotrait-monotrait (HTMT) ratio technique, 81–82

I

ICT skills, 28–29
Information and communication technologies (ICTs), 5–6, 21, 53, 103–104, 187
Information systems (IS) infrastructure, 55
Information technology (IT), 55
Innovative systems, 188
Intelligent Cities Challenge (ICC), 10
Intelligent city, 22, 190
Intercultural competence (ICC), 38
International Organization for Standardization (ISO), 26
International Telecommunication Union (ITU), 26
Internet of Things (IoT), 124, 191, 315, 316
Internet of Vehicles, 309
Internet protocol hourglass model, 108
Interoperability models
  components of, 274
  ecosystem, 273
  functions, 276
  institutional interoperability, 273
  intelligent systems, 274
  resource tree structure, 275
  RESTful API, 274
  semantic interoperability, 273
  technical interoperability, 273
INVEST collaboration, 282

J

Job roles, smart city, 10–12

K

Key performance indicators (KPIs), 26
K-nearest neighbors (KNN) algorithm, 322
Knowledge-based system, 317
Knowledge discovery from databases (KDD) process, 266

L
Language Model for Dialogue Applications (LaMDA), 253
Ledger, 297

M
Machine learning
  approach, 318
  clustering, 321, 323
  definition, 320
  interpretation, 321
Markov decision process (MDP), 321
Massive Open Online Courses (MOOCs), 17
Megacities, 259
Meta-design literacy, 220
Mobile applications, 248
Motor themes, 160
Multiple-country publications (MCP), 191
Multisystem construct, 124–125
Municipality of Patras, 268
  GIS application, 271
  [illegible] application, 270–271
  lighting management application, 270
  Sense.City application, 268–269
  vehicle fleet platform, 271–272

N
Natural resources, 188
Nested Dependencies model, 110
Network map, 192, 195
No-code
  applications
    electronic governments, 251–252
    features, 249
    managing public transportation, 252
    public records database, 251
    shopping and services, 252
  city API, 249–250
  methodology and principles, 249
  security and DevOps, 250–251
  smart assistants, 252–253
Non-smart city, 127

O
Oman Smart Ambassadors program, 206
Ontology 101 development process
  class hierarchy, 288
  create actual instances, 289
  data consolidation and integration, 288
  datatype property, 289
  object properties, 289
  relevant terms of the domain, 288
  reusing existing ontologies, 288
  sematic interoperability, 288

P
Partial least square-structural equation modelling (PLS-SEM), 80
Participation, 190
Participatory design, 190
Preferred Reporting Items for Systematic Reviews and Meta-Analyses (PRISMA) guideline, 125, 135
Problem-based learning
  didactic process structure, 56, 57
  DRs, 56–58
Project complexity model
  assessment question, 180
  concept of, 170
  descriptive complexity, 171
  ethic challenges, 178
  future research, 180
  identified challenges, 177
  management complexity, 176
  managerial and organizational challenges, 178
  project management, 171
  proposed framework, 179–180
  research methodology, 173
  significant challenges, 178
  smart cities' project challenges, 173–174
  special characteristics, 179
  stakeholder challenges, 177
  typology, 179
Project management skills, 129
Public deliberation, 230
Publicly funded services, 251
Public On-line Deliberation System (PODS), 239–243
  acceptability phase, 242
  citizens, 240
  comments integration index, 241
  conceptual architecture, 240
  deliberation authority, 239, 240
  electronic forum, 239
  evaluation phase, 242
  evaluation process, 239, 241
  information base, 239
  initialization, 241
  proposal plasticity index, 240
  proposal post, 241
  proposers, 240
  publication phase, 242–243
  rating phase, 242

theme post, 241
working flow, 243
Public participation, 188–190
Public-private-people partnership (4P), 24
Public records database, 251
Public sector, 29–30
Public sector modernization, 188
Public transportation, 252

Q

Quality of life, 188

R

R-based Biblioshiny app, 159, 190
Renewable energy (RE), 303
Representative democracy, 229
Representatives form, 229
Rescue MODES, 287
Responsibility, conception of, 188
RES-Q data consolidation, 293
RES-Q data layer (RDL), 293
RES-Q Living Lab (LL), 282
assessments, 286
comprehensiveness and applicability, 285
COSIMMA, 288
COVID-19 pandemic, 282
decision-making model, 287
DFO, 287
formalization of recovery actions, 284
knowledge discovery and evolution, 285
maintenance, 284
multitude and magnitude, 286
POLARISC ontology, 287
resilience, 285
risk/disaster management domain, 283
semantic interoperability, 284
synergistic approach, 287
RES-Q semantic model, 290
disasters, 290
FOAF, 291
infrastructure, 292
key performance indicators, 292
process, 292
risks, 291
top-level ontology, 292
upper ontology, 291
RES-Q software environment, 282
Rice's theorem, 319
Risk
clustering, 127
cyberattacks, 135
defined, 127
management, 125, 135
management skills, 129, 133
terminology, 127, 128
Risk Breakdown Structure (RBS), 129–133, 136
Risk response strategies, 133–135
Rule-based systems, 318
Rules classification, 322

S

Sankey diagram, 191
Science, technology, engineering, and mathematics (STEM), 29
Semantic web rule language (SWRL), 294
Sendai Framework UNISDR, 147
Sharing economy, 306
Signal spatiotemporal logic (SSTL), 319
Siloed architecture, 106, 107
Single-country publications (SCP), 191
Skills demand
in smart cities, 6–8
Smart assistants, 252–253
Smart cities, 3, 5–6, 21
academic community, 188
AI, 134–135
articles published per year, 190, 192
bibliometric analysis, 23
blockchain, 297–302, 306
centralized control, 311
citizen, 247–249
citizenship, 308
civic engagement (*see* Civic engagement)
competence, 38–39
concept, 129, 188
continuous improvement, 226
copy and paste generation, 222
data analysis
keyword clustering and classification, 42–45
data literacy, 222
development of, 187
domains, 45–48
dual education, 224
economy, 309
energy exchange, 308, 309
evaluation framework of citizen participation, 188, 189
and evolution, 22–24
features
actors and interactions, 24–25
components of, 25–26
indicators, 26–27
in Greece, 77–78

Smart cities (*cont.*)
healthcare, 306–308
human centered, 218
ICT and innovative solutions, 143
job roles, 10–12, 14, 15
knowledge and skills, 218
learning ecosystems, 223
lifecycle, 135
LIFE skills framework, 219
literature review, 142
mapping skills, job roles, 11, 14–15
measuring performance, 144–145
meta representation, 220
methodology, 8–10
multisystem construct, 124–125
no-code, 249–251
objects of knowledge, 220
people centered, 218
proficient employees for, 27–29
rapid urbanization/developments, 297
RBS, 129–133
research method
data extraction, 41–42
data source, 40
searching, identifying, and selecting articles, procedure for, 40
search summary, 40–41
resilience, 218
resilient city, 143
risk clustering, 127
risk management skills, 129, 133
risk response strategies, 133–135
road map, 87–88
security, 305, 306
selecting articles, 126–127
simulation of innovation processes, 224
skills, 11–15, 27–33
demand in, 6–8
formal educational system, 31
stakeholders involvement, 29–32
type of, 27–29
SLR, 125–127
smart governance, 188
smartness, 219
soft-digital competencies, 220
space of competences, 219–221
stakeholders, 125
taxonomic table, 221
transversal competences, 222
urban policies, 188
vision of, 217
workforce, 4, 6, 17
Smart cities reference architecture (SCRA), 103
Smart City Ambassadors Professional Community, 209
Smart City Ambassadors program, 201
ambassador database, 210–211
ambassadors, 211–212
COVID-19, 209–210
digital transformation, 212–213
ecosystem, 212
future research, 214
key success factors, 210
knowledge sharing, 208, 211
replication and scalability, 213
skills and learning
community skills, 203
data skills, 203
development, 204
employees and leaders, 204
formal learning programs, 204
innovation ecosystems, 205
knowledge workers, 204
Research Council, 205
technology skills, 202–203
structure and activities, 208
theoretical framing and methods, 207
Smart City IT Manager (SCM), 11
Smart City IT Officer (SCO), 11
Smart City Planner (SCP), 11
Smart economy, 25, 188, 217
applications, 303, 304
Smart environment, 25, 188, 217
careful, and gradual initiatives, 76
challenges, 71–72
digital strategy
Greek municipalities, 82, 83
planning and implementing, 85–87
digital transformation, 76
evaluation model
measurement model verification, 81–82
structural model verification, 82, 83
Greek municipality, 73
and mobility, 89
multidimensional definitions, 73
municipality forms, 72
research methodology
data specifications, 79–80
PLS-SEM, 80
research model, 78
smart cities
in Greece, 77–78
road map, 87–88

smart specialization, 72–74
urban environment, 75
Smart governance, 25, 188, 191, 197
Smart government, 217
applications, 307
Smart home sensors, 315
Smart living, 26, 188, 217
Smart mobility, 25, 188, 217
Smartness, 148, 223
Smart people, 26, 188, 217
Smart resilient city, 149, 150
Smart specialization, 72–74
Smart Sustainable Cities (SSC), 26
Smart traffic management platform (STMP), 323
Smart urban ecosystems, 73–74
Smart working, 224
Social and human capital, 188
Social Sciences and Humanities (SSH), 29
Soft domains, 5
Soft skills, 129
Software engineering, 248
Spatial temporal logic, 319
Spatial temporal reach and escape logic (STREL), 319
Special aggregation signal temporal logic (SaSTL), 319
Standards development organizations (SDOs), 97–98
Stewardship, 189
Structural equation modelling (SEM), 80
Student feedback, 62
Supervised learning, 320, 321
Supply chain management, 309
Sustainability, 191
Sustainable development, 26
Sustainable development goals (SDGs), 224
Symbolic logic- and rule-based approach, 318
Systematic literature review (SLR), 125–127, 135

T

Taxonomy framework, 302
Technical decisions, 237
Technical skills, 129
Technocracy, 188
Technology-pushed solutions, 30–31
Territorial pact, 223
Thessaloniki metropolitan area, 193–197
The Tragedy of the Commons, 259
Transportation/mobility applications, 304
Transport systems, 188
Triple helix model, 217

U

United for Smart Sustainable Cities (U4SSC), 26
Unsupervised learning, 321
Urban environment, 75
Urban landscape, 96, 100
capacity building, 112–118
competencies, 112–118
concerns and challenges, 104–105
core characteristics, 116
design paradigm, 114–115
design thinking, 116
and digitalization, 99–100
digital pedagogy
adaptive learning environments, 119
smart cities ecosystem, 119
tackling achievement gaps, 120
disruptive technologies, 97
eco-centric approach
circular cities, 111
circular city strategies, 111–112
circular economy strategies, 110–111
crucial imperative, 112
Nested Dependencies model, 110
imperatives
citizens' needs, 101
Covid-19 pandemic, 102
deal changes, 101
DX, 103–104
ICT, 103–104
90–90 formulae in design, 115
siloed architecture, 106, 107
Classic Saucer Champagne Glass, 108–110
internet protocol hourglass model, 108
from vertical to horizontal, 106–107
skills, 112–118
smart city, 96
standardization activities, 98–99
strategy framework, 113
sustainable engineering, 118
systems, 117
approach, 117–118
design, 116
engineering, 117
thinking, 116
systems-level approach, 105–106
Urban policies, 188
User access levels
department stakeholders, 276
directorate level, 276
interoperability center level, 277

V
VBA (Visual Basic for Application) script, 60
Vehicle networking, 315
Vehicular ad hoc networks (VANETs), 301
Vehicular energy network (VEN), 303
Virtual education concept, 66
Vocational and Educational Training (VET), 4, 33
VOSviewer (VoS), 38

W
Web of Science database, 190

The manufacturer's authorised representative in the EU is Springer Nature Customer Service Centre GmbH, Europaplatz 3, 69115 Heidelberg, Germany. If you have any concerns regarding our products, please contact ProductSafety@springernature.com

Printed and bound by CPI Group (UK) Ltd, Croydon, CR0 4YY
30/09/2026
02225250-0010